Global Television

How to Create Effective Television for the Future

Tony Verna

Edited by William T. Bode

Focal Press
Boston   London

Other Books by Tony Verna

Playback

LIVE TV: An Inside Look at Directing and Producing

ISBN 0-240-80134-2

Butterworth-Heinemann
80 Montvale Avenue
Stoneham, MA 02180

10 9 8 7 6 5 4 3 2 1

Printed in the United States of America

To my son, Eric Verna, and the future he represents.

Contents

Foreword

The key word in the title of Tony's book is, of course, *global*. Before satellites, fiber optic cabling, and digital compression, television was a local outlet sometimes attached to a national programming network. No more. Television—like all major forms of electronic communication—has gone global. Careers in the medium have reflected that. Tony went from working for a local TV station to directing for network television, to producing/directing for his own production company, to helming many of the biggest worldwide telecasts. I went from being president of a local O&O (WBBM-TV, Chicago) to being president of CBS-TV to being executive vice president of TBS (with its world-spanning CNN) to being president/CEO of Comsat with total immersion in global interaction.

But the extent to which TV programming and TV careers has been global in the past will seem almost insignificant to what lies ahead. Regardless of minor roadblocks, the European Community is forging a new giant coalition in Europe. Asian TV is leapfrogging the interim steps taken by other nations. Satellite transmission is catapulting Asian television into twenty-first century techniques virtually overnight. And with expanding services like PanAmSat, South America is on the brink of a similar electronic communications revolution. While it still faces major hurdles, Africa will not linger that far behind. Low earth orbit systems will revolutionize telephonic communications within only a few years. Anyone pretending an interest in a communications career must be aware of all of these interactions.

Tony Verna's career in network sports and news gave him unique preparation for his role as a global pioneer in TV. By the time satellite TV transmission became a reality, Tony's career had made him a world traveler, conversant with television systems worldwide. It is this unique experience that he shares in *Global Television,* an amassing of knowledge and techniques that have taken decades to develop, all summarized in easily accessible form.

Television Present is already global. Television Future will be increasingly so. I heartily recommend *Global Television* as a primer for anyone who aspires to be part of the future of television—or any other phase of electronic communications. It is a handbook for the twenty-first century and belongs on the bookshelf of everyone working in or allied with television.

Robert Wussler

Preface

To stay competitive in the 1990s, most major industries have been forced to function in the global market. Television is no exception. In fact, it is in large part the very tools of communication, symbolized by television, that have driven national economies into becoming world economies. We have not yet reached the state of Marshall McLuhan's famous Global Village. Television and telephone communication have flooded the so-called developed world, but a totally wired world—a very costly investment—is still decades away. During the 1990s, radio will remain the sole link to some parts of the world. The newer carriers of communication—satellites and fiber optic cable—promise change that will ultimately extend audio-visual communication literally around the world, but that change will be gradual, not dramatic.

The significance of these new channels of communication is more revolutionary in television, however, than in any of our other human intercourses. The reason is that as the new media of communication have developed, they have converged and have become new modes of signal transmission not only for outbound television signals but also for inbound signals. In short, satellites and fiber optics have become as much tools of television production as of television transmission. Combined with the development of charge-coupled devices (CCDs) and the use of digital compression to transmit analog signals, these new transmission tools—and their convergence—have revolutionized the television industry.

Modern technology has so altered all major communications media that using the word *global* to describe 1990s television is almost redundant. Like major manufacturing industries, communication industries have been forced to ignore national boundaries. Stock markets have been forced into round-the-clock operation. Electronic industries, like automobile industries, long ago went global. In television, satellites and fiber optic cabling make cable television a match for long-standing networks. Satellite feeds let local television news shows integrate world events with local electronic news gathering (ENG) reports—and forced the reshaping of television network news. Syndication, which had been global for years, found economic necessity dictating a faster, more aggressive pace of multinational coproduction, increasing the worldwide nature of non-network programming. Direct broadcast by satellite (DBS) hovers over it all. Like fax machines and cellular phones, satellite television is shrinking the world. Because of television, newscasts, Super Bowls, boxing matches, soccer matches, wrestling matches, rock videos, and rocket wars know no boundaries. We live in a television-saturated world, but we often appear more dazzled than educated by its electronic wizardry and ubiquity. And the technologies generating the changes in communication are themselves changing so rapidly that professionals working in television—and viewers watching television—need a sharp perspective to understand where we are and where we are going.

By the time you have finished reading this book, you should have a grasp of two things: (1) how a globalcast is put together—conceived, developed, produced, distributed, and viewed; (2) how the media are converging and how that convergence is affecting and will affect not only global television but all forms of communication.

My start in globalcasting was *Live Aid*. Since then I have executive-produced and executive-directed series of very different international telecasts. *Sport Aid* connected hundreds of countries with more than a billion people "Racing Against Time." *Prayer For World Peace* linked a billion and a half people around the world, including hundreds of millions of Catholics, allowing them to share prayer with Pope John Paul II. *Our Common Future* and *Earth 90* focused on the ecology and inspired viewers to become involved in rescuing planet earth. *The 1990 Goodwill Games* used international sports competition and interactive global television as a way of unifying the world and its citizens.

Although each of these shows was different in concept and execution, certain underlying patterns evolved. International satellite links made each show possible. Advancing technology let us devise new techniques and approaches. The methods of organizing, managing, and briefing the hundreds, sometimes thousands, of television professionals involved in such globalcasts became defined and refined. Alternative budgeting patterns emerged.

Paralleling this developing pattern of methods was the influence of these globalcasts on both national and local telecasts. News, in particular, was affected, but all television production has been touched in one way or another by pioneering advances in globalcasting. This convergence of patterns and influence prompted me to reevaluate the television industry, specifically its methods of operating, its influence, and its responsibilities.

As a result of globalcasting, television production and distribution and television's effect on the world will continue to change at an ever-increasing pace. What will these changes be? How can they be harnessed to improve all television programming, local, national, and global? How will work methods, budgets, and jobs be affected? This book is in response to these questions.

This book is presented in five parts. In Part I, Anatomy of a Globalcast, we dissect the pioneer globalcast, *Live Aid*, laying out the foundation for analyzing current techniques and for forecasting the future.

In Part II, The Spine of Global Television, we outline technical factors that are critical to television production: hardware—satellites, fiber optic cable, and charge-coupled devices—and software—analog/digital technology.

In Part III, Creating a Globalcast, we look at current techniques—techniques that didn't exist ten or even five years ago.

We look at some of the unique features involved in financing international television, because at the same time that you are being visionary with a globalcast, you have to be practical. We'll discuss the bottom line and how the word compromise joins other *C* words—like *CD, computers, convenience, convergence, coproduction*—that will be so much a part of our future.

We'll see how one thousand people or two thousand people can be brought together in a unified, functioning team. But most of all we'll diagram the steps globalcasters have to take to create a successful show. You'll go behind the scenes, inside productions, to get the *feel* of global-casting and to share the excitement and promise of this phenomenon that foreshadows the twenty-first century.

The most-viewed worldwide programs have been done by independent production companies and producers, such as myself, tired of mainstream programming and eager to embrace concepts centered in today's technological advances and reaching out to their hairy limits. (More people watched *Live Aid* than watched all of the Super Bowls and Kentucky Derbys I've directed.) The young technology of globalcasting has already had a substantial impact, enough to be judged, to be measured, and evaluated, even to be copied.

Part III was written with two main purposes in mind. First, I am aiming this book at professionals: those working in television who want to move into globalcasting; those who support production—administration, sales, public relations, and so on; and those who manufacture and supply equipment. Second, I want to aim at young people considering globalcasting as a career. As coproducer and executive director of *Live Aid* and *Sport Aid* and executive producer/director of Pope John Paul II's *Prayer For World Peace* and Ted Turner's *The 1990 Goodwill Games,* I have worked with literally thousands of people. I have worked with experienced professionals who thought they were ready for a massive production like a globalcast, only to discover they hadn't been given either the knowledge or skills necessary to handle the job. There was nothing comparable in their background that they could draw on. And I've lectured and trained young people just starting their careers and trying to understand the complexities so as to position themselves for the future. This book is designed to help make young people and television professionals and television viewers ready for the future.

Over a thousand people worked on *Live Aid* in 1985. Nearly two thousand worked on *The 1990 Goodwill Games.* I could never share with you the input of all these people doing all of these jobs. But I can share with you enough so that you can grasp the structure, the form. I will do this in much the same way mathemeticians invoke fractal geometry to abstract music. I will show you enough of the small pieces so that you will see the whole structure. By eliminating much of the detail, we will reveal the overall design.

We will reduce a television program's global components to a skeletal form. We will sketch the patterns that are repeated on productions of any size, in an effort to project future production techniques. From pieces of the past: the whole of the future.

This book is not aimed at any one phase of production. I am not suggesting that every young person starting a career in television should try to become an executive producer/executive director, a producer, or a director. We all have to choose our own careers and life paths. But regardless of what job anyone does in a globalcast, he or she will have to be familiar with the overall structures and all of the responsibilities involved. Further, many working in allied fields will find much here to draw on and apply to other media.

This book addresses all of the jobs—and the jobs in globalcasting often differ, in both scope and title, from those in local and national telecasting. This books shows how communication between and among all of the jobs can be accomplished in order to provide you with the necessary knowledge to be part of the team, to be able to work your job and to interlink with others.

To be successful at any job you have to want to do it, to understand it, to be dedicated to it and to work at it. In the days when people walked a field holding a plow behind a horse, one person could accomplish all of those career goals alone. In today's world that's not possible. And in the complicated world of global telecasting, everyone's job is so interfaced with everyone else's that people working in each job category have to understand the work the others are doing and how their jobs interrelate to produce the shared goal—a great interactive world telecast, a globalcast. To help understand the crossovers among the converging media, there is a glossary.

In Part IV, Transmission Technology, we address getting the message there. Television transmission used to be relatively simple. As the shift from analog to digital accelerates, however, it becomes increasingly important to understand the technology and how its existence has dictated the also accelerating convergence of the media. Transmission by satellite, by fiber optic cable, by copper cable is influencing not just globalcasting but all of television, all of radio, all of recorded music and sound, all of film production. The fundamentals applicable to any of these media are now becoming applicable to all. We'll look at the fusion of satellite transmission with photonic transmission via fiber optic cable. We'll focus on global television, but you'll find that understanding globalcasting will guide you to an understanding of virtually all of the communications media.

In Part V, Global Trends, we put globalcasting back together for you, to put it in perspective. After the transitional 1990s, what will globalcasts be like? What will the world globalcasting has influenced be like? What are the trends and how does new technology reflect those trends? Again: From the past will come the future.

Writing Part VI, Looking at a Crystal Ball, would be easier ten years from now, and it would be a lot easier for you to read. Why? Because for all the past changes and advances we've witnessed, the 1990s is a decade of transition. Yes, this is the "Age of Information." Yes, this is the "Age of the Image." But when historians look back on us a hundred years from now, they will almost invariably dub ours "The Age of Convergence"—the age when dozens of inventions came to fruition and joined to create unprecedented possibilities in the world of communication.

The electronic developments you'll read about in these pages are astounding: the digital audio we've already experienced in CDs; the fiber optic cables that will carry flawless digital audio and digital video any-where—and back; the same fiber optic cables that are linking computers with television sets with telephones with satellites.

Each of these phenomena is exciting in itself. The true excitement, however, will come when they all join in the predicted wired world and wired home. Gradually, as the decade progresses, walls in new homes will become television screens. Our phones will be linked to our television sets, which will be linked to our fax machines, which will be linked to our offices and to our banks and to our schools and to our doctors and to every-one and everyplace else in the world.

The question is not whether these communications media will be linked, but when and how. How will globalcasts be sent out to the world? By satel-lites to television stations, as they have historically gone out? By satellite to window-sized dishes, via direct broadcast systems? By fiber optic cable, which allows any program to be "interactive," with viewers able not just to respond to questions but actually to appear on the screen—from their home or from anywhere (as they have already on some quiz shows)?

No one can answer these questions now. Time will, as politics, econom-ics, and television audiences dictate how we communicate with each other. But in the meantime, we can't sit and do nothing for a decade. There are audiences all over the world watching television now. To deal with the future intelligently, we need to understand where we stand now, where we want to go, and how we want to get there. Knowledge, they say, is power. This book is designed to give you some of that power, to give you an edge, the knowledge you will need to stay ahead of the confusion in this incredible age of fusion.

Even after experiencing the worldwide response to telecasts of President Kennedy's funeral and of the Vietnam War, no one could have foreseen the power of television as reflected in the reactions to Tiananmen Square, or to the collapse of the Berlin Wall, or to the Persian Gulf War reporting, or in the role of television in the political upheaval in Europe and Asia and Africa. Television has erased boundaries and turned once secret diplomacy into nightly newscasts—teleplomacy.

Global Television will show you how our past will affect our future and how we can use the lessons of the past to forge a better future in television for the 1990s and beyond. This book is not a production manual addressing mike stands and camera angles. It is an analysis that addresses the concepts that underlie programming, the technical advances that will shape future production and distribution, and the budgeting and management required to translate theory into successful practicality.

And last, but certainly not least, this book will address the role and responsibility of television production as the last decade of the twentieth century carries us and our world into a new era. The medium has already become more than the message. It has helped reshape the world and the world's response to its environment. If you watch television, if you work in television, globalcasting is in your future, ready or not. This book is designed to make you ready.

Acknowledgments

Like television, publishing a book is very much a cooperative effort. The following acknowledgments are in no way inclusive. Many colleagues, associates, and friends have suffered with me through the years required to bring this work together. Perhaps the thanks it is most impossible to express are to my family, especially my wife Carol and daughter Jenny, for putting up with endless phone calls, proofreadings, and so on. Had they not been accustomed to the normal craziness of the television business, I'm sure this chaos would have disturbed them more than it did. Retrospective thanks, nevertheless, for their patience and interest.

All of the following have in one way or another contributed to bringing this book to fruition. My thanks to Robert Wussler, with whom I have worked on national telecasts and globalcasts, and who provided the foreword for this book. My thanks, too, to Michael McLees (he who has guided the technical and budgetary elements of most of my productions); Barry O'Donnell, former vice president, operations, of TBS's *Goodwill Games;* Hal Uplinger, with whom I have worked since our joint days at CBS Sports; Robin Barty King; Jim Tuverson, Jr. and Dave Herman, top professionals in the world of satellites and satellite television; Linda Wendell, who has served personally in so many capacities on so many globalcasts and whose company W-5 has booked talent for us on more than one occasion; Mike Kastner, CPA and friend extraordinaire who keeps track of so many details that I might otherwise lose track of; JHT Productions, Inc., the company that worked with me to isolate the video prints from our globalcasts; Sandi Fullerton and Lou Horvitz, two of the DGA's top concert directors who have worked with us through hours upon hours of global television from *Live Aid* forward; *Live Aid* executive producer Mike Mitchell; Harvey Goldsmith, the tireless concert promoter in the UK; the libraries in Pacific Palisades and Los Angeles, California, and the University of Central Florida Library.

At the Vatican I extend special thanks to Mother Teresa and Archbishop John P. Foley, President of the Pontifical Council for Social Communications of the Vatican. Special thanks, too, for the cooperation I received from The White House and from Band Aid and Bob Geldof.

On different continents for different shows, my thanks to Larry King, Paul Beckham, Terry McGuirk, Paul Temple, Ian Watson, Jim Slater, Paul Dietrich, Arthur Taylor, Simon Dring, Peter Derkson, and Graham Lacey.

Others I would like to thank are Jerry Adler, Richard Bazzy, Lloyd Bloom, Philippe Caland, Alexander Cohen, Joseph Dispenza, Pete Dodd, Claudette Doran, Bob Finkel, Carl Hahn, Bob Halloran, Mark Itkin, Terry Lanni, Dr. Robert Mendez, Lance Robbins, Hiam Saban, Arthur Taylor, Rudy Tellez, Tom Van Sandt, Tracy Verna, Michele Wallerstein, Bruce Walton, Murray Weissman, Eric Weissmann, Jeff Wohler.

In Japan, my thanks especially to Akira Yoshigi, Keiji Shima, Momoko Ito, and Joichi Ito.

In terms of the creation, construction, and content of *Global Television,* my thanks go to Phil Sutherland, development editor at Focal Press, and to my editor William T. Bode. I have known and worked with Bill since our days at WCAU-TV in Philadelphia in the 1950s, when television was in its infancy. His qualifications for editing this book include his stint as one of the two directors of CBS's *Action in the Afternoon,* the only live outdoor western ever on network television, scripting for *Studio One Summer Theater,* as well as writing and/or producing for a number of my globalcasts.

Special thanks go to Dr. Jose Maunez, Associate Professor, School of Communication, University of Central Florida, Orlando, who was kind enough to read the text for us in manuscript form.

A note for our readers in the United Kingdom and several other countries: The terms *billion* and *trillion* recur throughout the text. Rather than trying to translate USA-billions into UK-milliards or USA-trillions into UK-billions, the parenthetical (USA) has been inserted with each use. If you find a reference not so noted, please make your own notification that the USA terms have been used throughout this book.

Part *One*

Anatomy of a Globalcast

Chapter *1*

Defining Live Television: *Live Aid*

Globalcasting can be described as television transmission from multiple points on earth to multiple points on earth, usually live or live with tape inserts, employing all of the technical facilities that allow a maximum number of viewers/listeners around the world to share the television experience. But one sentence cannot define globalcasting. There are degrees of globalcasting. The most basic level is global transmission: a telecast from a single source, sending its signal around the world. The next step, from global transmission to globalcasting, implies that there are multiple sources (or injects) coming from around the world and being televised to the world. Newscasts and programs like ABCs *Nightline* and, occasionally, *The Today Show* and *Good Morning America* are examples. But these programs tend to inject only talking heads and interviews. One step beyond that, at the highest level of globalcasting, are programs with multiple inject points, each providing full production. The Academy Award telecasts, the *Miss Universe Contest,* and similar programs have been expanding into live overseas locations. And, of course, a show like *Live Aid,* with sixteen hours of programming in multiple locations and multiple simultaneous broadcasts using multiple satellites, defines globalcasting at its extreme.

But programs like *Live Aid, Sport Aid,* and *Prayer For World Peace* are really not one single program. Within each is a series of different programs delivered simultaneously to different countries. In the case of *Live Aid,* depending on where in the world viewers were, they saw different versions of the show. Some countries prohibited telethons, some prohibited commercials; some contracts demanded performances by certain artists at specified times to specified audiences; and because there are twenty-four time zones around the world, what was shown at prime time differed in different parts of the world. For many reasons, public service announcements were transmitted according to the spokesperson's appeal in the receiving country.

To accomplish all of these variations on a globalcast, you need a control room with a myriad of monitors and audio consoles to match the incoming and outgoing transmissions. You face thousands of decisions that cannot be completely planned or rehearsed. You have to use technology, production techniques, and language not employed in regular television.

Globalcasts like those we'll discuss have been relatively few in number. They'll doubtless continue to be few in number, due to the limited scope of causes and the high costs involved. But the structure of globalcasts stretches global television to its limits. When inputting from and outputting to multiple countries, different languages, different TV standards, and different equipment are involved. By dissecting these gigantic solo efforts, we can anticipate every possible element that might be encountered in a global television effort.

And I'm not talking just about the executive director position on these shows. I'm talking about everyone. Everyone doing every job has to understand what is going on, or the shows just don't work. To remove whatever mystery may lie behind the process, I will dissect *Live Aid* and indicate the technical structures involving satellites and fiber optic cable. Then I will start at the beginning of a globalcast—with an idea and a budget—and trace the production from start to finish, the finish being a debriefing to prepare for the next globalcast.

With this structure in mind, then, let me take you back to the day that made history—and made globalcasting a household word around the world.

Hiding the Seams: The Many Productions Called *Live Aid*

At 7:00 a.m. Eastern Daylight Time on July 13, 1985, a global television program began that eventually would be seen and heard by more than 1.5 billion people. *Live Aid* surprised and thrilled the world: almost half of the people on earth saw and listened to the same television/radio program. Satellites had been used before to televise news and sports events. But *Live Aid* astounded the world. It was celebrated in newspapers and magazines and, yes, on television. Globalcasting news told the world about the monumental globalcast that signalled a new era in worldwide communication.

Had Thomas Edison, Alexander Graham Bell, and other scientists whose imaginations made globalcasting possible been watching, they would have been amazed. One of those scientists—Arthur C. Clarke, whose idea it was to orbit satellites so we could use them around the clock to send signals around the world—was watching. This book celebrates Clarke's and the other scientists' contributions to globalcasting, to the now

Figure 1–1 Satellite consultant Jim Tuverson, Jr., worked with Synsat when he designed this satellite schematic for *Live Aid*. This show used twenty-two video channels on thirteen satellites. Multiple satellite connections were needed to support not only the six separate outgoing feeds involved, but also many incoming feeds that aired and/or were taped for later playback.

Courtesy Westinghouse Broadcasting and Cable, Inc. and Jim Tuverson, Jr.

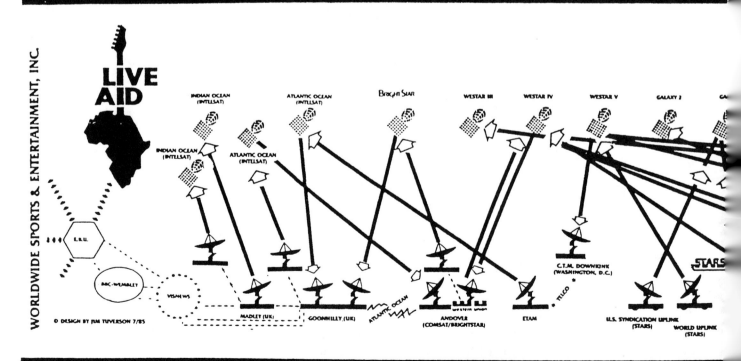

casually accepted live, instant communication that is helping to reshape politics, wars, the world economy—our lives. This celebration begins with an account of the program that surprised everyone—even those of us who worked on it.

Hiding the Sleight of Hand

The goal of any television production is to communicate the program's content to the viewing audience as effectively as possible. That job requires the skills of a magician—hiding the sleight of hand from the viewer. The production team has to know where the seams are in order to hide them. If you're working on a film or taped show, the assignment is easier. With skillful preplanning, most unforeseen problems can be solved in "post." If you've read my earlier book, *Live TV,* you know that the obligatory question for live television is "What if?" When it's live, you have to conjure up all of the possible "unforeseens" and have a standby course of action to control every situation so that the audience never knows a problem existed.

This same set of assignments applies to programs telecast worldwide. With globalcasts, however, you are dealing with many more elements, with many more seams, many more challenges. *Live Aid* was the first non-news/sports globalcast—and the first truly interactive globalcast, and it provides a clear-cut example of the steps a production team has to take to assimilate multinational inputs with multinational outputs.

Multiple Audiences Require Multiple Feeds

Anyone watching or listening to any part of *Live Aid* experienced a fully produced program. There were a lot of people watching and listening— more than 160 nations, 1.5 billion people, 85 percent of television viewers in the world, 75 percent of radio listeners. And each individual viewer/ listener was concerned only with what he or she was experiencing. For the

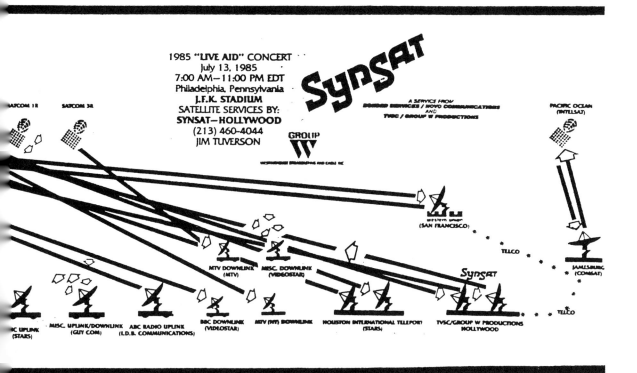

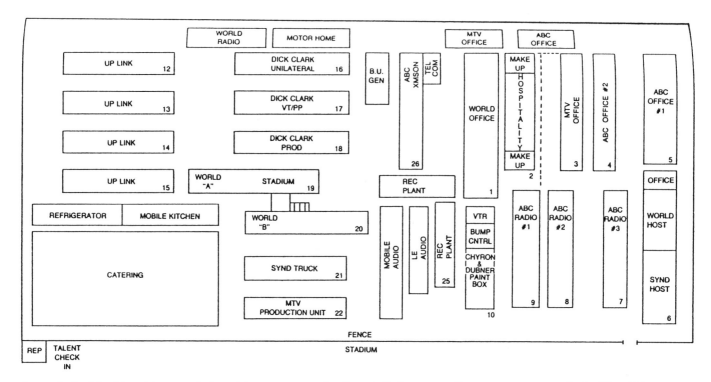

Figure 1–2 Matching this ground plan with Figure 1–3 offers a concrete impression of the complexity of the multiple productions of *Live Aid*. Multiple production companies all had to function separately and yet in coordination with the world feed coming out of the two world trucks. Note how the two trucks were linked so I could shift instantly from one control post to another.

production staff, however, there was not one single *Live Aid* program. There were seven basic program units that came out of my control room in Philadelphia's John F. Kennedy Stadium. (There were actually several more feeds, including some edited versions the next day in countries that could not air the live telecast, but I will deal here with the seven main live feeds.) There were six television feeds and one radio feed carrying the show:

World Radio: Stereo and mono
MTV: Full sixteen-hour coverage
ABC-TV: Prime time only, live and tape
Syndicated (USA) Feed: With commercials; ended 8:00 p.m.
USA World Feed: Commercial-free, in English
UK World Feed: Commercial-free, in English
DiamondVision Feed: In-stadium, London and Philadelphia

Figure 1–3 This photo shows a portion of the "Van City" created to produce *Live Aid*. The satellite uplink antennas are visible in the background, as is the catering tent. Compare this with the ground plan in Figure 1–2.

Each program had a separate producer, a separate director—a separate production crew, each tied into me. And for each of these crews to carry out my vision, I had to devise a format that provided each unit with its own unique needs.

World Radio

The radio feed was basically problem-free. There were no restrictions on performances, so whatever performance was live could go out. North America and Europe were fed stereo by the world radio feed and (in the USA) by the ABC Radio Network. The mono feed to the rest of the world was supplemented by the Voice of America. We used four international satellites for the audiocast, with five on-air broadcasters providing commentary and interviews.

The television feeds were vastly more complicated than the radio feed. Not only were there different audiences involved, there were also restrictions on some of the content.

MTV

MTV's was the easiest feed to coordinate. MTV incorporated portions of the entire telecast into their own format, using their regular MTV hosts, weaving in and out of the stadium feed. The main challenge was providing MTV with the concert talent they wanted for interviews. Figure 1–10 shows one of our talent-coordination teams escorting Bette Midler from one performance/interview to another.

You can imagine the massive task it was to coordinate all of the talent on the two continents. We had a huge staff to organize transportation, lodging, and equipment for all of the stars. Their performances were donated, but the cost and organization of the mechanics were the responsibility of the production unit. A walkie-talkie network was critical to the timing involved in getting talent not only from hotel to stadium to performances but also to World Feed, Syndicated Feed, MTV Feed interviews, photo sessions, and so on.

ABC-TV

ABC-TV aired in prime time only, from 8:00 to 11:00 p.m., EDT. This coverage started live with Phil Collins, newly arrived via the Concorde from his earlier appearance at Wembley Stadium in London. Other live coverage included Robert Plant; Crosby, Stills, Nash and Young; Duran Duran; Patti LaBelle; Hall and Oates; Mick Jagger and Tina Turner and the USA "We Are the World" finale with Lionel Richie, Harry Belafonte, et al.

That was all live and no problem. It was ABC's taped coverage that posed one of the major production complexities. To get ABC's approval to televise portions of the show in prime time, we had to promise non-cable USA exclusive coverage of some acts, including Madonna, Elton John, Prince, the reunion of the Who, and Paul McCartney. There were complications in terms of providing ABC-TV with the coverage: it meant taping the exclusive acts and making the tape available to ABC. The complexity arose with the USA Syndicated Feed. When the exclusive acts were on stage live, those performances had to be blacked out and replaced by alternative programming. The replacement had to be done in such a way that it didn't bother the viewers. No visible seams.

Figure 1–4 This video print from an off-the-air tape shows the massive *Live Aid* audience in Wembley Stadium in London, England. Note the dual DiamondVision screens, which allowed integration for the audience of close-up shots of UK performers, live performances from the USA and elsewhere, and taped performances.

Figure 1–5 As in Figure 1–4, the DiamondVision screen is very visible here in John F. Kennedy Stadium in Philadelphia. In addition to the seated spectators, many fans stood or sat on the stadium grass. One major function of the big screen was to provide clear coverage to those whose view was blocked by production equipment. See Figure 1–11 to sense the size of the equipment.

Figure 1–6 Global stars like Madonna contribute their talent to worthy causes like Bob Geldof's *Live Aid*, but there are production expenses—and jobs—connected with coordinating such stars.

Figure 1–7 Technical limitations blocked an intercontinental duet between Mick Jagger and David Bowie, but nothing limited the show-stopping magic between Tina Turner and Mick Jagger, live and together, in Philadelphia.

The Syndicated Feed

To replace the blacked-out performances, the syndicated production team had to have adequate alternative programming available. They had to know what it was, where it was, and when it would be available. And it had to fit into their format, which was constructed to include commercials. (The funds we raised through the telethon effort were for African Relief. We had to cover production costs up front.)

The syndicated team had four major concerns: to provide comprehensive coverage of the concerts; to insert commercials without interrupting the flow of the concert; to insert alternative program segments so that they would appear to be a natural part of the concert; to restate the show's theme by use of selective interviews supplied by the mother source.

World Feeds

Although much of the programming on the USA and UK World Feeds was shared, the two were separate productions for a number of reasons. The UK World Feed went to the UK and most of Europe and had its own host/presenters. Because certain performers were more popular there than in the USA and elsewhere, some artists who were featured on the UK feed were not seen in the USA, and some seen in both places were given more time on the UK feed. Similarly, the UK production staff felt that certain of the public service announcements (PSAs) would be more effective in the UK and Europe, so the UK feed frequently aired different PSAs. Also, to improve the technical quality of the UK feed, I often ordered them to air certain acts performing in London without beaming them back to the USA for retransmission and subsequent loss of quality. To coordinate all of this, I had the UK producer on a separate headset so that we could communicate without distraction.

There were also language considerations in the World Feeds. Music is called the international language; comedy is not. It is virtually impossible to translate verbal comedy, and it's even harder for people speaking different languages to understand—let alone enjoy—the many idioms and double meanings on which comedy is often based. Because of the language conflicts, the two World Feeds could not use some of the comedy sequences with Chevy Chase and Joe Piscopo, as well as some of the interviews. When I wished to air these segments for the USA audience, the production team for both World Feeds went through the same process of substitution as did the Syndication team. These substitutions, of course, occurred at different times for different lengths of time. They had to be preplanned, timed, and controlled in the same manner as those for the Syndicated Feed. I controlled the "live" time of the comedy bits in the same way I worked with the concert promoters to control the timing of the musical acts.

In chapter 9 you'll find an extended discussion of the internal communications needed on a globalcast. Here, however, you can sense the complexity involved, with the need for my control room to be in constant communication with each of the production teams, with the backstage crews on both continents controlling the flow of talent, with the individual camera operators, with the satellite operations, with the crew coordinating travel between locations, and so on.

The World Feeds had one other critical assignment: providing consistent outcues so that nations around the world could cut out to insert their own local segments. Many of the nations inserted native-language pleas for the Live Aid Fund. Some countries that do not permit such fund raising inserted their own hosts, over whom we had no control.

DiamondVision

The DiamondVision Feed went to giant screens in the USA and UK stadiums. This feed provided coverage for stadium audience members whose view of the stage was blocked by equipment; provided close-ups for the stadium audiences; and provided the UK feed for the USA Stadium audience and vice versa, so that "the show went on" in each stadium even while acts were changing equipment on stage. It also provided me with a production tool on the air.

In retrospect, one of my proudest moments during the sixteen hours of *Live Aid* occurred during Autograf's live performance from the USSR (see Figure 1–8). The DiamondVision screens allowed me to feed Autograf's performance, live, to the USA and UK stadium audiences. When I saw the young people in those countries reacting to the music from the USSR, I realized that a unique moment had been created. By having my switcher split the screen three ways, I was able to show live audiences in the UK and the USA reacting to live music from the USSR. This triple-nation interaction was a magical moment, created with the use of satellites and made possible because we had DiamondVision to carry the satellite images to our stadium audiences.

The Grid

To communicate alternative programming and other needed information, I developed a tool I have used on most of my shows since *Live Aid*. Figure 1–9 shows a page of the grid that I created to block out, minute-by-minute, the programming for each of the individual shows-within-a-show.

Anyone who has worked on any television program knows that communication among members of the production staff is critical to the success of the program. No live show rundown (or format) is final and written in stone. But it is a necessary guide, the foundation upon which all the variations are built.

If you study Figure 1–9, you'll see that it lists:

- The live action taking place on both the UK and USA stages
- The planned feed for each of the five video outputs (DiamondVision, World, Syndicated, ABC, MTV)
- Instructions for the tape operators
- The output from the stadium truck, the master feed for live programming

You can imagine how long it took to plot out this grid covering all sixteen hours of the *Live Aid* production. Each page contained one half-hour of programming, so there was a total of thirty-two pages. I went through many versions, changing it up to a half-hour before we went on the air. The biggest concern in planning was whether the concert promoters, Bill Graham and Harvey Goldsmith, could keep the acts running on time.

In chapters 7 and 8 we discuss in detail the roles of directors and producers. *Live Aid* offers an example of some of the on-air functions of these roles. Each of the production teams responsible for each of the outgoing feeds had its own control room and control room staff. They functioned as any production team would. Using my grid as the basis of available programming, each team had its own rundown and was connected by headset to me.

Figure 1–8 The USSR did not air *Live Aid* domestically, but it let us transmit via satellite a live performance by the Russian group Autograf. All "live" supers were inserted in the upper left-hand corner to avoid competition with IDs in the lower third and supers in the upper right-hand corner, reserved for local insertions.

MASTER SCHEDULE/"LIVE AID CONCERT"/6:30 AM REVISION—TONY VERNA

TIME	ENGLAND STAGE	USA STAGE	DIAMOND VISION	STADIUM TRUCK	WORLD TRUCK	SYNDICATION	FREEZE & VTR	ABC FEED	MTV FEED
AM 11:31	BBC Diamond	Judas Priest	Judas Priest	Judas Priest	Judas Priest	TV Host	VTR:	VTR:	Judas Priest
:32	Germany/	▼	▼	▼	▼	Judas Priest	Germany	FOR	
:33	▼	▼	▼	▼	▼	▼	7:24 PM	8 PM	
:34	▼	▼	▼	▼	▼	▼	World VTPB	TRANS-	
:35	▼	▼	▼	▼	▼	▼	2:41 PM	MISSION	
:36	▼	▼	▼	▼	▼	▼	SYNDIE		
:37	▼	▼	▼	▼	▼	▼	VTPB		
:38	Paul Young	Judas Priest/	Judas Priest/	Judas Priest/	Judas Priest/	Judas Priest/	Allison Moyet		
:39	Allison Moyet	Chevy into	Chevy into	Chevy into	A Enrico	A Keevama	TAPE FOR		Moyet
:40	▼	Vignette #5	Vignette #5	Vignette #5	B	B	DIAMONDVISION		
:41	▼	Mark Gastineau	WATCH	WATCH	Logo (:05)	C	▼		
:42	▼	Tracy Austin	▼	▼	TV Hosts:	Adv #48	▼		
:43	▼	Steve Lundquist	▼	▼	Rolland Smith	Adv #49	▼		
:44	▼	Joe Piscopo	WATCH	WATCH	Sheena Easton	TV Hosts	▼		
:45	▼	lead into	Adv #10		lead into	Dionne Warwick	▼		
:46	▼	Allison Moyet	Adv #11		Allison Moyet	George Segal	▼		Iglesias
:47	▼		Adv #12		▼	Allison Moyet	▼		A
:48	▼		▼		▼	▼	▼		B
:49	▼		▼		▼	▼	▼		Host: Goodman
:50	▼		▼		▼	▼	▼		Quinn into:
:51	▼		▼		▼	▼	▼		Moyet
:52	▼		▼		▼	▼	▼		
:53	▼	Open	▼		▼	▼	▼		
:54	▼	Bill Graham	▼		▼	▼	▼		
:55	▼	Everybody	▼	Open	▼	▼	▼		
:56	▼	Stretch	▼		▼	Moyet	▼		
:57	Young/Moyet/		▼		Moyet	Adv			ADV #30
:58	Geldof welcomes	London/USA	London/USA		MCM #4 (45 sec)	Adv			ADV #31
:59 12:00	Philly	into Jack Nicholson	into Jack Nicholson		TV Host	Adv			

Figure 1–9 To coordinate the six television productions emanating from the World Feed, I devised this minute-by-minute grid from which all producers worked. Because of multiple restrictions on which performers could air when, each producer and tape room operator had to know what was available live and on tape. The grid was supplemented by constant live communication.

Inserting Commercials

By using the grid, I maintained control over the insertion of the commercials. The commercials had to stay in their assigned sectors—they could not slide past that point. I always called down on my alert box to the syndicated producer to tell him when and what to insert. Further, my control room was intimately involved in the Syndicated Feed's getting into and out of commercials. I had to coordinate their needs with the clean (commercial-free) World Feed.

First, I had to program and time live or taped acts to fill out all the time requirements. Second, I had to have electronic control over all of the inputs—both audio and video—to prevent upcutting or cutting out of a performance. To provide that control, the master control switcher needed multiple outputs—clean and dirty, in terms of graphics and sound. My master audio booth had to provide for natural sound and abbreviated natural sound from each feed. For example, after Sting sang "Every Breath You Take," the World Feed might have continued with Sting performing "Roxanne" from his remote inject. But at the same time, I was cuing the Syndicated Feed to cut away to a commercial as soon as "Every Breath You Take" ended.

The audio from Sting's performance was coming in on the only program transmission audio line, the one in my control room. Therefore, that line had to be split to accommodate the mixed-media feeds. In this case, the

World Feed would continue to air Sting singing "Roxanne," but the Syndicated Feed needed us to bleed down the Sting audio so they could mix that live sound level with taped applause (from their control room) for an on-camera lead into commercial. If we couldn't fade the Sting audio cleanly, the syndicated audience would think we were cutting away from Sting, robbing them of a performance to "get in a commercial."

Inserting Taped Performances

To fill the holes left where the Syndicated Feed blacked out restricted performances, videotape playbacks (VTPBs) had to be planned—and timed—to fit. As an example, let me review one sequence, the hour from 2:20 p.m. to 3:20 p.m.

At 2:20 p.m., David Bowie was scheduled to appear live on the stage at Wembley Stadium in London. By contract, that performance was to be restricted in the USA for ABC. Therefore, while the world watched Bowie, the syndicated truck had a VTPB of the performer Sade. Sade had made her appearance at Wembley earlier (9:50 to 10:10 a.m.). During that time, however, Syndicated had carried Black Sabbath, performing live in the USA. The tape room had taped all of Sade's performance for this later playback, and I selected the portion to play back.

At 2:35 the Sade tape ended and Syndicated aired, in sequence, a taped pitch for the Live Aid Fund, the start of the live performance by The Pretenders on the USA stage, two commercials, and more of The Pretenders, live, until the 2:57 break for commercials and station break.

At 3:00 p.m. The Who began their live reunion performance in London—another blacked-out performance. So while the tape room was taping The Who for later playback by ABC, they were also playing back a tape of Spandau Ballet, which aired on the Syndicated Feed from 3:00 to 3:08 p.m. The performance of Spandau Ballet had been pretaped from Japan for all of its appearances on the show. The entire tape had aired earlier (8:45 to 9:03 a.m.) in the UK and on DiamondVision in the USA.

At 3:08 p.m. the Syndicated Feed aired another pitch for the Live Aid Fund (3:00), followed by a commercial pod (2:30).

At 3:13:30 the Syndicated Feed aired a live performance from Germany. It ran until 3:22 p.m. and was followed by another commercial pod.

At 3:23:30 the Syndicated Feed joined-in-progress the performance of Santana, appearing live on stage in the USA.

That summary covers one hour of just the Syndicated Feed. Multiply that by sixteen and you get some sense of the pressure on the production team in the syndication truck. Then multiply that by the other four USA feeds, plus the UK feed, and you get some sense of the number of split-second decisions I had to make in master control.

You can also get a sense of the tremendous demands on the technical staff. First they had to understand all of the electronic paths I needed to maintain the necessary controls. Then they had to design and build those paths, using miles of cable. Even if we hadn't been dealing with a multinational show using thirteen satellites, *Live Aid* would have been a massive technical challenge.

Production Complexity Reflected in Technical Complexity

You can see from this broad sketch of *Live Aid* that it was not an easy assignment. Figure 1–2 shows the "city" of thirty-seven vans and trailers that I designed to handle the technical and production elements. We used

Figure 1–10 One of the many critical job areas in globalcasting is coordinating talent. This video print from a tape of *Live Aid* shows walkie-talkie-armed coordinators easing performer Bette Midler's path to an upcoming interview.

Figure 1–11 Chevy Chase and Joe Piscopo acted as hosts, introducing acts as well as performing. I briefed them, as well as the technicians, on the technical complexity of the multiple feeds. Language barriers restricted airing some of their comedy material in non-English-speaking nations.

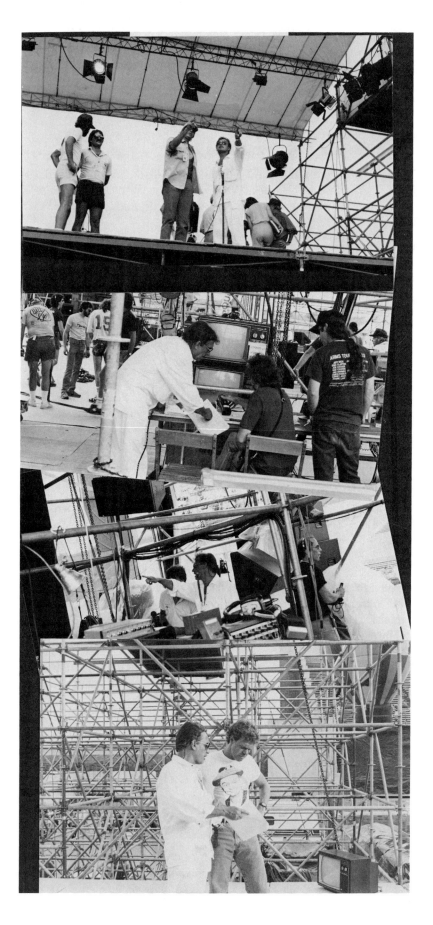

seventeen cameras, including one on the Goodyear Blimp and one capable of traveling on a high-wire over JFK Stadium. To cover all of the elements in the telecast, in the USA I worked with six producers, eight directors, and a dozen assistant directors (ADs) and tape ADs. In the UK there were seven producers and three directors. And I was in touch, of course, with the on-site directors at our remote locations in Australia, the Netherlands, and the USSR.

I had thirty-four monitors in the World Truck and another forty in the USA truck that was attached to it. No single truck contained enough equipment to handle the vast number of inputs and outputs required, so I had two giant control trucks latched together, facing opposite each other, so I could move in a matter of seconds from one control post to the other. Even if the technicians had been able to put all seventy-plus monitors and all of the communications lines into a single truck, I would have had them divide the equipment for psychological reasons. I found that the physical move from one truck to the other helped me create a change of mind and attitude. In those few seconds it took to drop a headset, walk across a platform, look at a new array of monitors, and pick up a new headset, I could shift mental gears, in a way, leaving one show and going to the other. It worked every time, for all sixteen hours.

The entire program was beamed to thirteen satellites. Eleven channels on five Intelsat satellites were among the six satellites used for the international transmission of the event, which was fed to nine locations: Australia, England, Ghana, Hong Kong, Japan, Korea, New Zealand, Taiwan, and the USA. The grid reflected this scope, but the complexity wasn't simply in the seven independent productions. With the two clean World Feeds possibly airing two separate performances, while the dirty Syndicated Feed aired another, while MTV might be airing a fourth and DiamondVision a fifth, there were also moments when one act fed North America live while I fed another act to the Asian market and replayed a tape to the European market. In essence, my switching input to the Indian Ocean or Pacific Ocean satellite created a third World Feed.

As the show went on, it became harder to think out. Here's a sample of my thoughts over just a few minutes.

Do I cut out of Paul Young with Allison Moyet? Let's see, it's 11:55 a.m. MTV had gone into their commercials and an on-camera plea by Julio Iglesias, and they should be back at 11:58 with their announcers [Martha Quinn and Mark Goodman]. Syndicated TV has rejoined after a plea by Preston Keevama and a commercial block with an album cover wipe [a presold item to raise money for the Live Aid Fund] and has aired their announcers [George Segal and Dionne Warwick]. Now comes the "World." The European nations are now being fed Paul Young with the EBU land line, but on the other World transmission over the Pacific and Indian Ocean satellites, we are just coming out of a pledge by Ted Kennedy, Jr., and an on-camera with their anchors [Rolland Smith and Sheena Easton]. Just in the nick of time, Young and Moyet finish. Okay, get Jack Nicholson standing by. We'll split screen him with Bob Geldof in London.

Also, in the last-minute technical rush, I got input from all of the sources I needed, but I wasn't able to receive it on a single headset. As it turned out, that was a blessing. Rather than possibly being overwhelmed by input from dozens of others simultaneously, I was able to control communication by my choice of headset. Outgoing communication was no problem: I had "alert boxes" as well as ring-downs to each of the other trucks. The

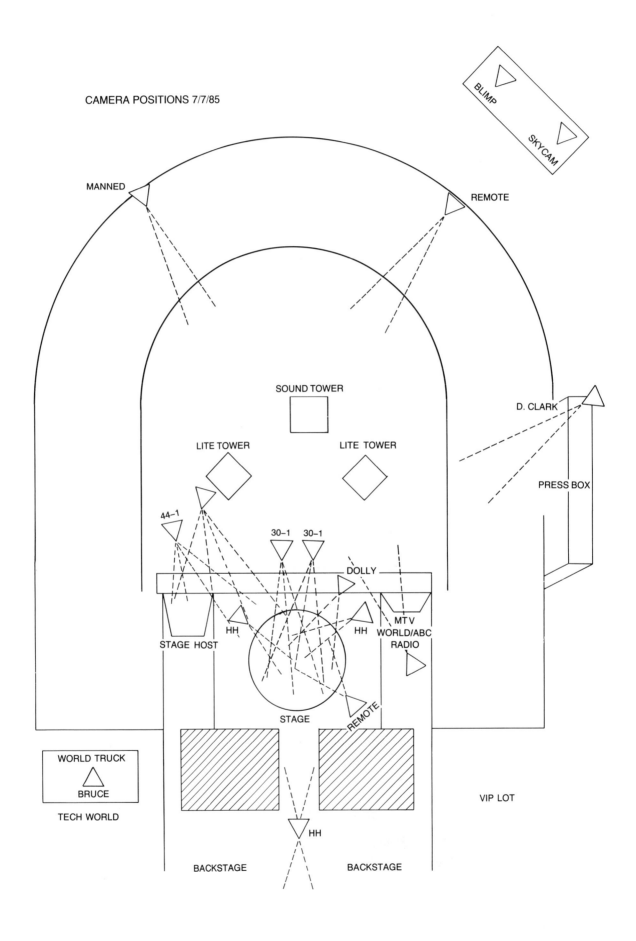

CAMERA POSITIONS 7/7/85

ring-downs let me establish instant two-way communication, a normal phone call. The alert boxes were one-way—I talk, you listen; you can't talk back—a necessary tool on shows of this magnitude.

Putting *Live Aid* Together

Perhaps most astounding about *Live Aid,* aside from its worldwide impact, is that it was created in only ten weeks. From the moment Bob Geldof approached Mike Mitchell, president of Worldwide Sports and Entertainment, on May 5, we had a ten-week window in which to prepare for the July 13 globalcast. Co-executive producer with Mitchell, and my co-producer, was Hal Uplinger, executive vice-president of Worldwide. Through Hal Uplinger the world was contacted. There was initial reluctance in Europe, where the first reaction was to play the show an hour, perhaps two. Uplinger told them, "You don't understand. This is a *live* concert for sixteen hours." Europe was skeptical, but eventually eleven countries agreed to do telethons to raise funds for the Live Aid Fund.

Live Aid was as expansive and complex as you could imagine in 1985, but you can't allow the complexity of a production to overwhelm either the production team or the audience. On the contrary, the more complex the telecast, the simpler the communication must be.

The goal in telecasting *Live Aid* wasn't the mere presentation of music and performances, but rather to fashion the sixteen hours of music and stars so that when the songs ended, the message would linger on. The rest of this book is dedicated to outlining the tools and techniques used on *Live Aid* and subsequent globalcasts to accomplish maximum worldwide communication. The single most powerful tool we have—the one that made globalcasting a reality—is satellite transmission. In chapter 2 we discuss satellites, fiber optic cabling, charge-coupled devices, and the shift from analog to digital technology.

Figure 1–12 This preproduction sketch of camera positions for J.F.K. Stadium in Philadelphia details many of the positions discussed in the text. Bruce camera refers to a stationary, unmanned camera position.

Part *Two*

The Spine of Global Television

Satellites as Production Tools

Communication Satellites in Geosynchronous Orbit

Live Aid could not have happened without the battery of communication satellites in geosynchronous orbit. It was these astounding electronic "birds" soaring some 22,300 miles above the equator that made global-casting possible. This unique satellite technology was born forty years before *Live Aid,* during World War II.

World War II underwear, World War II rockets, and 1990s television may not appear to be related, but war and television communication have been, and continue to be, inexorably linked. During World War II German soldiers in the field dried their white underwear on tree limbs. It was visible on allied military intelligence aerial photographs. During the same war, the Japanese sent aloft 9,300 unmanned Fu-90 balloons, each armed with an explosive payload. The jet stream carried the balloons over the 6,500-mile stretch of the Pacific Ocean into North America. The balloons peppered eighteen states, from Hawaii to Michigan.

How do wartime balloons and underwear relate to communication satellites? In 1945, Arthur C. Clarke, a twenty-eight-year-old radar officer in the Royal Air Force, wrote a letter to a British magazine, *Wireless World,* outlining the value of rockets for research in space. He detailed how three repeater space stations, 120 degrees apart and in the correct orbit, could provide television and microwave coverage for the entire planet. These "artificial satellites," wrote Clarke, would have to be placed in a specific orbit so they could make one revolution every twenty-four hours and thereby remain in the same relationship to earth and be within visual reach of approximately half the earth.

Others before him had imagined satellites in space, but it was Clarke who suggested putting them in geosynchronous (or geostationary, or Clarke) orbit. Others after Clarke, such as his friend Eric Burgess, suggested unmanned—as opposed to manned—geostationary satellites. A satellite put in this orbit, 22,240 miles (35,780 kilometers) above the equator and revolving at the same rate of speed as the earth's rotation, will appear to remain stationary. Because its orbital speed matches the rotational speed of the earth, it will always be in the same spot in relation to earth and will, therefore, always be available to receive and send back signals transmitted to it. (Actually gravitational forces from the sun, moon, and earth cause man-made satellites to drift slightly, but the drift is both predictable and controllable.)

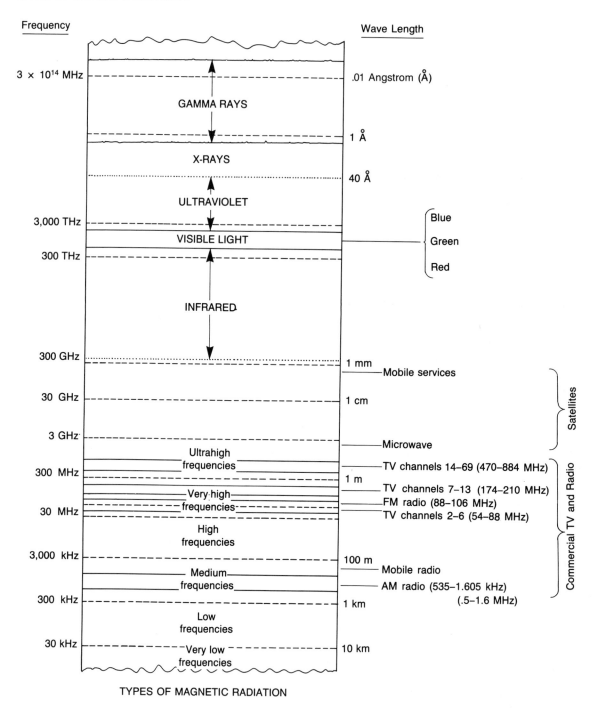

TYPES OF MAGNETIC RADIATION

Figure 2–1 The technical term *radio* encompasses both audio (radio) and video (television) and means transmitting/receiving electromagnetic radiation waves in the radio frequency. Information can be encoded on a radio wave by interrupting its transmission (telegraphy) or by modulation. Amplitude modulation (AM) maintains a constant frequency but varies the intensity or amplitude. Frequency modulation (FM) maintains a constant amplitude but varies the frequency of the signal. Using these methods, the audio and video are converted into electrical signals by microphones and cameras, amplified, then transmitted by an antenna that converts the electrical signals into electromagnetic waves that travel through space at the speed of light.

The wavelength of an electromagnetic wave is the distance between successive crests (or successive troughs). One full wavelength (one crest and one trough) is a *cycle*. The number of cycles is measured in Hertz, 1 Hertz representing one cycle per second. One kiloHertz is 1,000 cycles per second; 1 megaHertz is 1 million cycles per second; 1 gigaHertz is 1 billion cycles per second; and 1 teraHertz is 1 trillion cycles per second.

When producers became aware of the powers of geosynchronous satellite feeds, they became an integral part of program planning, as we saw in the integration of live and taped *Live Aid* segments from around the world. The term applied to these integrated feeds, regardless of the medium of transmission, is *backhaul.* Technically, backhaul means transmitting a signal from point-of-origination to a second terrestrial point—by cable, satellite, or microwave—before broadcasting it for distribution. An example is backhauling action from a sports event game site to a studio. Usually the purpose of backhauling is to add something to the signal before distribution, for example, a studio host or commercials. In the section on news we discuss how the USA networks' budget-cutting plans included keeping their news anchors in their New York studios for election coverage. In this example, convention coverage would be backhauled from the convention site to the studio, where the anchor would be added to the outgoing video.

Satellites were used effectively before *Live Aid*. Newscasts began using satellite feeds as early as the 1960s. For some time each broadcast by satellite carried a "super" identifying it as "Live by Satellite," but before long the supers disappeared. The same restrictive process occurred when I created the instant replay. In the beginning the network insisted that a "Videotape Replay" super be added to each replay, but gradually the super disappeared. Similarly, live satellite news feeds not only became acceptable, they became expected. During coverage of global stories like summit meetings, hostage releases, plane hijackings and bombings, and Operations Desert Storm and Restore Hope network camera operators were supplemented by hundreds of local news reporters feeding back live satellite coverage.

To understand the use of satellites in global—or local—television, you have to understand: how satellites work, what territory each satellite covers (its footprint), how to plan satellite back-ups, how to book time on satellites, the cost of satellite time, and time delay and mix-minus.

Chapter 13 deals with satellites as transmission tools—getting the message out. Here we deal with them solely as production tools—getting parts of the message in.

How Satellites Work: Bands, Uplinks, Downlinks

The magic of satellites, of course, is that they allow instant video to accompany long-established instant audio. The power of the picture is indisputable, as has been proven time and time again in commercial, military, diplomatic, and other applications. But satellites are not philosophical tools. They are mechanical transmission tools. They are "repeaters." They pass along whatever message we as communicators send. It's our use that determines their value.

Much of the strength of satellite communication comes from our use of it as a production tool. Just as I turned the recording device of videotape into the production tool we call instant replay, so television production quickly incorporated the power of satellite transmission into the producer's bag of tricks. I had the privilege of directing the first international use of a satellite for television (never aired) in 1962. Our satellite carrier, Telstar I, had been launched into low earth orbit for emergency testing and provided the first video service across the Atlantic Ocean.

Let me emphasize the need to keep the power of satellites in

APPROXIMATE WIDTH OF ELECTROMAGNETIC WAVELENGTH BANDS IN GIGAHERTZ (see note below)	LETTER DESIGNATION
22.2 to 60.0	E-Band
10.7 to 22.2	K-Band
10.95 to 12.75	Ku-band Downlink
14.0 to 14.5	Ku-band Uplink
18.3 to 22.2	Ka-band
6.5 to 10.7	X-Band
3.4 to 6.425	C-Band
3.7 to 4.2	C-Band Uplink
5.925 to 6.425	C-Band Downlink
1.7 to 3.0	S-Band
1.0 to 1.17	L-Band
0.3 to 1.0	U-Band

Figure 2–1 *(continued)*

To keep these signals separated so they don't interrupt and distort each other, the wavelength frequencies are divided into bandwidths, which are assigned letter designations. Most communication satellites currently operate in the C-Band and/or Ku-Band.

These bandwidths are measured in Ångstroms (Å), for Swedish physicist Anders Jons Ångstrom. One Ångstrom unit equals 1 hundred-millionth of a centimeter (10 to -10 m) 10^{-10}, now more commonly referred to in nanometers (nm). One nanometer equals 10 Ångstrom. Both charts show low frequences and wide bandwidths at the bottom.

L-Band and U-Band frequencies are limited generally to Inmarsat communications with ships at sea. The two Voyager interplanetary satellites were designed to use S-Band on uplink signals, both S-Band and X-Band on downlink. For commercial comsats, C-Band and K-Band dominate.

Two notes: (1) different sources use different numbers for the same bandwidths. From reliable sources I have compiled as many as three alternate sets of numbers for some bands. *Always* check with a satellite specialist. (2) As the FCC and worldwide agencies shift bandwidths for HDTV and other new technologies, these numbers may change. There is no "final answer." Always seek current, professional information.

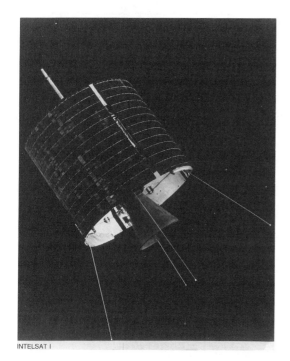

INTELSAT I

Figure 2–2 Intelsat I (Early Bird shown) was launched April 6, 1965, and began service June 28, 1965, in the Atlantic Ocean Region (AOR). The first Intelsat II Satellite was deployed in early 1967 for service in the Pacific Ocean Region (POR). Intelsat II was twice as powerful as Intelsat I and introduced multipoint communications capability between earth stations in the area of coverage, allowing several earth stations to access the satellite simultaneously. Intelsat III had five times the capacity of Intelsat I and II. The first Intelsat III was launched 1968 in the AOR. The second became Intelsat's first Indian Ocean Region (IOR) bird.

perspective. Like the other production facilities, satellites do not create good programs. No matter how ingeniously established or connected, cameras, cables, and satellites do one thing—transmit messages. The messages are only as good as the programming provided. Of course, in some cases talent triumphs over transmission, as in the show-stopping Mick Jagger–Tina Turner duet on *Live Aid,* which was staged basically to a single camera.

In Chapter 5 we look in depth at electromagnetic wavelengths, analog and digital signals, and how digital compression is greatly expanding the capacity for satellite—and cable—transmission. Here we are concerned only with the broad picture: how satellite signals work and how they get from one place to another. Also, we deal here only with communication satellites involved in television transmission. There are hundreds of other satellites used for business, military, weather, and other transmissions. These satellites often operate on different bandwidths. Most communication satellites operate on C-Band or Ku-Band.

The Satellite Bands: What You Need to Know

Figure 2–1B illustrates the approximate bandwidths assigned to various bands for satellite signal transmission. (You will find that different experts cite different numbers.) Orbital positions for satellites are assigned to individual nations by agencies within the United Nation's International Telecommunications Union (ITU). Until all the new technologies are available, it's important for globalcasters to be aware of certain characteristics of the satellite bandwidths.

C-Band and Ku-Band are the bands currently used by virtually all communication satellites (comsats), although the Ka-Band is increasingly used. Weather and military satellites use different bands. (It was wartime uses for radar that led to microwave radio transmission, and the need for secrecy that gave rise to digital transmission formats, which formed the basis for the digital revolution taking place in civilian—specifically television—communication.)

However, the importance of the difference between C-Band and K-Band to globalcasters is the potential effect of outside influences on signals transmitted by satellites using those bands.

Communication Satellites: C-Band, Ku-Band

C-Band and K-Band are different portions of the radio frequency spectrum allotted to satellite transmission, and it is the variations created by the differences in spectrum that can affect globalcasts.

The C-Band was the band used for the earliest comsats and is still the most-used for that purpose. Each C-Band comsat is given 500 megahertz (MHz) of space of the total allocated space on the spectrum, but because the C-Band operates in the same region of the spectrum as microwave, there are regulations that limit C-Band use. To limit interference with terrestrial microwave systems, the downlink power density of C-Band satellites is limited to between 38 and 48 decibels/watt of power (dbW), depending on the elevation angles. In practical terms this limitation means that a large antenna receiver is necessary to pull in C-Band transmission.

K-Band satellites, on the other hand, do not operate in the same region as microwave systems, so there is no limit on their downlink power density. The wider bandwidth and uncontrolled power density on the downlink make signals from K-Band satellites easier to receive, which means the earth antennas receiving Ku-Band or Ka-Band signals can be smaller. This ability to use smaller antennas, in addition to the lack of interference from microwave systems, makes the K-Band the overwhelming choice for satellite news gathering (SNG) services. And, it is the combination of smaller home satellite dishes and channel expansion via digital signal compression that will make direct broadcast by satellite (DBS) a serious factor in television's future. Another predictor of expanded Ku-Band usage was WARC's 1992 addition of an extra 250 mega Hertz (MHz) to the Ku-Band, increasing the amount of available bandwidth by fifty percent.

There are, however, a couple of drawbacks to K-Band satellites. Because of their higher power, charges for renting K-Band transponders are higher, often more than double the rates for C-Band. The other drawback is the effect of weather. Whereas, rain has little effect on C-Band transmission, it can cause serious degradation of K-Band transmission (see the discussion on fiber optics in Chapter 3). If a globalcast is using K-Band transmission, especially in a tropical area subject to intense electrical rainstorms, it would be wise to back up any critical feeds in case of a storm during the telecast. This is true for both the Ku-Band and Ka-Band satellites, being used increasingly for communication. It is expected that technology now in development will soon provide automatic rain-fade compensation for signals transmitted in the K-Band. Sensors comparable to those that adjust light levels for video on the Jumbitron in Times Square, New York, will react to signal-fade caused by rain and kick in added power to compensate. But for the foreseeable future globalcasters have to be aware of potential rain-fade problems in K-Band transmission.

There is a further weather phenomenon—a problem for both C-Band and K-Band satellites—that globalcasters need to be aware of: Sunspots. There are periodic and predictable times, which vary by geographic location, when sunspot activity occurs and causes outages in satellite signal transmission. Satellite specialist Dave Herman, of On Call Communication, describes the situation as a point at which the sun is "looking over the shoulder" of the satellite, aiming into the dish antenna base. Although the problem cannot be prevented, Comsat and satellite manufacturers publish timetables of expected outages, which can last as long as five to fifteen minutes. Predicted sunspot outages can be a determining factor in deciding when to air a globalcast.

There are also occasional massive occurrences of sunspots, known as *solar storms*.

One further note on these two bands: there are already in orbit several communication satellites that house transponders with multiple bands. With these satellites, called *hybrids* or *tribrids*, it is possible to use a different band for uplinking and downlinking of signals. The process of interconnecting the different bands is called *cross-strapping*.

What and Where the Satellites Are

There are many satellites available for use in television production—more than two hundred in geostationary orbit and a number in other orbits—all of which function on the same basic principles and provide the same services.

Figure 2–3 Intelsat IV (F-2) was launched in the AOR in January 1971, with capacity for 4,000 telephone circuits plus simultaneous television. The Intelsat IV-A series, launched beginning in the mid-1970s, provided 6,000 circuit capacity. Six IV-A satellites were launched before 1980, when the first of the Series V was launched. Fifteen V (shown) and V-A satellites were launched (two unsuccessfully) prior to 1989, when Intelsat celebrated its twenty-fifth anniversary. Ku-Band spot beams became available for the first time with the Intelsat V series. Note the widely extended solar power panels.

Figure 2–4 Intelsat VI closed the 1980s, and the Intelsat VII (shown) series began the 1990s, both providing nearly 200 times the capacity provided by Intelsat I. Intelsat expanded its capacity in the 1980s, with lease capability (as opposed to occasional use) becoming available. The 1990s and Intelsat VII brought spacecraft that could be configured to meet client needs. Increased power on the newer satellites permitted use of small fly-away uplink antennas, permitting live television coverage anywhere in the world.

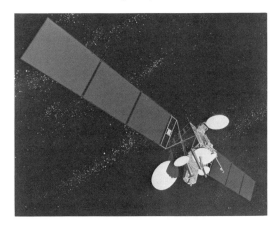

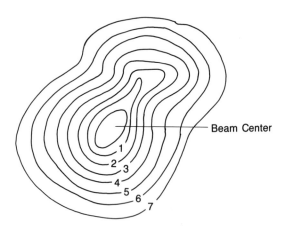

— Beam Center

Figure 2–5 The contour lines of a satellite footprint map represent the graduated strengths of the signal coming from a satellite transponder. These strengths are expressed in decibels as they relate to a single watt of power (dbWs).

The signal strength is called the satellite's effective isotropic radiated power (EIRP), derived by mathematical calculations involving the gain of the satellite's transmitting antenna, the power of its amplification, and loss factors connected with equipment and/or distance.

On this simulated footprint, the beam center's EIRP strength might be 54.7 dbW. The first contour strength might drop to 53.7 dbW. Each succeeding contour, farther from the center, might diminish by one point until the seventh contour's strength is only 47.7 dbW.

The Intelsat Telecommunications Satellite Organization is a nonprofit coalition of nations providing satellite services. Combined, Intelsat and the East European/Asian Intersputnik system provide global coverage. Figures 2–2 through 2–4 illustrate the progress in the development of Intelsat's satellites.

Satellite Construction

The major elements of a communications satellite are the housing and the payload.

The Housing The most visible parts of the housing are the solar power cells. These frequently wing-like contraptions capture and convert the sun's power into energy (some of which is stored in batteries) to run the satellite's mechanisms. The space vehicle, or bus, houses the hardware used by the earth station tracking, telemetry, and control systems to keep the satellite on course and functioning.

The Payload The heart of the communications satellite, the payload consists of antennas (two sets), receivers, and transponders. One set of antennas receives signals from earth (the uplink signal). These are separate from the space vehicle's antenna, used for housekeeping purposes, and from the transmitting antennas.

The receivers relay the signals from the receiving antennas to the transponders. The number of transponders may vary, and the number of signals a transponder can handle can be greatly increased by a number of techniques, including digital compression of the signal. The transponders perform two functions: they shift (lower) the frequency of the signal to prevent interference between the uplink signal coming from earth and the downlink signal going back to earth; and they amplify the signal, increasing its power for its trip back to earth.

The second set of antennas, the transmitting antennas, beam the shifted, amplified signal to its downlink station on earth.

Signal Transmission

This sequence dictates that you have four concerns regarding satellite transmission as part of a TV production: getting the program signal from its source to an uplink station; sending that signal up to the satellite; receiving the signal back from the satellite; and getting the signal from the downlink station to the control room for integration into the overall production.

Uplinks The first step in getting your signal to an uplink station is contracting for the facilities agreed to in your budget/production planning. The common carrier closest to the control room is where all of the input signals must arrive and normally where all of the output signals originate. Usually the common carrier is the local telephone company, but there are companies that specialize in booking satellite and uplink/downlink facilities. Satellite transponder time is booked in fifteen-minute segments, at costs ranging from 200 to 800 USA dollars per hour, depending on the time of day (night rates being lower).

Once satellite transponder time is booked, the second step is to arrange how to get your program signal to the designated uplink station. This is done in the least expensive, easiest way, whether by cable (coaxial or fiber

LONDON UNIVERSAL COORDINATED TIME (UCT)	NEW YORK/ WASHINGTON	CHICAGO/ DALLAS HOUSTON MEXICO CITY	LOS ANGELES	SYDNEY/ MELBOURNE	TOKYO SEOUL	HONG KONG SINGAPORE	NEW DELHI	MOSCOW	JOHANNESBURG ATHENS HELSINKI ISTANBUL	MADRID/PARIS ROME/VIENNA BRUSSELS BERLIN	BUENOS AIRES RIO
100	2000	1900	1700	1100	1000	900	650	400	300	200	2200
200	2100	2000	1800	1200	1100	1000	750	500	400	300	2300
300	2200	2100	1900	1300	1200	1100	850	600	500	400	2400
400	2300	2200	2000	1400	1300	1200	950	700	600	500	100
500	2400	2300	2100	1500	1400	1300	1050	800	700	600	200
600	100	2400	2200	1600	1500	1400	1150	900	800	700	300
700	200	100	2300	1700	1600	1500	1250	1000	900	800	400
800	300	200	2400	1800	1700	1600	1350	1100	1000	900	500
900	400	300	100	1900	1800	1700	1450	1200	1100	1000	600
1000	500	400	200	2000	1900	1800	1550	1300	1200	1100	700
1100	600	500	300	2100	2000	1900	1650	1400	1300	1200	800
1200	700	600	400	2200	2100	2000	1750	1500	1400	1300	900
1300	800	700	500	2300	2200	2100	1850	1600	1500	1400	1000
1400	900	800	600	2400	2300	2200	1950	1700	1600	1500	1100
1500	1000	900	700	100	2400	2300	2050	1800	1700	1600	1200
1600	1100	1000	800	200	100	2400	2150	1900	1800	1700	1300
1700	1200	1100	900	300	200	100	2250	2000	1900	1800	1400
1800	1300	1200	1000	400	300	200	2350	2100	2000	1900	1500
1900	1400	1300	1100	500	400	300	50	2200	2100	2000	1600
2000	1500	1400	1200	600	500	400	150	2300	2200	2100	1700
2100	1600	1500	1300	700	600	500	250	2400	2300	2200	1800
2200	1700	1600	1400	800	700	600	350	100	2400	2300	1900
2300	1800	1700	1500	900	800	700	450	200	100	2400	2000
2400	1900	1800	1600	1000	900	800	550	300	200	100	2100

Figure 2–6 The earth's twenty-four time zones are determined by how far distant a particular part of the the world is, east or west, from Greenwich, England. Adjacent time zones are differentiated by one hour, but local legislation may dictate different times within countries and/or states within countries. International satellite services are ordered using Universal Coordinated Time (UCT), formerly called Greenwich Mean Time.

The chart lists time differentials in standard, versus daylight savings, time. The latter globally shifts times.

Globalcasters often must coordinate inputs from many different time zones and frequently must work around the clock, making such information vital.

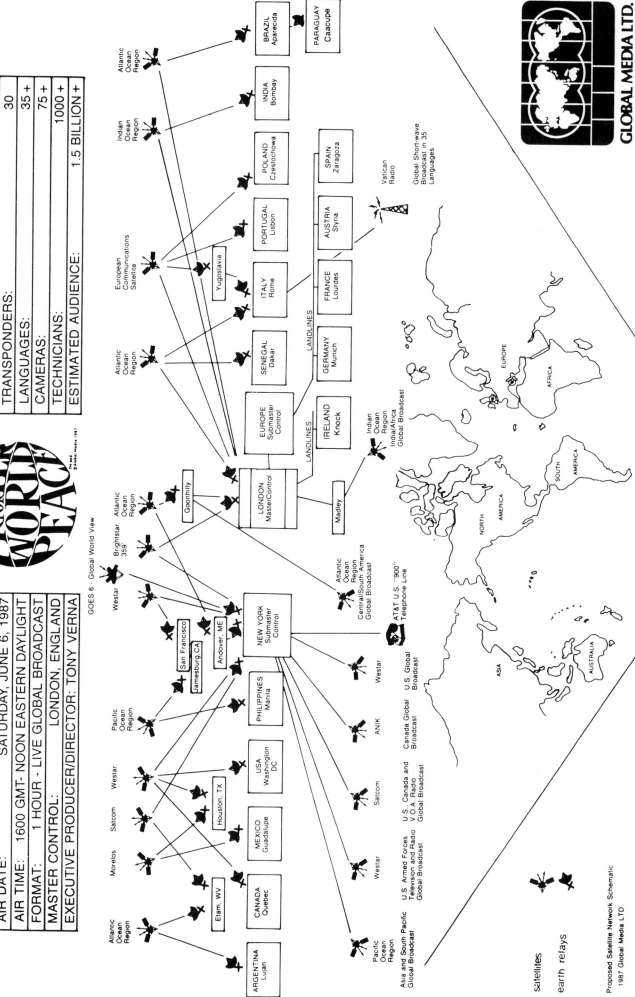

PROGRAM:	"PRAYER FOR WORLD PEACE"
AIR DATE:	SATURDAY, JUNE 6, 1987
AIR TIME:	1600 GMT- NOON EASTERN DAYLIGHT
FORMAT:	1 HOUR - LIVE GLOBAL BROADCAST
MASTER CONTROL:	LONDON, ENGLAND
EXECUTIVE PRODUCER/DIRECTOR:	TONY VERNA

SATELLITES:	18
TRANSPONDERS:	30
LANGUAGES:	35 +
CAMERAS:	75 +
TECHNICIANS:	1000 +
ESTIMATED AUDIENCE:	1.5 BILLION+

GOES 6 - Global World View

satellites

earth relays

Proposed Satellite Network Schematic
1987 Global Media LTD

GLOBAL MEDIA LTD.

optic, whatever is available), by microwave, or by another satellite hookup. As you will see, we used all of these methods for *The 1990 Goodwill Games*.

Next, an uplink facility is identified. It may be as large as a building or as small as a suitcase and may be a mobile satellite news gathering (SNG) truck. The uplink station for global transmission is called an international gateway. Whatever their size, all uplink stations perform the same basic functions. Figures 13–2 and 13–3 diagram the pattern for the uplink and downlink stations. In essence, these stations process and amplify the television signal, the uplink to make the signal strong enough to reach and activate the satellite transponder, the downlink to make the signal ready for reception. The uplink station adds *pre-emphasis* to the signal, creates a subcarrier for the color data, and separates the audio portion of the signal onto a subcarrier frequency. Usually, it also creates other audio subcarriers to transmit other data, such as cues. The satellite transponder receives the uplink signal, downshifts its frequency, and amplifies it again for the trip back to earth.

Downlink The downlink station, whether professional or a small television receive only (TVRO) station for home reception, reverses the uplink process. It collects the signal from the satellite, amplifies it (it has been weakened by its 22,300-mile journey), and removes the pre-emphasis placed on it at the uplink station. While it is readjusting the signal's frequency, it also filters out any extraneous signals.

Note, however, that satellite signals, like television signals, travel in straight lines. They can be deflected by intervening objects, such as buildings or trees. They also weaken as they travel over distance. In addition, signals are weaker at the fringes than at the center. Because of these characteristics, satellite signals have to be aimed at downlink stations (and vice versa) for proper reception. This relationship between downlink stations and the satellite will become clearer when we discuss satellite footprints.

Figure 2–7 (opposite) Details coordinating this satellite schematic with the video paths used on *Prayer For World Peace* appear in chapter 9. Note that two separate satellites, one for back-up, carried the live signal from Rome. I was able to cut from a wide shot on one satellite to a close-up on the second satellite as easily as cutting from one studio camera to another.

Courtesy Global Media Ltd.

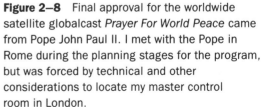

Figure 2–8 Final approval for the worldwide satellite globalcast *Prayer For World Peace* came from Pope John Paul II. I met with the Pope in Rome during the planning stages for the program, but was forced by technical and other considerations to locate my master control room in London.

Figure 2–9 This video print shows one of the wide shots I used on *Prayer For World Peace* to cover the recitation of the Rosary from St. Mary Major in Rome. This was a regularly scheduled ceremony, occurring the first Saturday of each month, normally broadcast only by Vatican radio.

Figure 2–10 The Pope recited his portion of the Rosary in English, French, German, Portuguese, and Spanish and was able to see as well as hear live responses from parishioners on five continents around the world.

Master Control Getting the signal to master control is the final stage. Using the best and cheapest routing via cable, microwave, or satellite, you arrange to have the signal sent to your master control for incorporation into the production and eventual retransmission to the audience. Unfortunately, routing paths are not always this neat, as you will see in the analyses of *Live Aid, Sport Aid,* and *Prayer For World Peace,* the most complex globalcasts produced. But the basic pattern is the same for any production. Molding the equipment and paths to a show's needs is one of the jobs of the production team.

How Satellites Are Managed

To determine which satellite(s) you need for a production, you have to know the location(s) involved in the production and which satellite(s) can receive from and send to these areas. Satellite contractors plot the areas their satellites cover on maps and call them footprints. Figure 2–5 is an example of such a plot. Before describing the footprints in detail, let's look at how satellite systems are organized.

Earlier we noted that Arthur Clarke determined that by spacing a small number of geostationary satellites in the proper orbits, signals could be received and sent around-the-clock from anywhere on earth (with an uplink station) to anywhere on earth (with a downlink station). Now that there are hundreds of satellites in orbit, coverage can be pinpointed with great accuracy, the only limitations being budget and availability of rental time on the satellites you need.

Organization: Trunks and Branches

When planning a production using satellites, you will turn to professionals to determine the available time/space. But it's critical to understand the overall structure of satellite transmission, for two reasons. First, you can't plan a show intelligently unless you know the possibilities, just as you can't plan camera coverage without knowing how much floor space is available, how many hand-helds are available, and so on. Second, if a satellite quits functioning while the show is on the air, there is no time to study satellite paths to conjure up alternate routings. The redesigned switching pattern will come from a satellite expert, but you need to understand the redesigned patterns.

A convenient way to describe the existing satellite networks is to compare them to telephone systems. The big satellite systems—Intelsat and Intersputnik—are used for trunking purposes. Reaching out from them are many branches. With proper planning and availability, these trunks and branches can be interconnected to produce virtually any pathing a show's format demands and budget will allow.

International Satellite Organizations: Intelsat and Intersputnik

The Intelsat trunk is based on Clarke's 1945 proposal. By placing satellites over the Atlantic Ocean Region (AOR, in 1965), the Pacific Ocean Region (POR, in 1967), and the Indian Ocean Region (IOR, in 1969), Intelsat was the first to establish true *global* audio/visual communication. Intelsat now has multiple satellites over each of these regions. By tying into this trunk, any television production, or any other satellite system, can reach virtually any downlink earth station in the world.

At the time of its formal creation in 1971, Intersputnik was kind of an Eastern Bloc response to Intelsat. Having also launched a series of geostationary satellites, Intersputnik provided similar global coverage. It established a geostationary system with the Raduga, Ekran, and Gorizont satellites, but, for geographical reasons, it needed more than that. As noted earlier, satellite signals travel in a straight line. But the earth's surface curves, and the farther a downlink earth station is from the equator the weaker the signal it will receive from a geostationary satellite. Because some of its vast territory stretched north, out of direct-line contact with the geostationary satellites, the USSR orbited an additional group of satellites (the Molniya series) into nongeostationary, elliptical, north/south orbit. These were and are included in Intersputnik's coverage and extend coverage as far north as the Arctic Circle.

Established as an international satellite organization open to all, Intersputnik swelled from an original nine nations (highlighted with an *) to fourteen: Afghanistan, Bulgaria*, Cuba*, Czechoslovakia*, German Democratic Republic*, Hungary*, Korean People's Democratic Republic (North Korea)*, Laos, Mongolia*, People's Republic of Yemen, Poland*, Rumania, Socialist Republic of Vietnam, and the USSR*. The realignment of political boundaries will change the role of Intersputnik gradually over the next few years starting with the thirteen-member broadcast group in Eastern Europe (OIRT), absorbed in 1993 by the European Broadcast Union (EBU), a thirty-two-member counterpart in Western Europe.

Figure 2–11 Fatima, Portugal, was one of the sixteen cities linked to Rome by satellite audio and video circuits. See Figures 2–12 and 2–13 to see how Fatima was worked into the quads and super quads used during the program.

Regional and National Satellite Systems

One organizational level below Intelsat and Intersputnik, and interlocking with them, is a series of regional satellite organizations, some public and formal, some private and informal. Below these regional organizations come satellite systems operated by countries or private companies for domestic use.

Other Satellite Organizations: Comsat and Inmarsat

One gradually becomes aware, when considering the global satellite system, that despite international cooperation and planning much of the system grew with little preplanning. There *is* a structure to the system. It developed gradually and was eventually administered by the United Nations (UN).

The International Telecommunications Union (ITU), an agency of the UN, is the overall governing unit for worldwide services. Two groups within the ITU have been authorized to develop and control policy-making decisions. These groups, the World Administrative Radio Conference (WARC) and the Regional Administrative Radio Conference (RARC), collectively allocate use of the limited space, both physical space and bandwidth, for orbiting satellites. By treaty, each member nation is assigned a number of orbital slots. Each individual country, in turn, has set up a national body (or bodies) to regulate/distribute its slots. In the USA that organization is the Federal Communications Commission (FCC). But

Figure 2–12 To control and use the seventeen live television signals inputting to London, I used quads—four separate signals transmitted as one. See the text for detailed explanation.

Figure 2–13 This super quad, showing sixteen live signals simultaneously, was used at the end of the globalcast. Its impact was enhanced by the sound of bells ringing at all of the churches.

there are quasi-governmental groups involved also: Comsat, a national organization, and Inmarsat, a global nonmilitary, mobile satellite communications system.

Comsat A good example of how satellite organizations "just grew" is Comsat (COMmunications SATellite General Corporation), which was created in 1962 when the US Congress passed and President Kennedy signed into law the Communications Satellite Act. This act established the USA's policy on the creation of Intelsat and authorized the formation of Comsat as a private company—but with obligatory government representation—to represent the USA to Intelsat and to manage earth stations to provide access to the Intelsat system. Comsat still represents the USA on the Intelsat board and the Inmarsat council, but its powers have gradually eroded, even as its nongovernmental activities have swollen.

In the 1980s, as part of President Reagan's administration's emphasis on deregulation, private companies were given limited permission to apply for satellites to handle global television traffic in competition with Intelsat. At the same time, Comsat sold its satellite earth stations to a number of private USA common carriers. As part of the sales agreements, a series of *gateways* was set up to provide access to Intelsat's global facilities. It is through a common carrier, through these gateways, that global television productions arrange for satellite transponder space.

While its quasi-governmental functions were shrinking, however, Comsat's commercial enterprises were expanding. It launched a series of Comstar satellites that were leased to AT&T before AT&T launched its own satellites. In the early 1980s a Comsat subsidiary built satellites to operate a direct broadcast by satellite (DBS) system, but the financing fell through and the Ku-Band satellites were never launched. During the tenure of president and CEO Robert Wussler, Comsat showed renewed interest in DBS transmission in addition to its other commercial enterprises, but Wussler's resignation in January 1992 put Comsat's future interest in DBS in question.

Inmarsat Inmarsat (INternational MARitime SATellite Organization) is a global satellite communications network run by forty-eight member nations and headquartered in London. It is a nonmilitary, mobile satellite communications system that serves all ships, remote oil rigs, remote land sites, and so on with data, facsimile, telephony, and telex transmissions. It uses both C-Band and L-Band for shore-to-ship and ship-to-shore transmissions. Frequencies for these and other bands are shown in Figure 2–1.

Comsat is not only the USA's representative to Inmarsat; it helped create the system. Prior to the establishment of Inmarsat in 1979, Marisat, a multinational group brought together by Comsat, operated a series of satellites. When Inmarsat was formed, it took over the Marisat satellites.

Inmarsat operates on both the C-Band and L-Band. Although it is not normally connected to television transmission, it is important to know about this system. In 1985, when a massive earthquake knocked out all terrestrial telecommunication systems in Mexico City, it was a combination of suitcase satellite uplinks and Inmarsat's C-Band that brought detailed news of the city's disaster to the world. CNN's audio line out of Baghdad in 1991 traveled via an Inmarsat terminal. In 1992 a flyaway

Ku-Band uplink used an Inmarsat terminal to carry out news from Sarajevo in Bosnia-Herzegovina. A one-person news reporter with carry-on equipment often has easier passage through customs, roadblocks, etc.

Note, too, that although the USSR/Commonwealth of Independent States has long been a member of Inmarsat, it has operated a parallel system, called Morya ("Seaman").

Satellite Footprints

Now that we have looked at the structure of communication satellite systems and the electronic paths necessary to access them, let's turn to the actual coverage provided by the satellites—their footprints.

Depending on their location, dish owners can receive signals from multiple countries. Living in the Los Angeles area, I can receive signals from Mexican and Canadian satellites, in addition to a large number of CONUS satellites. And although I can't receive signals from Atlantic Ocean satellites, I often receive European signals apparently being relayed from an east coast downlink to a west coast station. For example, I can watch RAI's news in Italian. It is both fun and educational to tune into multinational sources; newscasters on Morelos (Mexican) and Anik (Canadian) satellites do not always share the same political views as US newscasters. But beyond that, for globalcasters footprints are significant because they can influence the focus of the rented transponder.

Because satellite beam patterns arrive at the earth in a straight path, they are strongest at the center, weakest at the fringes. It is therefore logical that the narrower the coverage area of a satellite signal, the stronger its beam will be (see Figure 2–5).

For international satellites, such as Intelsat, beam patterns can be adapted to deliver specific coverage. For example, *global*-beam transmission from a single transponder can cover more than 40 percent of the earth's surface. This kind of transoceanic coverage is critical for data such as stock markets, TV network news coverage, radio, and telephony.

Hemispheric footprints, as the name implies, are aimed at a more restricted area, covering only about half as much as a global-beam transmission. Because the coverage area is smaller, the signal levels from hemispheric antennas are stronger. This kind of coverage is especially important for domestic transmission, such as for the CIS's Gorizont satellites or those aimed at South America.

Domestic and regional satellites not serving areas as vast as those of the CIS or the Pan-American regions often have even more restricted beam patterns, called *zone* and *spot* beams. These beams provide concentrated, and thus even stronger, coverage to areas smaller than 10 percent of the earth's surface. An example of the use of such restricted areas was the creation of Live Aid's "third World Feed." By designating separate signals to be sent to a transponder on a satellite over the Indian Ocean, focused on the Asian market only, I was able to feed them individualized programming simultaneously with, and without disturbing, the other satellite feeds going to the rest of the world.

Again, the requirements of the globalcast will dictate the necessary coverage. This can be restricted by budget, by sponsor demand, by governmental regulation—by whatever factors are involved in the programming. The production and business staffs have to reach a consensus before approaching the common carrier before reaching a final decision on satellite booking. The ultimate decisions on coverage area also

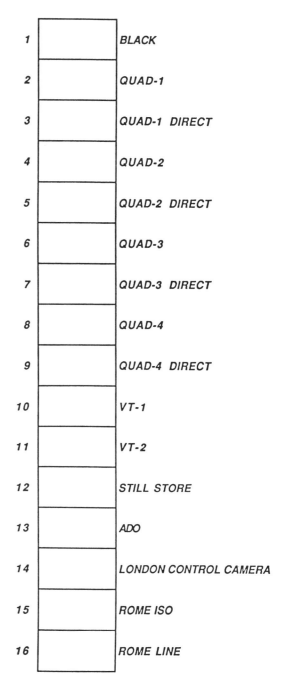

PRAYER FOR WORLD PEACE
MASTERCONTROL SWITCHER

1 BLACK
2 QUAD-1
3 QUAD-1 DIRECT
4 QUAD-2
5 QUAD-2 DIRECT
6 QUAD-3
7 QUAD-3 DIRECT
8 QUAD-4
9 QUAD-4 DIRECT
10 VT-1
11 VT-2
12 STILL STORE
13 ADO
14 LONDON CONTROL CAMERA
15 ROME ISO
16 ROME LINE

Figure 2–14 This chart is a reproduction of the actual chart used for the globalcast. This was the switcher's lineup, the parts of which are broken down in detail in Figures 2–15 and 2–16.

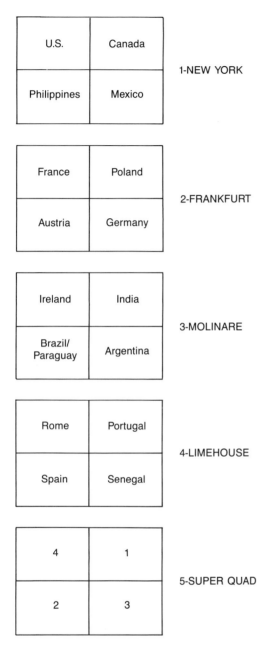

Figure 2–15 This chart indicates which countries were located in each of the quads listed in Figure 2–14. Technical routing required us to add the hot-switch capability between Brazil and Paraguay.

dictate the television standards with which you will work and which conversion(s) will have to be arranged, for both incoming and outgoing signals. (See the Appendix for television standards.)

How to Plan Satellite Back-ups

As with all the other elements of production, satellite back-ups are dictated by program content and budget. *Prayer For World Peace* serves as a good example about determining satellite back-ups.

Prayer For World Peace aired June 6, 1987. Figure 2–7 illustrates the satellite schematic: eighteen satellites, thirty transponders, more than thirty-five languages, more than a thousand technicians. The basis of the hour-long globalcast was Pope John Paul II's recitation of the Rosary from the Basilica of St. Mary Major in Rome, with parishioners in seventeen countries on all five continents responding—live. While portions of the ceremony were conducted in a total of eleven languages, the Our Father and Hail Mary's were recited in only five languages: English, French, German, Portuguese, and Spanish. The Pope was able to see and hear the parishioners, and the parishioners were able to see and hear the Pope. The production called for live injects from eighteen cities.

Because of certain union rules in Italy, it was impossible to have the control room in Rome. There was also the possibility that a strike would take place during the time of the telecast. As a result, we set up the master control room in the Limehouse Studios in London. Feeds arrived there by both satellite and land lines. Not having the control room near the action required having three submaster controls, one in New York City, one in Frankfurt, Germany, and one in a production house, Molinare, in London. All of these fed into my master control at Limehouse.

Coordinating eighteen live remote, multilingual injects was a technical challenge. Each inject was important, but only one was critical in terms of satellite back-up. The responses from the parishioners did not require satellite back-up because they were backed up by carefully planned alternate locations. For example, when the Pope prayed in Portuguese, if the satellite feed from Fatima, Portugal, had gone down, we could have switched instantaneously to Rio de Janeiro, Brazil, where churchgoers were also praying in Portuguese. If signals from both these satellites feeds had gone down, a portion of the audience in St. Mary Major in Rome was praying in Portuguese. Their responses could have been picked up on camera and on specially placed microphones to complete the prayer cycle. Similar back-ups were arranged for each of the other languages. French input came from Lourdes, France; Dakar, Senegal; and Quebec, Canada. Spanish input came from Guadalupe, Mexico; Zaragoza, Spain; Lujan, Argentina; and Caacupe, Paraguay. German input came from Frankfurt, Germany; and Mariazelle, Austria. English came from Manila, Philippines; Washington, D.C., USA; Knock, Ireland; and Bombay, India.

Each of these remotes was important, but in terms of program content each was sufficiently backed up. The only satellite feed we could not replace was the one from Rome. St. Mary Major and the Pope were the core of the telecast. If we lost Rome, we lost the program.

To provide back-up for Rome, therefore, we incurred the expense of a second feed. And we booked the back-up feed not just on a separate transponder, but on a separate transponder on a separate satellite. Providing that back-up was a nonnegotiable item on the budget. Without it, the program could not have been planned intelligently—and we could not have acquired insurance to cover our production costs.

Decisions on back-up transponders are made by the producers. Common carrier staffs can advise on costs and coverage, but the ultimate decisions on what back-up is needed lie wholly with the production unit. This is one reason we stress in the section on budget that the unit/business manager has to work with the executive producer(s) from the outset. A show should be designed originally without considering the budget. The wildest, most wonderful ideas should be conceived and explored. But once a show reaches the production stage, reality—back-ups and budgets—has to determine the final form.

How to Book Time on Satellite Transponders

To summarize, the first steps in booking time on satellite transponders are determining the desired locations and feeds for the program, based on program planning and the budget, and establishing a location for the master control room (point of origin). The common carrier nearest the master control site is the starting point for booking telephone communication facilities, land lines, including fiber optic cabling, and satellite transponder facilities. Since all signals normally originate from master control, that is the logical focal point. If satellite footprints of continents/countries from which injects may come and to which broadcast signals may go are extremely complex, as they were for *Sport Aid* and *Prayer For World Peace,* it may be decided, for technical or budgeting reasons, to use signal compression.

There are four ways to gain access to transponder time on a communications satellite: petition for space and launch a satellite; buy the use of transponder space from the owner of a satellite; lease space for the life of the satellite or another set amount of time; or rent use on a one-time-only basis.

Text continued on page 41.

4 Rome	Portugal	1 U. S.	Canada
Poland	Senegal	Philippines	Mexico
2 France	Spain	3 Ireland	India
Austria	Germany	Brazil	Argentina

Figure 2–16 Note that this super quad has quad 4 in the upper left, to keep Rome in the same upper left-hand position. Where Brazil is listed, I had the capability of hot-switching to Paraguay while airing the super quad.

```
                              4TH JUNE 1987

FROM: TONY VERNA

TO: SUB CONTROL STATIONS

HEREWITH AN EXPLANATION OF TONY VERNA'S COMMANDS
DURING TRANSMISSION:
```

TONY VERNA COMMAND	INSTRUCTION
1. STAND BY *'COUNTRY NAME'*	CUT *'COUNTRY NAME'* TO PRE-SELECT CHANNEL
2. TAKE *'COUNTRY NAME'*	(LIMEHOUSE VISION MIXER) CUT LIVE TO *'COUNTRY NAME'*
3. IF PICTURE IN SAME QUAD, TONY WILL SAY "SWITCH" (TO DISTINGUISH "TAKE" COMMAND TO VISION MIXER)	(AT SUB CONTROL) CUT LIVE TO *'COUNTRY NAME'*
4. CHANGE	STILL STORE OPERATOR MIX TO NEXT STILL
5. GO TO	VISION MIXER MIX TO NEXT SHOT UNLESS HE PREFERS CUT.

Figure 2–17 This printed communique on camera commands supplemented verbal communication of the same information. Dealing with multiple languages among the crews requires added concentration in preparing and executing a globalcast.

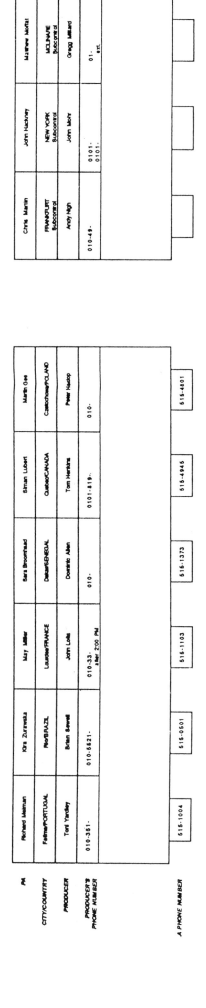

Figure 2–18 This chart shows the ground plan of the production assistants' (PAs) room behind my master control room. A glass window between gave visual communication to supplement headset communication to the producer there. Each PA had a phone line linked to a producer at one of the sixteen remote locations. Contact with Rome was via a four-wire staffed by a producer directly behind me in the control room.

Guidelines For Control Room PAs In London
for
"Prayer For World Peace"

Welcome to our production. As you're already aware, the live broadcast of our "Prayer For World Peace" will most definitely become the largest television event of its kind in history, and we appreciate the addition of your professional abilities and skills to its success. To aid you in your work on the program, we've created this brief guide to your responsibilities.

First and foremost, you are our constant lines of communications to our "troops" in the field. At each of the 16 inject locations from which our program will be fed to us here in London, as well as at our subcontrol points in Frankfurt and New York, we have placed special "Inject Field Producers" who are our direct liaisons with the local broadcast company or network which is handling the coverage technically. These are the people on the other ends of your phone lines, and with whom you will connect as a tightly knit team of two, relaying information to and from our Executive Director, Tony Verna.

Your communications with these locations should be efficient, rapid, and clear, as the program is a totally live broadcast and there are no "second takes" possible. Supervision of your activities will be by Mr. Bill Bode, who will be in the window to the main control room and will relay information from that room to your room in the client gallery immediately behind it.

There are a few basic facts that must be stressed again and again to the people in the locations around the world:

(1) No matter where they are listed in the official script or time grid format, Tony Verna MUST have the artistic and technical option of switching to ANY location LIVE at ANY time during the hour of the broadcast. He must, therefore, count on getting clean shots and audio from each location at all times. Do not let the crew at your location "rest" between "takes" on the grounds that, "Oh, we're French and they're doing Spanish now." This is vital.

(2) Should there be any breakdown in the satellite audio or visual signal, it is most important that we know about it in advance. For example, our audio plans include having someone "pre-hear" each location to be sure that, should a signal be about to go down or have lots of static, we can substitute another audio from a different place speaking the same language. The same policy holds for picture. Therefore, if you hear from our location producer that his signal is going to go down in five minutes because the backup generator is failing, OR conversely, you are told in London to inform your location that their signal is getting fuzzy and they should increase the gain on their transmission, it is IMPERATIVE that such information/instructions be given to the appropriate party quickly, and that the results of such activities be made known to Mr. Bode so that Tony Verna can plan his next few shots accordingly. Often in such circumstances, you may only be operating a minute or less ahead of Tony's line cut, but that minute of warning and pre-planning will make all the difference in the world to the final look of the program.

3. The whole key to your success in this broadcast and to its success thanks to your efforts, is concentration and control. During your training and rehearsal period, you will have occasion to see the "work reel" which is a mock-up of the actual broadcast made from previously filmed footage of the actual locations. Study the architecture, language, clothing styles, and customs of your assigned location as seen in the tape. Get a "feel" for the place and its portion of the broadcast. Take advantage of all of the learning opportunities available to you in this brief rehearsal period, and we're sure they'll pay off on the broadcast day.

4. Fourthly, please pay close attention to all security regulations and practices. They are here for the safety and protection of us all, and that of the broadcast.

Again, our thanks in advance for your expert assistance. We hope that you'll look back on your participation in "Prayer For World Peace" as a milestone in your own professional career as much as it is an historic milestone for the broadcast television medium.

Figure 2-19 Note the instructions to the production assistants in these guidelines to be ready for an audible at any time. A live globalcast never runs exactly as formatted, so each member of the crew must be ready for any change at any time.

```
                P R A Y E R   F O R   W O R L D   P E A C E

                          (P.A. OUTLINE)

      A) Wide shot for each location                          06:01:33

         Sequence:

            1.  Fatima, Portugal  (Live or still store (film))

            2.  Lourdes, France

            3.  Zaragoza, Spain  (Still store)

            4.  Mariazell,  Austria

            5.  Frankfurt,  Germany (Still store)

            6.  Czestochowa, Poland

            7.  Knock, Ireland

            8.  Dakar, Senegal (Still store)

            9.  Bombay, India  (Still store)

           10.  Manila, Philippines

           11.  Washington D.C.

           12.  Quebec, Canada

           13.  Guadalupe, Mexico

           14.  Rio, Brazil

           15.  Lujan, Argentina

           16.  Caacupe, Paraguay


      B) ROME - THE POPE                                      06:10:19


      C) FATIMA (PORTUGAL)                                    01:10:19


    (1) D) 15. The Our Father in Portuguese

            1) Rome

            2) Fatima
```

Figure 2–20 These are the first two pages and the final page of the outline (or rundown or running order) for the globalcast. The designation P-1 on page 2 indicates Portuguese, first recitation. E-9 on the last page denotes English, ninth recitation. See the text for why and how this written outline had to be supplemented.

Figure 2—20 *(continued)*

E) 16: Hail Mary (Portuguese)

  1) Rome (P-1)

  2) Fatima (Portugal) (P-1)

  3) Rome (P-2)

  4) Rio (Brazil) (P-2) ✳

  5) Rome (P-3)

  6) Fatima (Portugal) (P-3)

  7) Rome (P-4)

  8) Rio (Brazil) (P-4) ✳

  9) Rome (P-5)

10) Fatima (Portugal) (P-5)

11) Rome (P-6)

12) Rio (Brazil) (P-6) ✳

13) Rome (P-7)

14) Fatima (P-7)

15) Rome (P-8)

16) Rio (Brazil) (P-8) ✳

17) Rome (P-9)

18) Fatima (P-9)

19) Rome (P-10)

20) Rio (Brazil) (P-10) ✳

```
P)  Hail Mary (English) Continued.....


    17)  Rome (E-9)

    18)  Washington D.C. (USA) (E-9)


    19)  Rome (E-10)

    20)  Manila (Philippines) (E-10)



Q)  38: CHOIR (GLORIA PATRI)                            06:36:03

    39: THE POPES'S MESSAGE (in English)



    40: POPE INCENSES THE ICON OF THE VIRGIN MARY       06:42:56

    41: ETHNIC CHILDREN PRESENT FLORAL TRIBUTE

    42: POPE MAKES APOSTOLIC BENEDICTION                06:44:55

    43: CHOIR: SALVE REGINA

    44:  a)  POPE BOWS, PROCEEDS TOUR MAIN AISLE         06:44:57

         b)  ALL SITES: (CUE BELLS) at:                  06:48:01

         c)  INJECT FOR ALL SITES OF WAVING HANDKERCHIEFS/
             CHURCH BELLS

    45: SATELLITE PICTURE OF THE WORLD

   (46: USA OUTCUE)

    47: ADD CREDITS: NEW YORK VTR -                      06:58:50

    48: ALL SITES: INJECTS PLUS CATHEDRAL STILL STORE    06:59:13

    49: WORLD OUTCUE ON SITES

   (50: VTR: LONDON CONTROL ROOM - ADDS CREDIT FOR VTR)  06:59:26
```

Figure 2–20 *(continued)*

TIME SCHEDULE

6 MAY, 1987

16:00-17:00

16:00:00   *   All inject producers and P.A.'s prepare for test.

          *   Audio/Video circuits inbound to Limehouse and subcontrols activated. The order of activation is as follows:

1. Visnews subcontrol
2. Frankfurt subcontrol
3. Molinare subcontrol
4. Rome
5. India
6. Senegal
7. Philippines
8. Portugal
9. France
10. Austria
11. Spain
12. Ireland
13. Poland
14. Argentina
15. Brazil/ Paraguay
16. Mexico
17. Washington D.C.
18. Quebec

          *   Communication circuit activation
4 wires (two way talk circuit)
2 wires (one way talk circuit) This is a one way directorial circut featuring Tony Verna
10 KH2 audio lines (music circuits)
phone lines (business phone at inject producers position minimum of 2)

16:05:00   * System check and tuning of all inbound circuits and out bound circuits by engineers.

          *   Inject producers standby for system check. Systems to be checked in the following orders:

1. Wait for phone call from P.A. in London on your phone in the O.B. Van
2. Send bars and tone and an audio loop identification on outbound audio/video circuit; repeating.ie: This is Fatima ...This is Fatima... This is Fatima. Wait for response verification.
4. Confirm all other 4 wire and KH2 audio lines, wide screen
5. Await orders from London for audio check.

16:15:00   *   In order to confirm the status of the audio 4 wire circuits the In Rome a designated person will begin counting numbers into the outbound audio circuit to London. London will then send this

Figure 2—21   These three pages detail the schedule of technical events for the hour preceding the globalcast. Technical complications telescoped these events into less than thirty minutes.

Figure 2—21 (continued)

audio signal to the injects via the 4 wire circuit. Each will confirm reception of the audio signal with their designated P.A. and stands by to start numerical voice count thru a microphone in the church (altar mic) on the outbound audio circuit from the inject to London. London will pass this audio inject to Rome for check. The The following is the order for audio system check. (Note: if an inject cannot establish circuit integrity by 16:15 London Time, London will bypass the inject and reconfirm at the end of the test.

Rome begins count: Rome 1, Rome 2, Rome 3, Rome 4 and Rome 5. Injects in the following order will confirm with their P.A. that they are receiving the audio from Rome.

1. Fatima, Portugal
2. Lourdes, France
3. Zaragoza, Spain
4. Mariazell, Austria
5. Frankfurt, Germany
6. Czestochowa, Poland
7. Knock, Ireland
8. Dakar, Senegal
9. Bombay, India
10. Manila, Philippines
11. Washington, D.C.
12. Quebec, Canada
13. Guadalupe, Mexico
14. Rio de Janiero, Brazil
15. Lujan, Argentina
16. Caacupe, Paraguay

Following confirmation of reception, the injects will respond by counting to Rome in the following order:

1. Fatima, Portugal
2. Lourdes, France
3. Zaragoza, Spain
4. Mariazell, Austria
5. Frankfurt, Germany
6. Czestochowa, Poland
7. Knock, Ireland
8. Dakar, Senegal
9. Bombay, India
11. Manila, Philippines
12. Washington, D.C.
13. Quebec, Canada
14. Guadalupe, Mexico
15. Rio de Janiero, Brazil
16. Lujan, Argentina
17. Caacupe, Paraguay

At the conclusion of this test, Rome will call for immediate responses from individual injects in the following order.

1.. Fatima, Portugal
2. Lourdes, France
3. Zaragoza, Spain
4. Mariazell, Austria

Figure 2–21 *(continued)*

```
           5.   Frankfurt, Germany
           6.   Czestochowa, Poland
           7.   Knock, Ireland
           8.   Dakar, Senegal
           9.   Bombay, India
          10.   Manila, Philippines
          11.   Washington, D.C.
          12.   Quebec, Canada
          13.    Guadalupe, Mexico
          14.   Rio de Janiero, Brazil
          15.   Lujan, Argentina
          16..       Caacupe, Paraguay

16:40:00  *  Inject producers will standby for the camera direction
             from London in the following order: ( give Wide Shot )

           1.   Rome, Italy
           2..  Fatima, Portugal
           3.   Lourdes, France
           4.   Zaragoza, Spain
           5.   Mariazell, Austria
           6.   Frankfurt, Germany
           7.   Czestochowa, Poland
           8.   Knock, Ireland
           9.   Dakar, Senegal
          10.   Bombay, India
          11.   Manila, Philippines
          12.   Washington, D.C.
          13.   Quebec, Canada
          14.    Guadalupe, Mexico
          15.   Rio de Janiero, Brazil
          16.   Lujan, Argentina
          17.   Caacupe, Paraguay

16:55:00     Standby for program air.
```

The Cost of Satellite Time

Insuring and orbiting a communications satellite may cost as much as USA $300 million. Launch failures for USA companies for the past decade have ranged from about 9 to 30 percent. Therefore, except for the occasional billionaire, only governments, coalitions of governments, and large corporations have the financing to function in this area. Most television networks, superstations, and local stations buy or lease transponder space with long-term planning. Virtually all producers of one-time programs rent only the time needed for that specific program.

Transponder time is rented in fifteen-minute segments. When budgeting, you always have to be sure the figures include booking sufficient time for making technical checks, for "rehearsal," and for possible run-over. Many of these mega-events are like sports events. They may run over their scheduled length—with prior approval of the stations/networks feeding the show. Also note that "rehearsal" on a globalcast is a relative term. There is no such thing as a full top-to-bottom rehearsal. Aside from the impossible physical coordination of people and sets, the cost of satellite time alone would rule out full rehearsal. You have to provide local, non-satellite fax-time to rehearse musical numbers, etc., at each location. But a full run-through could never happen. You do,

Country: _____

Mix-minus feeds: _____

Production Co-ord: _____

Technical Co-ord: _____

Directors open mic. circuit: _____

Figure 2–22   This checklist was given to each production assistant. The importance of the mix-minus is indicated by its position on the list. See the text for an explanation of mix-minus.

however, have to book enough satellite time to make the many technical checks required and to let your production staff coordinate with the technicians and with each remote location. And that final check-in for a globalcast is impossible without hot satellite hookups. Occasions may arise when governmental restrictions require your using more expensive satellite paths than the cheapest available paths. You learn all of these details from the satellite specialists who work for the common carrier.

Current hour rates for communication satellites range from about USA$200 to $800, depending on the time of day. USA domestic rates are based on Eastern (Standard or Daylight) Time. Prime time runs from 4:00 p.m. to 2:00 a.m. Eastern time and ranges from about USA$400 to $800 per hour. Non-prime time runs from 2:00 a.m. to 4:00 p.m. Eastern time with rates of about USA$250 to $300 per hour.

Global times are based on Greenwich Mean Time. *Domestic* satellite services are ordered based on Eastern Standard Time. All *international* services are ordered based on Universal Coordinated Time (formerly Greenwich Mean Time).

Satellite transponders are used primarily for telephonic and business data transmission, and normally these uses, not television transmission, present the major competition for time. But there are times, such as when a president or the Pope plans a trip, when the networks deliberately overbook to ensure that they have enough time to transmit reports. For program time, especially runover time, satellite specialists might bargain with a network to clear time.

Although there are hundreds of communication satellites in the Clarke orbit, there may be times when the common carrier has to bargain for the time slots you need. Transponder time is finite: occasionally the needed time slots cannot be cleared, and a show has to replan its air time. For this reason, it is critical to begin booking satellite time as early as possible.

A major advantage of the proliferation of satellites in orbit is the competition inspired by that proliferation: the cost of booking time on a satellite transponder has decreased markedly over the past decade, and the costs will tend to become even lower with increased competition from fiber optic cabling.

Two Problems: Time-Delay and Mix-Minus

Time Delay

Time delay is a problem every global television production faces, whether it uses satellite feeds or fiber optic feeds. The electromagnetic waves that carry television signals travel at the speed of light, but there is always *propagational delay,* a phenomenon caused by (1) resistance to the waves as they travel through the air or cable; and (2) capture and redirection of the signals by the antennas and amplifiers in the transponders and/or fiber cables. The delay is less with fiber optic cables, not because of the medium, but because the signals travel a shorter distance.

You've probably noticed the few-second delay on newscasts in conversations between continents. While this delay is acceptable in conversation, it may be critical in other situations. Let me give you an example.

In the planning stages of *Live Aid,* Mick Jagger and David Bowie wanted to do a live duet, with Bowie in London and Jagger in Philadelphia. When the request reached me, I reluctantly had to say no. Why? Because of the

```
PRAYER FOR WORLD PEACE
TONY VERNA
6-4-87
CONFIDENTIAL WORKSHEET
```

LONDON	1.	VTR	GLOBE TURNING GLOBAL MEDIA PRESENTS PROGRAM TITLE. (Voice over announce from multi-audio tape in all languages). MUSIC 1/4″ TRACK DISSOLVE EXTERIOR ST. MARY MAJOR BASILICA - TRAFFIC SOUNDS UNDER	:14	06:14:00
			LONDON FEEDS ROME PRE-RECORDED REGINA CAELI	:07	06:21:00
ROME	2.	LIVE	INTERIOR ST. MARY MAJOR MUSIC ANNOUNCER COMMENTARY (V.O. LONDON) OPENING:		
			"Live from St. Mary Major in Rome, today, June 6, 1987, Pope John Paul II will lead the recitation of the Rosary.		
			The Pope recites the Rosary with the faithful on the first Saturday, of each month for broadcast by Vatican Radio, but today the recitation of the Rosary, from the most famous Roman Basilica dedicated to the Mother of God, assumes a special significance, because it is the vigil of the feast of Pentecost, the opening of the "Year of Mary".		
			On this Saturday, responses to the prayers of the Holy Father will come not only from the people present in the Basilica of St. Mary Major, but from the faithful gathered in the most famous Marian Shrines and other churches on all five continents, linked directly with Rome by television and radio.	:47	06:01:08
LONDON CONTROL ROOM	4.	LIVE	ANNOUNCER COMMENTARY "This London Control Room will coordinate the live pictures from Rome and 16 other cities. The telecast will use 20 satellites and 30 transponders. The program will also be broadcast by radio and transmitted by telephone lines to a potential audience of more than 1 billion people in an experience of instant and interactive global communication and of a world visibly united in prayer.	:25	06:01:33
LONDON	5.	LIVE	ANNOUNCER COMMENTARY IN EUROPE		
	1.		FATIMA, PORTUGAL, where the Blessed Virgin Mary appeared to three children with a message of prayer as the key to peace;		
	2.		LOURDES, FRANCE, where Bernadette Soubirous saw the Virgin Mary in a stone grotto;		
	3.		ZARAGOZA, SPAIN, where Christopher Columbus visited the Shrine of Our Lady of Pilar before departing for the new world;		

Figure 2–23 These two opening pages of the copy read by Archbishop John Foley can be compared to the production assistants' outline/rundown (Figure 2–20). Similar copy was read in other languages by narrators located in the Vatican. I was in headset communication to a producer working with them.

4.   MARIAZAELL, AUSTRIA, a popular pilgrimage location for German
speaking Catholics for more than 800 years;

5.   FRANKFURT, GERMANY, where local parishioners are gathered in
the Church of St. Sebastian;

6.   CZESTOCHOWA, POLAND, a world renowned place of pilgrimage which
Pope John Paul II will visit next week;

7.   KNOCK, IRELAND, where the Blessed Virgin Mary appeared more
than 100 years ago;

IN AFRICA:

8.   DAKAR, SENEGAL, in the beautiful Cathedral of Our Lady of Victory.

IN ASIA:

9.   BOMBAY, INDIA, where Hindus and Moslems often join Christians at
the Shrine of Our Lady of the Mount;

10.   MANILA, PHILIPPINES, where the president of that nation is
scheduled to join the Cardinal Archbishop in the Cathedral of Our Lady;

IN NORTH AMERICA:

11.   WASHINGTON, D.C. at the National Shrine of the Immaculate Conception;

12.   QUEBEC, CANADA, at the Shrine of Our Lady at the Cape de la Madelaine;

IN LATIN AMERICA:

13.   GUADALUPE, MEXICO, where the image left by the Virgin Mary on the
cloak of an Indian continues to attract millions of pilgrims;

14.   RIO DE JANIERO, BRAZIL, at the Shrine of Our Lady of Penha in the
shadow of Corovado.

15.   LUJAN, ARGENTINA, a national place of pilgrimage 40 miles from
Buenos Aries.

16.   CAACUPE, PARAGUAY, where hundreds of thousands have travelled many
miles to join millions around the world in prayer led by the Holy Father."

LONDON CONTROL ROOM			ANNOUNCER COMMENTARY "The Rosary will be recited in Latin, Portuguese, French, Spanish, German and English and selections of scripture will be read in Greek, Polish, German, Dutch, German and Arabic"		
			FADEOUT AUDIO TAPE MUSIC		
ROME	6.	LIVE	CUE-LIVE CHOIR POPE ENTERS BASILICA POPE & PROCESSION MOVE DOWN THE AISLE Pope will wear short red cape and red stole. Preceding him will be a cross bearer, two candle bearers and one incense holder. choir sings "PRAISE TO THE LORD ALMIGHTY"	3:27	00:05:00

Figure 2–23 *(continued)*

Italy	Italy	Germany	France	Austria	Spain	Portugal	Poland	India
AOR London	Cable Yugoslav ECS London	Cable Germany AOR London	Cable Germany AOR London	Cable Germany AOR London	Cable Germany AOR London	ECS London	ECS London	IOR London
Brazil	Paraguay	Argentina	Canada	Mexico	USA	Philippines	Ireland	Senegal
AOR London	AOR London	AOR West Va Satcom New York AOR London	Westar AOR London	Morelos Houston Westar AOR London	Westar BrightStar London	POR San Fran. Westar Maine New York BrightStar London	Cable London	AOR London

Quad #1 New York	Quad #2 Frankfurt	Quad #3 Molinare	Quad #4 Linehouse
USA Canada Philippines Mexico	France Poland Austria Germany	Ireland India Brazil/ Paraguay Argentina	Rome Portugal Spain Senegal

Superquad	
4	1
2	3

Figure 2–24 This chart of the video paths used during *Prayer For World Peace* shows the complexity of handling the multiple inputs and outputs inherent in any globalcast. Compare this with the chart outlining audio channels on *Sport Aid* (Figure 3–11) and compare both charts with their respective satellite schematics (Figures 2–7 and 3–3).

three- to four-second propagational delay in the transmission of sound via satellite. Bowie and Jagger could have sung together, but they would never have been in sync. It would have been a hopeless jumble. Fortunately for the *Live Aid* audience, Bowie and Jagger taped a duet that provided one of the highlights of the show. But it wasn't a live duet, and until fiber optics spread worldwide it is unlikely there will be a live duet involving multicontinental satellite feeds. There are suggestions that the use of low-earth-orbiting satellites (LEOs)—which reduce the distance the signal travels—will help solve this problem; but even if it is practical, this solution is many years in the future.

I confronted this problem again with *Earth 90,* a three-hour globalcast that aired June 2, 1990 (in the USA; June 3 in Europe and Asia). This was the day designated by the United Nations Environmental Programme as Worldwide Clean-up Day. The program was coproduced by NHK, and we had sister concerts in Tokyo (cohosted by John Denver and Yu Hayami) and New York City (cohosted by Debbie Gibson and Rolland Smith). We also had a series of live and tape pickups, including a live segment from Rio de Janeiro, hosted by Gilberto Gil. We had originally planned and scripted a conversation about some of Brazil's major conservation efforts between John Denver in Tokyo and Gil in Rio, but we had to switch that conversation to one between Rolland Smith in New York and Gilberto Gil in Rio. Why? Because of the available satellite links, the signal from Brazil was routed through the New York control room to Tokyo. This routing meant a *double* delay: the three- to four-second delay in the uplink/downlink signal from Rio to New York, plus another three- to four-second delay in the uplink/downlink signal from New York to Tokyo. Had we attempted the John Denver–Gilberto Gil conversation, there would have been a minimum six-second delay between the time each of them heard a question and could respond to it—an intolerable situation. By switching the conversation to New York/Rio, we cut the delay in half and experienced no more problems than those seen on a newscast. But the fact remains that the technical limitations required a change in planned programming.

The time delay problem does not disappear with the use of fiber, but it is shorter. According to Manuel Sierra, VideoTeleconferencing and Product Market Manager for AT&T, the propagational delay over 3,500 miles of fiber optic cable is one-half second. The satellite signals don't travel any slower than those on cable, they just have to travel farther—22,300 miles up, 22,300 miles back down. In this instance, intercontinental fiber wins the time-delay battle with satellites. If we had had the current transAtlantic fiber cable in 1985, we might have been able to trick our way through that live Bowie–Jagger duet.

Mix-Minus: A Key to Interactivity

This second problem concerns feedback in audio circuits; and while the problem can exist in any production, from local to national to international, it's a problem that is magnified on a global scale.

Feedback is the high-pitched squeal that results when audio from a program monitor is fed back into a live microphone. If you saw the

Figure 2–25   This monitor wall in the London control room for *Prayer For World Peace* shows the physical as well as electronic space saving of the quad feeds, with each of the bottom monitors providing video from multiple sources. See Chapter 9 for details on how I used these quad feeds and the monitor screens.

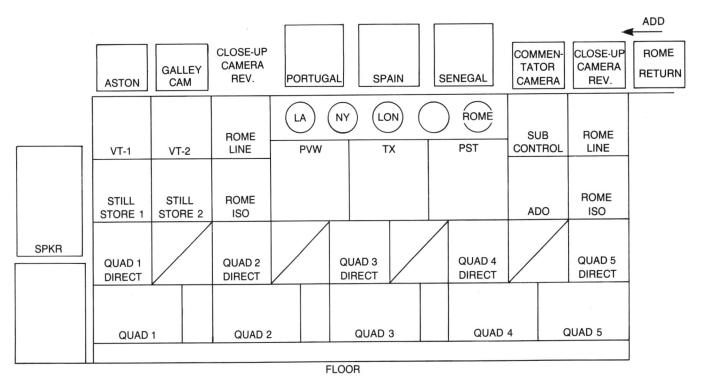

Academy Awards globalcast several years ago, with Jack Lemmon in Moscow, you know what feedback can do. The Moscow feed was almost destroyed by uncontrollable feedback that disconcerted both Lemmon and the audience. The feedback came because the Moscow audio was included in the mix. Each time Lemmon spoke, his audio came over the program system provided for the Moscow auditorium audience and into his microphone, competing with what he was saying. The way to avoid this problem is to use a mix-minus feed.

Mix-minus is created by setting up the audio lines so that the remote source gets all audio *except its own.* In *Prayer For World Peace,* the entire program was structured around the Pope reciting and the parishioners responding to prayers. The Pope had to hear the parishioners and the parishioners had to hear the Pope, in each of sixteen different locations. Therefore, in each of these locations we had to set up audio cabling so that each remote location could hear everyone's audio except its own.

It's much like a walkie-talkie setup: when the speaker from the remote location talks, all other sound is shut off. For example, parishioners in Fatima could hear the Pope and they could hear responses from any of the other fifteen remote injects *but not their own.* When the microphones in the church at Fatima were picking up their responses, there was no audio feeding into the speakers in their church.

This is not a simple system to set up. It requires conscientious cabling and checking out. And when you're intermixing seventeen remote locations it's a real challenge. It's also tantalizing, because you can only test such a set-up when everything is connected and when you are on satellite time—one very important reason to book ample technical check time.

For *Prayer For World Peace* we actually went one step beyond mix-minus. As I explained, portions of the audience in Rome were praying in each of the response languages, as back-up in the event that we lost satellite signals from all of the remotes. In addition, we synchronized their prayers with those of the remote parishioners—and so with the Pope.

We accomplished that essential synchronization by routing all of the audio signals from the remote locations through our London master control room *to* Rome. We received all of the signals in London, but we did not air them from London. The pattern was this:

> All of the feeds from the seventeen remote inject churches came in on separate studio lines to our London control room.
> We routed the audio feeds from the selected church directly to speakers in St. Mary Major in Rome.

This meant that the people praying in Rome heard those praying in the remote church and automatically prayed in sync with them. The combined synchronization was the only audio the Pope heard, so his prayers were in pace with the parishioners in the remote inject churches. The Pope knew when to begin speaking because he knew—simply by listening—when the parishioners stopped speaking. Their audio was as intimate to him as was that of the faithful praying in front of him in St. Mary Major.

The separate audio lines that provided this synchronization were the same lines that provided the electronic safety of mix-minus. If you look at the sequence called for in the *Prayer For World Peace* Time Schedule/PA checklist in Figure 2–22 you'll see that the mix-minus feeds were the first thing the PAs had to check, once the preliminary technical checks has been made.

It was the interactive nature of this telecast that made it such a challenge for the technical crews. One-way audio is not that tricky. When two or more locations start talking among themselves, the potential problems multiply rapidly. Mix-minus is one of the salvations for that complexity.

Chapter **3**

Fiber Optic Cabling

Why Use Fiber Optic Cable?

The introduction of satellite TV transmission has been visibly world-altering. But while we have been looking up at the sky, under ground there has been a subtler but constant expansion of optical fiber as a replacement for copper wiring—and as a tool for television production.

Fiber optic cabling is extremely lightweight cable that replaces copper wire with hair-thin strands of glass or plastic. Because plastic cable cannot transmit as much data as glass, its use is still in the experimental stage. Virtually all fiber optic cable in use today employs glass fiber.

The main reason optical fiber cable is making less expensive copper cable obsolete is that, as its name implies, optical fiber uses light to transmit signals—at the speed of light—and can transmit vastly more data than can copper wire, which historically has used electrical impulses for transmission.

Copper wire can transmit only about 10 to 16 million bits of information per second. This is vividly more data than the one bit per second transmission of Samuel Morse's first telegraph message in 1844, but in the 1990s it is slow. Glass fiber optic cable routinely sends 100 million bits of information per second. Telephone companies have successfully sent 300 million bits per second in tests. Other experimental reports claim speeds as high as a trillion (USA) bits-per-second. If transmission at this rate becomes practical for daily use, one fiber could carry fifty thousand voice channels on a cable as tiny as a strand of hair. And because video signals require vastly larger capacity than audio signals, glass fiber becomes a desirable medium to transmit TV signals.

There are other advantages to fiber optic cable that will continue to spur its growth. Transmission through fiber optic cables is static-free, immune to electromagnetic and radio frequency interference. There is no interference from fiber to fiber, no cross talk. In business settings, signals are immune to electric signals generated by office or factory equipment.

Transmission through fiber optic cables is not affected by the weather as transmission by satellite can be. And although fiber optic cable is lightweight and made of glass, it is strong. You can run over it with a truck.

Because light beams can carry digital signals through fiber, making it possible to transmit large quantities of information, fiber is a good candidate for transmitting high-definition signals. Whatever advanced television transmission standards are approved and used, fiber optics have the capacity to carry them.

Signals transmitted over fiber optic cables can be sent over much longer distances without repeaters (amplifiers) to reinforce them, whereas copper wire requires frequent repeaters. When there are high bandwidth requirements, as with high-definition (HDTV), signals through coaxial cable weaken after only one hundred feet or so. With existing technology, fiber optic cable can carry signals for miles without a loss of strength.

Optical fiber offers the option of two-way, interactive communication, in terms of both transmission and production. In terms of transmission, viewers on cable-TV or DBS systems can order pay-per-view (PPV) events conveniently, and the company can bill them conveniently. As a production tool, fiber optics can be used to involve viewers directly in a live telecast. Viewers calling in to a talk show no longer need be disembodied voices—video and audio can carry together, so the callers can appear on screen. When videophones advance beyond the sluggish rate of ten images per second, contestants on game shows will not have to be at the studio to participate; with two-way transmission they could be anywhere. Using virtual reality images, any person in any country on any continent in any setting could appear on-camera with any other person in any country on any continent. This two-way capability is sure to spark types of programming not possible in the past.

Another advantage of fiber optic cable is that, because it uses light, not electric signal, it is virtually impossible to tap without detection. (Think of all the spy and detective stories you won't see in the future!)

Figure 3–1 This map shows the availability for television signal switching via fiber optic cable of just one company, Vyvx NVN. The television control center for the entire system is located in Tulsa, Oklahoma.

Courtesy Vyvx, Inc.

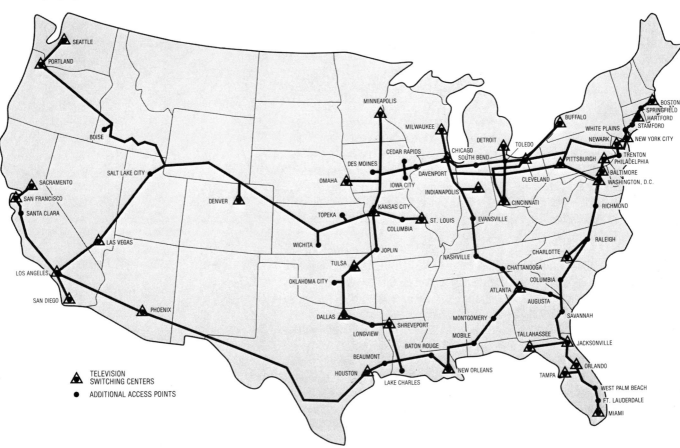

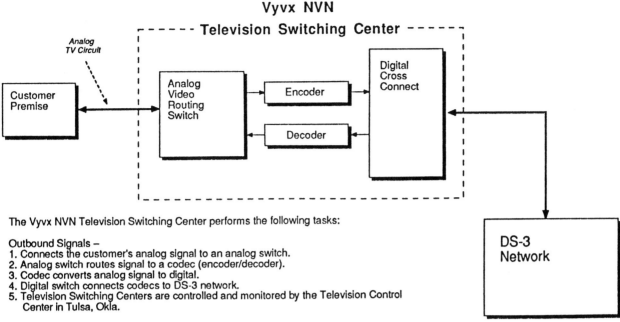

Vyvx NVN
Television Switching Center

The Vyvx NVN Television Switching Center performs the following tasks:

Outbound Signals –
1. Connects the customer's analog signal to an analog switch.
2. Analog switch routes signal to a codec (encoder/decoder).
3. Codec converts analog signal to digital.
4. Digital switch connects codecs to DS-3 network.
5. Television Switching Centers are controlled and monitored by the Television Control Center in Tulsa, Okla.

Inbound Signals –
Opposite of outbound signals.

Customers are responsible for bringing the video signal into the analog video routing switch located in the Vyvx NVN Television Switching Center. From that point, Vyvx NVN is responsible for the signal. The customer is again responsible for delivery of analog video output from the switch in Vyvx NVN's Television Switching Center to the customer premise.

Figure 3–2 This diagram shows the path of a television signal arriving at a Vyvx fiber optic switching center. The client has only to provide analog signal. The company uses its own codecs to encode and decode the signal to/from digital signals (DS)

Courtesy Vyvx, Inc.

Another reason many production units include fiber optic cable in their backhauling plans is the instant switching service available from companies like Vyvx National Video Network (NVN), a long-distance carrier whose basic coverage is limited, at least now, to the USA. Owned by Wiltel, Vyvx NVN uses Wiltel's 11,000-mile switched fiber network to some sixty-five USA metropolitan areas. Central Control is in Tulsa, Oklahoma, but there are switching centers in more than fifty cities around the system.

Vyvx's switching capability allows its Central Control to direct routing of TV signals just as voice and data are switched through traditional long-distance networks. Signals can be switched to match customer requirements. You can transmit from one point to another single point, from one point to multiple points, from multiple points to one point, from multiple points to multiple points—and you can change your mind at any time, with the flick of a switch. They multiplex digital signal-3s (DS-3s) of 45 million bits-per-second—the industry standard for broadcast quality—into systems that can carry from 565 million to 2.4 billion bits-per-second.

This rapid routing capability also builds in a solution to potential cable breaks. For example, if a cable is cut on a line between Washington, D.C. and Atlanta, Georgia, during transmission from Atlanta to New York, a digital switch in Washington would detect that the incoming signal was lost. Simultaneously, alarm signals would go out from Atlanta and New York to the Tulsa TV Control Center, which would call up preplanned alternate routings and perform the necessary switching routines to reconfigure the circuits. This all happens within two seconds, and the rerouted TV signals are once more being received in New York from Atlanta.

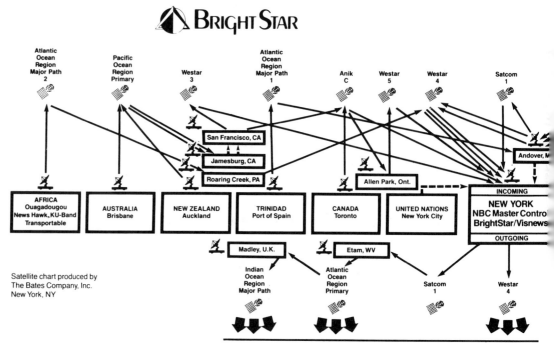

WORLDWIDE BROADCAST

Satellite chart produced by
The Bates Company, Inc.
New York, NY

LARGEST SATELLITE BROADCAST EVER. "Sport Aid" brought live television coverage of fundraising running events in 13 countries on six continents to a potential worldwide viewing audience of over 1.5 billion. Satellite coordination for the two-hour broadcast, which utilized 24 transponders on 14 satellites, was handled by BrightStar Communications and Visnews International (USA). Produced by Global Media Ltd., "Sport Aid" was a joint effort of UNICEF and "Live Aid" organizer Bob Geldof and reportedly raised more than $100 million for African famine relief.

Figure 3–3 The caption on this satellite schematic for *Sport Aid* proclaims "Largest Satellite Broadcast Ever," and it was to that date. However, *Prayer For World Peace,* which used thirty transponders on eighteen satellites, exceeded this scope. See Figure 3–4 for a visualization of the two quads shown in the schematic.

Courtesy BrightStar Communications

A major economic advantage Vyvx provides is its encoding and decoding equipment. They use their own codecs (COde-DECode) to convert from analog to digital at the input point and from digital back to analog at the output point(s). This means the company using the system doesn't have to invest in coding-decoding equipment. Deliver analog signal from your sending point; Vyvx delivers analog signal back to you at your receiving point(s). Your equipment has to handle only normal analog signals.

The system also provides convenience in making last-minute orders. While only full-time service guarantees exclusive use twenty-four hours a day, broadcasters can call to reserve available transmission lines on just a few hours' notice, less in a crisis situation. This is especially important for fast-breaking news events or unavoidable last-minute changes of plan. (Singers and actors have been known to join megaevents as late as the day of the broadcast.) In 1991 the average cost of this "occasional service" was about USA$400 per hour.

How Fiber Optic Cable Works

Two separate developments led to combining high capacity with the speed of light in fiber transmission. First was the development of laser beams, providing the light source, and second was the addition of cladding to the fiber core. In early fiber optic experiments the light signals tended to bounce and zigzag away from the thin fiber. This problem was solved by placing a reflective material called cladding around the core. Cladding contains mirror-like angles designed to return signals escaping from the

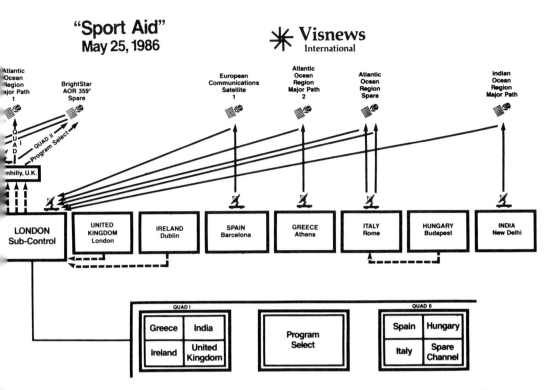

SPORT AID featured use of innovative QUAD FEEDS developed by BrightStar/Visnews to handle transmission of feeds from seven countries via three satellite signals from London to the U.S. Two signals, carrying four feeds each, enabled master control to preview the feeds simultaneously, select the desired segments, and receive them via the third signal for broadcast worldwide. *Land lines and connections are indicated by dotted lines.*

core, at different distances, back into the core at a consistent speed so that the resulting signal is not distorted. Laser beams and cladding made it possible to combine the fantastic speeds developed in computer digitalization with speed-of-light transmission. The brittleness of the hair-thin glass fiber created early problems with splicing sections of the cable together, but largely automated advanced splicing machines have solved that problem.

Another factor limiting the use of fiber has been the cost: glass fiber is still more expensive than copper wiring. A representative of General Instrument's Jerrold Division says that in just one year, 1989 to 1990, the cost of installing a transmitter/receiver for glass fiber cable was cut by one-third, and gradually the advantages have begun to outweigh the disadvantages of new installation. The only remaining block is the cost of replacing existing copper wire.

However long it takes to wire the world—and it will take a while—this convergence of transmission medium and method has put us on our way to twenty-first century communication that will make past efforts look like a horse and buggy.

How Fiber Optic Cable Is Used

Fiber optic technology was first used in broadcast TV during the 1980 Winter Olympic Games. It did not become the preferred method of transmitting TV signals until the early 1990s.

Figure 3–4 This is the master control room in Studio 8-H at NBC in New York City. Compare this with Figure 3–3. To highlight this new technology, I instructed the technical director to air the quads as soon as the lines became available.

Figure 3–5 This video print reflects the focus of the first hour of *Sport Aid*. We integrated live coverage of Sudanese world-class runner Omar Khalifa running through New York streets to the United Nations buildings with tape of his earlier symbolic runs through capital cities in Europe. Combining the past and present symbols both carried the program's message and led to the climactic moment when Khalifa's lighting a torch at the UN started millions of runners racing in 237 cities worldwide.

Figure 3–6 The sheer number of runners participating worldwide in *Sport Aid* made coverage of the Race Against Time exciting.

The main use of fiber optic cable for TV transmission has been the kind of feed described in the section on satellites—backhauling to bring in remote feeds to be inserted in a final production.

In the USA, in 1990, ABC-TV, CNN, and C-Span used fiber optic cable to backhaul feeds from the International Economic Summit held in Houston, Texas. CBS-TV used fiber lines to backhaul coverage of Super Bowl XXIV from New Orleans to its New York studios. Also in 1990 CNN signed a three-year agreement with Vyvx for two fiber optic links to and from Washington, D.C. One line links the Washington, D.C. studios with CNN headquarters in Atlanta, Georgia; the other ties Washington to New York City. CBS has signed with the same company for a permanent two-way line between Washington, D.C. and its New York studios.

A very significant use of fiber optic cable began in early 1991, when Vyvx and Hughes Television Network signed a multi-year nonexclusive agreement. Hughes is a division of IDB Broadcast Group and a leader in satellite distribution of sports events. The first use of their integrated systems was to backhaul coverage of a series of baseball games in the same way CBS used the system to backhaul the Super Bowl, NCAA basketball, and the All-Star Game.

This marriage of satellite and fiber technology is without question a foreshadowing of the future.

Available Fiber Optic Cable Services

There are four national fiber networks in the USA, plus numerous smaller local carriers. Three of the coast-to-coast networks—AT&T, MCI, and US Sprint—are better known because they provide national and international telephone communications, and they've been around a while. Vyvx, which provides predominantly video service, began offering its commercial services only in 1990 but has been expanding rapidly ever since.

Fiber optic cabling is not, by any means, limited to the USA. On the contrary, virtually all long-distance phone lines in Japan and the European Community (EC) consist of fiber optic cable, and the rest of the world is racing to catch up.

When broadcasting to multiple nations, however, there are two necessary precautions in planning. The first is that routing depends on the location from which the signal is sent and the location at which it's received. For example, USA to Tokyo is an all-fiber-optic connection. If you go to a city in Japan outside Tokyo, however, the connection may or may not be totally fiber. By the year 2015 Nippon Telegraph and Telephone (NTT) plans to have virtually every office and home in Japan on its fiber optic telecommunications system, a multibillion dollar investment. In 1992, AT&T completed installation of TPC-4, its first direct fiber optic link between the USA/Canada and Japan.

Fiber optic lines to Europe also vary by location. To Geneva, Switzerland, lines are almost all fiber optic. As of 1992, 92 percent of AT&T lines to Paris were fiber. Randy Berridge, District Manager of Public Relations for AT&T, stated that Switched Digital Intercom (SDI) is available, using one of several types of digital access circuits, to Australia, Belgium, Hong Kong, Jamaica, Japan, The Netherlands, Singapore, Sweden, and the UK. The charge for overseas lines is split between AT&T and the receiving nation.

The second precaution for planning global television backhauls is that with *any* fiber optic line you have to ascertain its transmission rate.

Telephone lines often operate in the range of fifty to seventy thousand bits-per-second, more than enough for voice transmission but far short of the forty-five million bits per second (megabits) needed for a television signal.

AT&T

According to Berridge, within the USA, including Alaska and Hawaii, and Canada, AT&T's long-distance system is all fiber optic, and by the end of 1990 75 percent of all of AT&T's domestic and international traffic was carried via fiber optic cable. Expanding globally, AT&T bought the UK information-service company ISTEL and formed alliances with Italy's leading telecom manufacturer (Italtel) and with a similar firm in Spain (AMPER).

MCI

According to Steve Fox, a public relations spokesperson for the Southern Division of MCI, MCI has partnership rights in the consortium-owned fiber backbones between the USA and Europe (TAT-8) and between the USA and the Pacific Rim. MCI entered the expanding international videoconferencing field in 1991, announcing a cooperative effort with PictureTel Corporation, one of the two largest USA manufacturers of videoconferencing equipment.

US Sprint

US Sprint's current and future systems offer a dramatic view of how fiber networks are expanding. US Sprint has the largest domestic fiber optic network in the USA. According to Vince Hovanec, National Media Manager for US Sprint, the system is co-owner, with the UK company Cable & Wireless, of PTAT-1, a private line to the UK. US Sprint also owns interest in the North Pacific Cable (USA to Japan); the Pacific Rim (PacRim) East and PacRim West, scheduled for completion in 1993; the Asia-Pacific line from Tokyo to Singapore, scheduled for completion in 1993; and Coqui-1, from San Juan to the British Virgin Islands.

The "Baby Bells"

There are other USA companies involved in fiber optic cabling around the world. In the UK, for example, USA cable operators and tele-communication companies ("Baby Bells") control 90 percent of cable franchises. NYNEX (New York/New England) has offices in Frankfurt, Geneva, Hong Kong, London, and Singapore. It owns half of Gibraltar Telephone Company, which has announced plans to install fiber optic services. In addition to UK ties, it has agreements with companies in France and Scandinavia. Pacific-Telesis (California/Nevada) is part of a group installing a digital system in Germany and is part owner of a Japanese long-distance phone company. Southwestern Bell owns part interest in a cable company in Israel. BellSouth is involved in joint ventures in Argentina, Australia, and France. Bell Atlantic is involved with data communications in Austria, France, Germany, Italy, and Switzerland. US West is working in France, Hong Kong, and Hungary and is part of a multicompany group scheduled to install the longest fiber optic cable system in the world, across the CIS.

Figure 3–7 Adding aerial coverage to *Sport Aid* introduced dramatic shots like this video print of the London race and allowed us to show the vastness of the numbers participating, which would have been impossible using just ground-level cameras.

Figure 3–8 Heat overcame the portable uplink station that fed the first live television pictures of Ougadougou to the rest of the world, but a hard-working British and Burkina Faso crew resurrected the uplink in time to reestablish contact later in the show.

Figure 3–9 My background in sports directing was helpful in airing video of the winner in each of the twelve races we covered live on *Sport Aid.* By asking the eternal question "What If?" I had arranged to tape all of the winners as they crossed their finish line. Two of the twelve crossings occurred simultaneously, and I had to show the second on tape. Note the contrasting night sky "Downunder" in Auckland. The difference in time zones is a constant challenge to globalcasting.

Figure 3–10 Conducting post-race interviews with race winners was a challenge due to the different languages.

This is not a complete listing of the Baby Bells overseas projects, and not all of these companies' links involve fiber optics. Many are geared toward cellular phone and other wireless services. But the potential for fiber optic networks now exists in all countries. And one striking fact is that fiber optic cable is reaching the point at which it is in direct competition with many satellite services. It is also very involved in transmitting expanded programming to USA cable-TV subscribers.

Fiber Optic Feeds Versus Satellite Feeds

Now that we've analyzed both satellite and fiber feeds as production tools, let's discuss which is better for use in global television. The battle breaks down into five categories: quality, reach, reliability, security, and cost.

Quality

Round one goes to fiber optic transmission, on at least two counts. Weather, including sun spots and excessive rain, can cause signal interference with satellite transmission, whereas fiber is not affected by weather. The second factor is the possibility of signal interference from the satellite equipment, from degradation through processing equipment. Satellite signals have a decibel signal-to-noise (S/N) ratio of 50 to 60 decibels, versus 64 decibels for fiber.

Reach

Round two goes to the satellites—and probably will for the foreseeable future. As indicated, most of the USA and Japan should be fully fiber-networked by 2015. But the so-called developed nations do not make a world. When we telecast *Sport Aid: The Race Against Time* in 1986, millions of people in more than 200 cities around the world, raced at the same time. The worldwide viewing audience was nearly two billion. And although there were 250 thousand people running in the streets of London and then United Nations Secretary-General Javier Perez de Cuellar officiated in Lisbon, Portugal, perhaps the highlight of the show was the signal we fed to the world from Ouagadougou, the capital of Burkina Faso, in Africa. *The Race Against Time,* produced by Simon Dring, was Bob Geldof's follow-up effort to *Live Aid,* and the goal of *Sport Aid* was to refocus attention on starvation in Africa and to raise funds to help alleviate that starvation. Burkina Faso was the only inject coming from Africa. It was also the first time in history that Ouagadougou had been linked live with the rest of the world. To make this event possible, Independent Television News (ITN) flew in a portable satellite station from London. We had a signal from Ouagadougou for the opening of the program, but lost it because the extreme heat (over 105 degrees Fahrenheit) burned out the portable satellite uplink. The technicians in Burkina Faso worked ceaselessly, using ice among other things, to reactivate the station. About two-thirds of the way through the program, the uplink reactivated and Ougadougou rejoined us live on global television. It was a thrilling moment, and it was made possible by a satellite feed. It would still be possible only by satellite feed today—and probably for the foreseeable future. Fiber optic transmission is possible only where the cables go, and it will be decades before the entire world is wired.

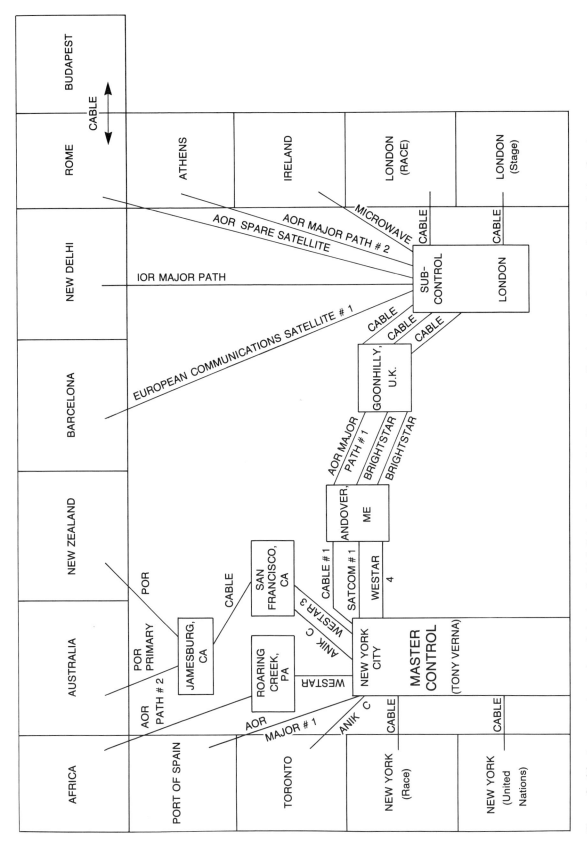

Figure 3–11 I prepared a series of charts, of which this is the first, detailing the audio channels of communication during *Sport Aid*. These charts convey a sense of the complex nature of the communications faced not just by the executive director, but by every member of the crew of a globalcast. Compare this to the chart of the video paths for *Prayer For World Peace* (Figure 2–24). See text description in Chapter 9.

Figure 3–12 These three lists detail the destinations for the thirty keys on my audio box for *Sport Aid*. The left hand column (keys 1 through 15) indicates connections to the on-air commentators. The middle column (keys 16 through 30) indicates connections to the same locations, but to the control rooms. The third column (blank keys 1 through 30) indicates lines used by others in the crew for business phones.

Commentators		Control Room		Business Phones	
				Production	Technical
1	NYC	16.	NYC	1	16
2	NYC	17.	NYC	2	17
3	Toronto	18.	Toronto	3	18
4	Port of Spain	19.	Port of Spain	4	19
5	Quagadougou	20.	Quagadougou	5	20
6	Auckland	21.	Auckland	6	21
7	Brisbane	22.	Brisbane	7	22
8	London	23.	London	8	23
9	London	24.	London	9	24
10	Dublin	25.	Dublin	10	25
11	Barcelona	26.	Barcelona	11	26
12	Athens, Greece	27.	Athens, Greece	12	27
13	New Delhi, India	28.	New Delhi, India	13	28
14	Rome, Italy	29.	Rome, Italy	14	29
15	Budapest, Hungary	30.	Budapest	15	30

There's another reason satellite transmission wins this round. In addition to the portability of satellite equipment, dramatically evidenced during the Persian Gulf War and Operation Restore Hope, there is the reach of the satellite footprint. Satellites can be orbited either to reach very specific portions of the globe or to cover larger areas. A series of satellites, properly positioned, can send a TV signal virtually anywhere in the world. It will be some time before fiber optic cable can duplicate that reach.

Reliability

Round three is a closer call, but fiber probably has a slight edge. Weather can affect satellite signals, especially the increasingly used Ku-Band and Ka-Band satellites, and no one has control over the sun and rain. And a much more ominous problem has begun to plague satellite users—the death of existing satellites, especially the older ones, before their replacements arrive. Technical innovations are gradually extending the lives of the newly manufactured satellites, but the problem will be around for some time. For example, the Japanese Satellite Broadcasting's WOWOW channel had to curtail part of its broadcast plans because the launch of a new satellite was aborted and the existing satellite on which WOWOW was able to borrow transponder time was weakening. CNN faced a similar problem in Asia. A satellite launch failed in 1990, and in mid-1991 the satellite that was to be replaced, Intelsat Vf8, began to falter. While CNN was able to switch to a temporary replacement satellite (Intelsat Vf3, also aging) the orbit of the replacement did not match that of the original. As a result, unless subscribers had a tracking device to move

Text continued on page 63.

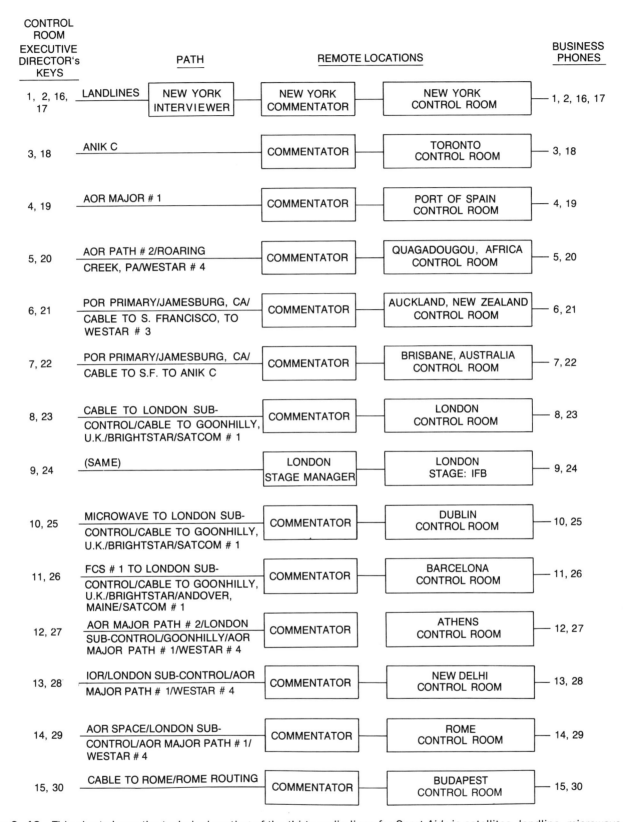

Figure 3–13 This chart shows the technical routing of the thirty audio lines for *Sport Aid* via satellites, landline, microwave, and so on.

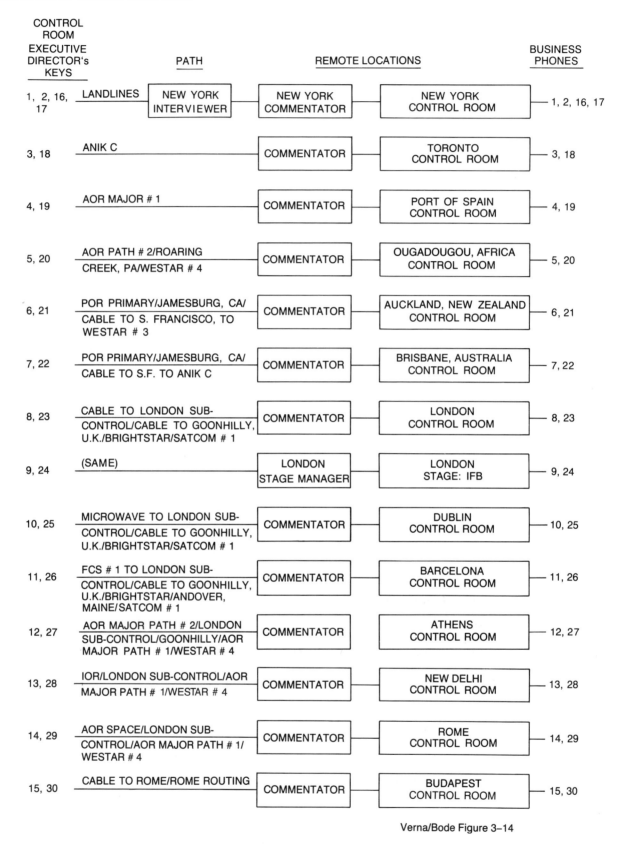

CONTROL ROOM EXECUTIVE DIRECTOR's KEYS	PATH	REMOTE LOCATIONS		BUSINESS PHONES
1, 2, 16, 17	LANDLINES — NEW YORK INTERVIEWER	NEW YORK COMMENTATOR	NEW YORK CONTROL ROOM	1, 2, 16, 17
3, 18	ANIK C	COMMENTATOR	TORONTO CONTROL ROOM	3, 18
4, 19	AOR MAJOR # 1	COMMENTATOR	PORT OF SPAIN CONTROL ROOM	4, 19
5, 20	AOR PATH # 2/ROARING CREEK, PA/WESTAR # 4	COMMENTATOR	OUGADOUGOU, AFRICA CONTROL ROOM	5, 20
6, 21	POR PRIMARY/JAMESBURG, CA/ CABLE TO S. FRANCISCO, TO WESTAR # 3	COMMENTATOR	AUCKLAND, NEW ZEALAND CONTROL ROOM	6, 21
7, 22	POR PRIMARY/JAMESBURG, CA/ CABLE TO S.F. TO ANIK C	COMMENTATOR	BRISBANE, AUSTRALIA CONTROL ROOM	7, 22
8, 23	CABLE TO LONDON SUB-CONTROL/CABLE TO GOONHILLY, U.K./BRIGHTSTAR/SATCOM # 1	COMMENTATOR	LONDON CONTROL ROOM	8, 23
9, 24	(SAME)	LONDON STAGE MANAGER	LONDON STAGE: IFB	9, 24
10, 25	MICROWAVE TO LONDON SUB-CONTROL/CABLE TO GOONHILLY, U.K./BRIGHTSTAR/SATCOM # 1	COMMENTATOR	DUBLIN CONTROL ROOM	10, 25
11, 26	FCS # 1 TO LONDON SUB-CONTROL/CABLE TO GOONHILLY, U.K./BRIGHTSTAR/ANDOVER, MAINE/SATCOM # 1	COMMENTATOR	BARCELONA CONTROL ROOM	11, 26
12, 27	AOR MAJOR PATH # 2/LONDON SUB-CONTROL/GOONHILLY/AOR MAJOR PATH # 1/WESTAR # 4	COMMENTATOR	ATHENS CONTROL ROOM	12, 27
13, 28	IOR/LONDON SUB-CONTROL/AOR MAJOR PATH # 1/WESTAR # 4	COMMENTATOR	NEW DELHI CONTROL ROOM	13, 28
14, 29	AOR SPACE/LONDON SUB-CONTROL/AOR MAJOR PATH # 1/ WESTAR # 4	COMMENTATOR	ROME CONTROL ROOM	14, 29
15, 30	CABLE TO ROME/ROME ROUTING	COMMENTATOR	BUDAPEST CONTROL ROOM	15, 30

Verna/Bode Figure 3–14

Figure 3–14 This chart combines the data from Figures 3–12 and 3–13, showing transmission lines plus the multiple destinations. In addition to maintaining an awareness of dozens of video images on multiple monitors, the executive director of a globalcast has to maintain constant awareness of the positions of the keys for audio communications. Mastering the key pattern is a little like learning a new typewriter keyboard for each globalcast.

SPORT AID
TONY VERNA
GLOBAL MEDIA

DATE:
TIME:

NOTE: THESE GRIDS CHANGE DAILY.
CHECK BILL BODE, TONA VERNA, WENDY ARMENDARIZ.

ITEM	TIMES (P.A.)	REMOTE NOTES	AUDIO	MUSIC & EFFECTS	STILL STORE	CHYRON	T.D. VIDEO EFFECTS	VTR TAPE A.D.'s	WRITTEN COPY	U.N. A.D.	MASTER CONTROL A.D.	COMMENTS
1 OPEN												
2 VTR SPOT												
3 SMITH (live)												
4 V.N. (live)												

Figure 3–15 I created this blank grid to serve as a rundown sheet for *Sport Aid* based on my experience with *Live Aid*. Combining the many production elements allowed everyone to share a single grid/rundown.

SPORT AID
TONY VERNA
GLOBAL MEDIA

DATE: 5/23
TIME: 5:30 p.m.

Note: These grids change several times a day. Check **Bill Bode, Wendy Armendaris,** or Tracy Verna for updates. AIR COPY WILL BE YELLOW.

ITEM	TIMES	REMOTE NOTES	AUDIO	MUSIC & EFFECTS	STILL STORE	CHYRON	TD/VIDEO EFFECTS	VTR TAPE AD'S	ANNOUNCER & WRITTEN COPY	U.S. AD & S.M.	MASTER CONTROL AD	COMMENTS
5	2:55	London New Zealand NYC, others if possible	R.S.: Mic +intl sound		Record when approp.		windshield wipe L to R		R.S. adlib re: cities around world		dispatching	
6	1:00		Track VTR PSA - 2 2 channel audio		Record when approp.			Roll VTR			Roll VTR	
7	:30	Beauty shot of UN with people on sideline	R.S.: mic		Record when approp.				R.S. V/O			
8	8:00	Prep cam-car and helicopter	feed both audio channels		Record when approp.					cue runner		VTR arrives Sunday 7 a.m.

Figure 3–16 This partially filled in grid/rundown for *Sport Aid* shows typed entries for all personnel to see, but there was space also for each individual to add instructions or communications unique to any specific job and/or time.

their antenna dishes to bring in the repositioned signal, reception deteriorated. CNN has nearly 100 thousand subscribers in Japan, in addition to feeding close to 50 thousand hotel rooms. Add to these numbers more than 50 thousand subscribers in Hong Kong and Taiwan (plus an additional 20 thousand hotel rooms), and you have a lot of potentially upset viewers.

While nothing lasts forever, fiber optic cable generally is expected to have a longer life than satellites. Transoceanic cable, with an estimated life of twenty-plus years, always runs the risk of disruption, and natural phenomena such as earthquakes could threaten some land lines. But these threats appear to be fewer than those facing satellites.

In the case of domestic feeds by Vyvx, the match goes hands-down to fiber. Much of Wiltel's 11 thousand miles of cable was buried in decommissioned oil and gas pipelines that belonged to its parent company.

Security

Round four tilts toward fiber, too. If it weren't for "Captain Midnight," the battle would be close, but industries allied with satellites will never forget the night of Sunday, April 27, 1986. General Instrument's Video Cipher II had become available to scramble satellite signals so that pay services like HBO could not be pirated by satellite dish owners. What had been free would be free no longer. The clear signal was to be scrambled, and dish owners would have to pay for a descrambler and the monthly subscription rate to receive the HBO signal. There was much dissent, of course, from dish owners. But the fears of all satellite users (including the government) were realized that April night when Captain Midnight went on the air, cutting into the HBO signal with a message protesting the newly imposed charges.

It turned out that Captain Midnight was a technician at a teleport, an earth station with access to a number of satellites. Although he had easy access to the HBO signal, the fact that *anyone* could, without detection, take over a nationwide satellite feed sobered many a mind. Issues aside, it was healthy for the industry—and government security forces—to be caught short, to be shown how vulnerable communications systems are.

(As an aside, when the USA was trying to prosecute students for hacking their way into government security files in computer banks, teachers in the Netherlands put the problem in a different perspective. Far from prosecuting students for gaining access to government files, instructors in the Netherlands *urged* their students to try to break all existing codes, government and other. The reason—quite properly—was that security is the responsibility of the people maintaining the files. If a student computer hacker could wend his or her way into "secure" files, someone less innocent could do the same.)

Captain Midnight is not the only blow against security of satellite feeds. A 1991 report by the Satellite Broadcasting and Communications Association estimated that as many as six of ten satellite antenna dish owners use illegal decoders to steal the signals of scrambled services like HBO and Showtime. This problem can be solved in part by new technology. General Instrument's Video Cipher II Plus RS (for renewable security) has changeable code slots so that if signal pirates break an existing code, a new one can be substituted. Scientific Atlanta's nonbroadcast B-MAC standard has a hard-scrambling technique that randomly inserts timing delays at the start of each and every television scan line. The key that decrypts the delay is also random and changes every one-quarter second. As Scientific Atlanta says, "That's secure." As we shall see in the section on sports (see Chapter 17), however, some events have been kept from DBS distribution because of the fear of piracy.

Even when the scrambling/descrambling system works, it remains an issue. Signals on fiber optics need no scrambling because they cannot be intercepted without detection.

Cost

Round five is probably a draw for domestic broadcasts, with satellites getting the edge for globalcasts. Hourly rates for one-way transmission via fiber are about USA$400 and for two-way (interactive) transmission, using two lines, are about $600. Satellite transponder time rates vary from about USA$200 to $800 per hour depending upon the time of day, night rates being cheaper.

Summary

Checking the score card, fiber wins for domestic broadcasts. For most of the 1990s, however, satellites will be the way to reach a global audience. But depending on the show's broadcast time and distribution patterns, it is conceivable, if the origination point is in the USA, that it will be cheaper to go by fiber to Europe or Asia and uplink from there. Each program is different. The rule of thumb is to approach each show as the unique effort it is and to compare distribution patterns and costs.

The good news for globalcasting is that there is no reason for fiber and satellites to fight it out. Use both: be aware of all of the facilities available for a globalcast and integrate their use in the most effective, most economical way.

Chapter **4**

Charge-Coupled Devices (CCDs)

How CCDs Are Used

Charge-coupled devices (CCDs) are not transmission tools like satellites and fiber optic cables, but combined with those transmission methods they have dramatically altered television production—local, national, and global—in terms of both equipment and coverage. When you combine the country-jumping, continent-jumping speed of satellites and light-weight fiber optic cable with the facility of solid-state CCD cameras, you can see just how much the world has opened up for global TV producers and directors. Where a decade ago you might have been limited by bulky equipment and aging coxial cable, now the main limitations are with human imagination.

A CCD is an image sensor, a type of semiconductor (chip) used, basically, in place of a camera tube. CCDs are also used in computers. In TV, CCDs are used in video cameras in place of picture tubes, especially in electronic news-gathering/satellite news-gathering (ENG/SNG) hand-helds. For nonbroadcast TV purposes they are used in sensor cameras on satellites and on military vehicles. First developed by Bell Laboratories in 1969, CCDs have been much researched and used by NASA. They are also used in telecine systems, in lieu of camera tubes, in film and tape editing systems, and in image-processing equipment—to translate images from analog to digital to be enhanced with computer graphics and to create the special effects that have raised image artists on many feature films to the same level of stardom as the actors. Is that Arnold Schwarzenegger or a digital Arnold Schwarzenegger?

How CCDs Work

A CCD is a solid-state semiconductor chip that contains light-sensitive elements. The terms *chip* and *semiconductor* are often used interchangeably, but they are different. Technically a *semiconductor* is a substance, such as silicon or gallium arsenide, with poor conductivity that can be improved with the application of heat, light, or voltage. A chip is a *microchip,* an electronic memory device that contains microscopically thin layers of a semiconductor configured into an integrated circuit. An *integrated circuit* is an electronic circuit containing multiple interconnected amplifiers and circuit devices. Currently most integrated circuits are printed onto semiconductor chips by passing light through a stencil that has transparent and nontransparent sections (much like a photographic negative). (For other techniques see *optical lithography* in the Glossary.)

CCDs use silicon chips with light-sensitive elements, each of which becomes independently charged by light coming through a camera lens. The chips convert the images into electrical signals for transmission.

The chips are sensitive to light just as a camera tube is, but they're not made with breakable glass tubes. They're very small, solid-state image sensors with so many size and weight and other advantages that they have helped revolutionize TV cameras and what TV cameras can do.

Why CCDs Are Used

Cameras made with CCDs are lighter, smaller, easier to handle, and more rugged. They require less power. (If you're working off batteries, they last longer.) And they produce better results under low-light conditions. The resolution and quality of pictures from solid-state cameras aren't as good as those from a tube camera, but except in a highly controlled studio setting no one seems to care. Other advantages are no burn-in, no comet-tailing after moving light sources, less maintenance and longer life, no need for warm-up, no magnetic-field interference from radio transmitters. But the main reason CCD cameras were teamed with microwave and satellite transmission to give us ENG/SNG news coverage is the combination of small size, light-weight, low-light sensitivity.

Electronic news gathering and satellite news gathering reports are the name of the game and have been for years. Reality programming has been soaring in the ratings in country after country. Worldwide, viewers took an immediate liking to the intimate up-close, in-your-face feel of the hand-helds. It's how you get the real news. It's how you get the gossip. It's how you get the home videos. It's how you get the power and intensity of a Mick Jagger–Tina Turner duet.

I've put CCD solid-state cameras on stage with musicians, in the field with ball players; I've mounted them on bicycles and on boats. The World League of American Football put one on top of a football quarterback. David Letterman put one on a monkey. One was planted in a bowling ball.

With satellites in geostationary orbit and planes and space vehicles carrying solid-state cameras into the wild blue yonder and beyond, the sky is literally the limit. Using satellites, fiber optic cable, and solid-state CCD equipment, you can transmit signals almost anywhere with equipment that can go almost anywhere and do almost anything. Add this hardware to the software synergy that is gradually revolutionizing all of the methods by which the hardware is used, and you sense the future.

Chapter 5

Analog/Digital Technology

Satellites, fiber optics, and CCDs are hardware—mechanical devices for recording and transmitting data. Analog and digital technology define the software—the forms the data take when stored, manipulated, and transmitted—the processes, not the machines. It may not be essential, but it is always helpful to understand how equipment works. You can seldom use equipment to its fullest potential if you don't understand the technology.

For example, when I developed the instant replay, it was necessary to have the videotape machine on the field, not in the studio. Unlike today's tape, in 1963 it took an inexact number of seconds for the video to lock to the point where it was synchronous and air-worthy. I couldn't use the crude footage counter because it was on the tape machine itself. I overcame this problem by having the tape operator put tones on the tape so I could identify each section: one tone for the offensive team huddle and two tones when the team broke huddle. When I heard the two tones, if the video locked in, I put the replay on the air.

Accomplishing this the first time required extremely close coordination with the technicians. It also meant conning a friendly technician into driving a truck from New York to Philadelphia so I could have the tape machine with us on the field and persuading other skeptical but dedicated technicians to work unpaid overtime to set up cabling from the isolated camera to the tape machine.

All of this didn't just "happen." It was created. And I couldn't have conceived the idea or persuaded the technicians to attempt "the impossible" if I hadn't understood enough about the equipment's potential to create the techniques.

Similarly, it's difficult to appreciate the new digital technologies—and the way in which the media will converge—without sufficient background to understand the fundamental physical processes. We are only at the *beginning* of this convergence. In the decades after the 1990s sight and sound will blend in ways we cannot now imagine. But you will be better equipped to imagine if you understand the science behind global TV.

"The Age of Information"

Basically, globalcasting—and the other dramatic shifts in the way we communicate with each other—have been made possible by the scientific explosion called The Age of Information. Four basic factors developed technically and converged to produce this explosion: (1) the form of

Figure 5-1 The instant replay is taken so much for granted now that it is hard to realize someone had to invent it. The original tape on which I recorded the action of the 1963 Army-Navy Game for the first instant replay was wiped, because the 2-inch tape used then was costly and was reused. This video print from the game film shows the 1963 touchdown action from the same angle as the camera I used to isolate the action for that first instant replay.

Courtesy United States Military Academy.

information itself; (2) the method of storing, manipulating, and sending information; (3) the channels through which information is sent; (4) the method by which information is displayed. Let's look at each of these more closely.

The Form of Information

A series of discoveries by physicists in the nineteenth century led to an awareness that sight and sound were different degrees of the same phenomenon, *electromagnetic energy*. These discoveries led to the later invention of radio and television communication.

The Method of Storing, Manipulating, and Sending Information

The major change came with the realization that analog information could be stored and manipulated in digital form and then reproduced in analog form. In this context *analog* refers to information you are familiar with: words and pictures. Analog transmission transmits images of these words and pictures—visual, not numerical, images—by varying the amplitude and the frequencies of electromagnetic waves. *Digital* refers to the use of a series of electric or electronic impulses to represent those words or pictures in the form of numbers. Analog images are "the real thing," images you can see. Digital images are a series of on-off impulses that contain and can reproduce the information in the images. Technically, analog equipment, including analog computers, reads and records light and sound waves using voltage levels, as opposed to the zero-one numbers of digital computers. The machines that let the two systems communicate with each other are called, logically, analog-to-digital (A/D) and digital-to-analog (D/A) converters. *Codec* (a contraction of COde and DECode) is also used to refer to this process. The manipulation, however complicated and complex, is all based on the fundamental digital binary system of zero/one, off/on.

The Channels Through Which Information Is Used

Until very recently, the wires through which information is transmitted were all made of copper. The shift from copper to fiber optic cable dramatically increased the amount of information that could be sent and the speed at which it could be transmitted.

The Method by Which Information Is Displayed

The development of liquid crystal displays (LCD) has led to very visible changes in the way clocks and watches look, the size and appearance of computer screens, and the method of displaying television pictures.

We discuss the channels through which information is sent in Chapter 3 and the method by which information is displayed in Chapter 15. Here we will look at the form of information and how it's stored, manipulated, and sent.

The Form of Information: Sight and Sound

The form of information has to do with the physical nature of electromagnetic radiation, the element we transmit in globalcasting, which is involved with waves charted on the electromagnetic radiation scale.

Electromagnetic radiation was first discovered in 1865, by Scottish physicist James Clerk Maxwell, and his theories were confirmed some twenty years later, when German physicist Heinrich Hertz verified the existence of radio waves. Figure 2–1 presents the measurements on the electromagnetic scale.

We are all familiar with most of the categories on the electromagnetic scale: radio waves that are beamed to our receivers; microwaves used in microwave ovens; x-rays used in doctors' and dentists' offices. All of these waves are forms of energy measured in terms of *wavelength* and *frequency*.

Scientists describe this energy as the result of an electric field interacting with a magnetic field, hence *electromagnetic*. Again, we are all familiar with magnets and magnetic fields: some particles are positively charged and some are negatively charged. The opposites attract each other and those with the same charge repel each other.

The current we tap into when we plug in the television is the flow of electric charges. In terms of globalcasting, it is important to know the relationship of the waves to each other. You will note on Figure 2–1 that radio waves are the longest waves, gamma waves are the shortest, and visible light waves are in the middle. *Photon* describes the energy that carries these waves. Photons have no mass and no charge, positive or negative. What they do have is momentum—energy. And they can travel through—in essence—nothing, meaning they can travel through a vacuum (or as near a perfect vacuum as we have achieved). This is important, of course, because vacuum tubes were a critical element in the development of the transmission of radio/television waves.

Radio (short for radiotelegraphy) was created by the transmission and reception of electromagnetic radiation from one place to another without wires (thus "the wireless"), and telegraphs are created by periodically interrupting radio waves in their transmission (sent first through wires, later wireless). Modulation is the process of varying the characteristics of radio waves. AM (amplitude modulation) radio is radio waves that are constant in frequency but vary in intensity (amplitude). FM (frequency modulation) radio is radio waves that are constant in amplitude, but vary in frequency. (See Figure 5–2.)

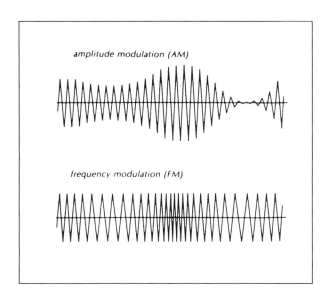

Figure 5–2 Modulation is a process of varying certain characteristics of an electromagnetic wave to carry data-containing signals. AM (amplitude modulation) transmits data by maintaining consistent frequency but varying the amplitude (the intensity) of the signal (A). FM (frequency modulation) maintains consistent amplitude but varies the frequency of the signal (B). Both AM and FM are familiar in radio use. FM is currently used for television transmission. A third type of modulation, pulse modulation, maintains consistent amplitude and width, with all signals equally spaced, but varies (by presence or absence) selected pulses in the carrier stream.

The Pace of Development and Convergence

As noted above, it was not until 1865 that humans became aware of the electromagnetic waves that made radio and television possible. Alexander Graham Bell invented the telephone in 1876, and Thomas Edison invented the microphone in 1877 and the incandescent lamp in 1879. Wireless telegraph came along in 1895, the cathode ray tube two years later, the first live signals over long-distance radio in 1900. In terms of history, we are dealing with very recent discoveries and inventions.

And that brief history highlights two important points. One is the discovery of the audio and video elements of broadcasting that are on the brink of converging with communication elements normally considered—incorrectly—outside the realm of broadcasting (e.g. computers, telephones). The second is the speed of the developments involved in globalcasting. The interval between the discovery of radio waves (1873) and the beginning of regularly scheduled radiocasts (1920) was almost fifty years. Compare that with the development of television. Within twenty years of John L. Baird's first public demonstration of television (1926), regularly scheduled television began in the USA (1946), delayed five years by World War II. Just ten years later (1956) videotape was introduced. Six years after that (1962) a satellite TV feed connected the USA and Europe.

The pace of electronic and photonic development has increased since the 1960s and will continue to increase during the 1990s.

Storing, Manipulating, and Sending Data: Bits of Information

Perhaps the most dramatic display of the accelerating pace of global communication is a comparison of the rate of sending data. When Samuel Morse sent his first telegraph message in 1844, the information was transmitted at the rate of one bit per second. In 1858, Queen Victoria cabled President James Buchanan—the first transatlantic telegraph cable—and it took more than sixteen hours to transmit her ninety-nine-word message. In the 1990s, glass fiber optic cable routinely carries 100 million bits of information per second, and in experiments fiber optics have transmitted one trillion (USA) bits per second.

How are such rates possible? The answer is that data are transmitted digitally, via computers. Why discuss computers in a book on globalcasting? Because computer technology and television technology are intimately connected. Charge-coupled devices, for example, are used in computers as well as in video cameras, digital telecine (film-to-video transfer) systems, and tape-editing and image-processing equipment.

A computer phenomenon like RAM (random access memory) is the reason the *electronic still store* replaced slides. When a sports director calls for the Chyron to produce supers for on-screen data, the process that gets that data on-screen instantaneously is random access. You can get to the information instantly, without having to go past anything. CD records players use the same technique. And think of all the digital video effects (DVE) you see in television today, with compressed images flying all over the screen. Think of the electronic switchers that make all the special effects images work. Increasingly, the media and the computer world are using the same technology.

Bits, Bytes, and Binary Code

An appreciation of the role fiber optics will play in globalcasting—and the reason convergence is such an inevitable theme for the 1990s—depends on an understanding of the words used in relation to transmitting information. *Bits* is a contraction of BInary digiTS. Digital computers use a binary system of numbers, *binary* meaning having two parts, in this case the digits zero and one. Using the analogy of a light switch, zero means off, one means on. Sequences of these bits, these zeros and one, are used to represent letters of the alphabet, musical notes, graphics, or other numbers. When you type on a computer keyboard, for example, each letter key you strike is recorded in the computer's memory in a sequence of bits. Historically, the on-off system of communication has been used in many ways: smoke signals, mirrors reflecting the sun's rays, ship-to-ship lights flashing on and off, the telegraph key. Under different guises, bits have been around for centuries.

Today, in most computer systems, eight bits are combined to form a *byte*. This system is called extended binary coded decimal interchange code. The letter A and the number one are represented in this code by the eight-digit bytes 11000001 and 11110001, respectively. You can see from these examples why it is important to have capacity in the millions, billions, or trillions. It is also important that these zero–one combinations be transferrable from the computer's memory to a permanent memory source— floppy disks, hard disks, or tape. What has been "memorized" can then later be called up and reproduced in its original form, *exactly* as memorized. Endless copies of documents, graphics, musical recordings, or motion picture films can be produced, each equal to the first. Numbers don't wear out the way needles, vinyl and tape do, especially not when they're read by laser light, which does not cause deterioration on the surface of a compact or laser disc.

Zero and One Add Up to Processed Images

Digital information, once stored, can be manipulated, and this makes possible the special effects so common in television production. Multiple images swirl about the screen. Pages turn, revealing new images. Images appear in one shape, are transformed into another and then another. In films like *The Abyss, Willow,* and *Total Recall,* images and effects were created through digital manipulation. The filmed image was fed into an image processor, transferred into digital data (zeros and ones), manipulated by computer graphics technology, and transferred back to film.

Originally this system was bulky: computers filled rooms and cost fortunes. As the systems were compressed and miniaturized, the simple off–on system made its way into homes and offices as desktop computers and video games. By 1990 IBM began pilot production of a 16 million-bit computer memory chip. Each of the 16 million bits represents an off or an on, a zero or a one. This tiny chip can store the equivalent of approximately 1,600 double-spaced typed pages. And scheduled for completion by the mid-1990s is a supercomputer, commissioned by a group of fifteen universities, designed to perform one trillion (USA) mathematic calculations per second, a rate equal to that of experimental fiber optics transmission.

And for those who "don't know anything about computers," future computers will adapt to your needs. There are already computers with touch-sensitive screens. You can write on them with an electronic pen. There are also computer programs you can talk to, such as "Talk and Draw," a program developed to help educate the physically handicapped. This program lets the user control the screen by talking—as a sportscaster might while describing an upcoming play or a replay. Computers will be so fast and so user-friendly that we will all need them and be able to use them. In fact, the day is near when a single machine will integrate the functions of computer, television, telephone, fax machine, and stereo. The word *computer* will disappear—we will all own Vidtalk-Stereofax machines. Such an appliance is already being developed jointly by Time Warner and IBM.

Another extremely critical factor in the converging of the media is cost. Almost as astounding as the changes computers have brought about in communication are the changes in the cost of a computer. When you buy a new car with many ballyhooed improvements, it costs you more than the old model. Computers are the opposite: as they expand in speed, capacity, and reliability, they shrink in price. Within a decade, the cost of exactly comparable computer systems dropped from $25,000 to $2,500. If automobiles had done the same, a car that cost you $5,000 ten years ago would cost $500 today.

Information: Stored, Manipulated, Sent, Displayed

I don't think anyone in either the television or computer industry doubts that there will be a marriage between the two. Whether it occurs through one of the current programs or one still to be invented, a new standard will merge voice, video, and graphics. One sign that industry leaders have begun to act on this inevitable convergence was formation of the twelve-company First Cities in October 1992. Microelectronics & Computer Technology (Texas) sparked the group, which announced its aim as creation of a nationwide multimedia network to provide home-consumers with everything from videophones to videogames, from film libraries to educational libraries. The group's outstanding feature: its comprehensiveness. Kaleida Labs—the joint venture of Apple and IBM, the research amalgam of the seven Baby Bells, Eastman Kodak, the investment firm of Bieber-Taki Associates, plus more, and more to come. Telecommunications, electronics, and financing comprise the shape of our multimedia future—not realized yet, but in formation. In the meantime, domestic and global telecasters have to stay alert to the shifting systems and standards and adapt programming to them, or them to programming. No doubt at all, the 1990s are a decade of transition.

Technically, as we've seen, the history of today's globalcasting began with the nineteenth century discovery of electromagnetic energy, through the development of radio and the visible radio we call television, to the computer-inspired emphasis on shifting from analog signals to digital signals, to the development of satellites and fiber optic cable capable of transmitting digital signals, to the development of high definition television and flat screens that provide portable TV and promise walls filled with whatever programming wonders are available on the hundreds of channels made possible by video compression.

And once again, we arrive at the inevitable underlying theme: To what use will we put the expanding, converging technology? What these technical miracles need is miraculous programming.

Part *Three*

Creating a Globalcast

Chapter **6**

Initiation and Budget

In these chapters on creating a globalcast, I will dissect a global production from conception to conclusion. Just as the minute-by-minute grid of each program varies, so do the details of production. But there are certain factors common to all productions. I've grouped fifteen steps in producing a globalcast into four arbitrary phases: getting an idea, getting a format, getting backers, and getting distribution. But there is much overlap. As I tried to show in my discussion of *Live Aid,* many things happen at the same time.

Getting an Idea

There is no way to predict where or when an idea will originate. Bob Geldof was driven by seeing starving people in Africa to conceive and create *Live Aid* and *Sport Aid.* Because the United Nations needed to warn of impending environmental disaster, *Our Common Future* and *Earth 90* were born. The idea I can speak of most personally is the one I created: *Prayer For World Peace.*

I grew up in South Philadelphia, a very Italian, very Catholic part of the city of Brotherly Love. My family's photography business/home stood next to an appliance store. I grew up with a still camera in my hand and pictures and video techniques in my head. When I saw images on TV sets in the appliance store window, I always got mad inside. The images I saw simply weren't as compelling as the technology. Moving pictures! Live! From all over! But most of the pictures on the tubes were dull, repetitive, the same, unimaginative. I wanted them to be more. I wanted the pictures to make me feel as excited and fascinated as the *fact* of the pictures made me feel. I was driven to make the meaning match the medium.

It was this constant striving that made me become a director, that made me force the video hardware into being a production tool. Videotape was too great a tool just to show programs later. I wanted it now—*instant* replay. And I wanted to take those pictures and make them do something more, to use the magic I saw in the immediacy of the medium.

After my experience with *Live Aid* and *Sport Aid,* I had my chance. I became president of the company Global Media, and when Pope John Paul II declared a fourteen-month period of prayer to the Virgin Mary dedicated to world peace, I saw the possibility of using the power of the satellites to link the world in prayer.

If there is a test for developing the basis for a globalcast, it would be that the concept of the program "travels"—that it applies not to a single nation or a single continent or a single group of people, but to the world. One of the buzzwords in film production today is travelability. Increasingly one nation's box office won't support the cost of producing a major motion picture. Similarly, global television has to span cultures and continents. The idea has to be large enough, universal enough to carry worldwide distribution, to justify the use of satellites. The message must justify the medium.

In the case of *Prayer For World Peace,* my company's board of directors approved my approaching the Vatican with the idea of making the Pope's prayer for world peace truly global with interactive response from seventeen different locations around the globe. It was a challenge and a dream I couldn't refuse. After weeks of discussion and negotiation, the concept was accepted and planning went forward.

What I envisioned—and what we were ultimately able to accomplish—was the program I've outlined: an hour-long globalcast of Pope John Paul II's recitation of the Rosary from the Basilica of St. Mary Major in Rome with parishioners in seventeen countries on all five continents responding—live—with the Pope able to see and hear the parishioners and the parishioners able to see and hear the Pope. Theologically it was a dream program. Technically it was a massive challenge. But we made the idea a reality.

In my earlier book, *Live TV,* Steve Allen tells the story of how one man's television work influenced him and two others, an author of books about farm workers and a producer/director of films about farm workers. All three had been inspired twenty years earlier by a single television program, Edward R. Murrow's 1952 *Harvest of Shame.* Murrow changed the lives of three people he never met. Those three, in turn, influenced others. An individual can make a difference. And an idea can come from any inspiration. Whether it is your own or someone else's idea, it is with a single thought that any creative work begins. It is so with globalcasts, as well.

Getting a Format

The second step in developing a globalcast is turning the abstract idea into a concrete proposal. People are starving in Africa. They need help immediately. What can you do about it? You can call on your friends. If you're a recording star like Bob Geldof, your friends are musicians and powerful promoters like Harvey Goldsmith. So you call Mick Jagger and Tina Turner and Hall and Oates and Phil Collins and Sting and a few dozen others. And when you realize that you have enough support, you reach out for someone who knows about producing huge television events.

Geldof reached out to Mike Mitchell at Worldwide Sports and Entertainment, and the *Live Aid* dream began to take shape. Performers in the UK, performers in the USA. Can't get them all together. Okay, two concerts: London and New York. Can't get New York. What can we get? What about Philadelphia? Okay, we have to raise money as well as consciousness. Do we use a tote board? I vetoed the idea of a tote board. We had to keep Geldof's vision. I opted for the global symbols—the sky, the clouds. Images. We could put them around the PSAs. We had a style. That was fine for the framework, the skeleton. What about a style for the sixteen-hour concert performances?

With a show that long you can't design a tight format the way you can with a half-hour show. But you have to establish an approach, an attitude, otherwise you'll have chaos. In writing, this attitude is called psychic distance range—how close you get to your material, how far you can stand back from it. The closest range you can establish between your subject and your audience with a camera is to use a subjective camera, making the lens the eyes of a participant, as director Robert Montgomery did, with mixed results, in *Lady in the Lake* (1946). Orson Welles reportedly considered filming *Citizen Kane* using a subjective camera but changed his mind. The elemental choice of your psychic distance range impacts on all subsequent decisions.

With *Live Aid* I chose to let the event happen at its own pace, in its own way—not to let our technical apparatus and techniques get in the way. I wanted the feelings and moods to develop minute by minute, hour by hour, song by song. When Phil Collins said he was willing to fly the Concorde to appear in both concerts, and we were able to promote the flight on the Concorde, I knew the psychic distance had gotten a big kick added. It was going to happen right before the viewers' eyes: "Never been done before . . . Watch it happen . . . See it happen . . . Grab it . . . It's unique! . . . Never before, never again!"

During the show, when I began intercutting shots of kids in the USA with kids in the UK watching Autograf performing in Moscow, I had my TD split the screen and put the shots on DiamondVision in the stadiums. The kids saw themselves on the screen and began waving, group to group. Though they were communicating by way of inanimate screens, each group felt that the other could see and read and feel their meaning. So I put them all together. We split the screen three ways and put the shots on the air so the young people on three continents could see their wish to communicate come true.

I wasn't superimposing my style on their show, I made it possible for their show to happen. I realized their wish for them by using the technology I'd amassed: satellites, DiamondVision screens, and split screen.

The moment couldn't have been planned. The kids created it, and the psychic distance I'd set up made it possible for their moment to circle the globe.

But you can't just randomly start a show and hope it will take some shape and end well. You need a running order, an outline. Planning the beginning, middle, end, and transitions (and commercials and PSAs) is what formatting is all about. Determining your psychic distance establishes your attitude. You then apply that to the material. For *Prayer For World Peace,* we constructed supporting explanations and celebrations before and after the prayer, but the formal structure of the Rosary dictated the format. With musical shows like *Live Aid,* you structure the format depending on available talent, satellites, and so on. These simultaneously predictable and unpredictable live programs force you to create a structure and to be able to alter that structure at any time.

An attitude of adaptability let me turn Paul McCartney's dead microphone during *Live Aid* into an advantage. When he started to sing "Let It Be" and his microphone didn't work, I tried to persuade the British producers to run a live mike in on-camera. I couldn't convince them to do that, so I called for shots of the audience on both continents singing along to help Paul. They knew the words and the tune, and we heard them loud and clear. The audiences rescued the song and created an historic musical moment on global TV—because we were ready for it. Our psychic distance attitude let it happen.

Okay, we have our psychic distance. We need more talent. Who'll book the rest of the talent? Who'll coordinate all the musicians, all the talent? We need music directors. Some countries don't allow televised fundraisers. We'll have to do a clean world feed and let the local countries make the pitch. More calls . . . Listen, MTV will carry it—the whole thing. We need another feed. Figure out how we do that. That's three separate feeds . . . Wait a minute! ABC wants three hours of prime time, but they want exclusives on some performers. It's getting complicated. We're up to four feeds, with some acts blocked out. More trucks, more coordinating producers, more coordinating directors, more switches, more patches—and more decisions to be made by one person, me. This could get out of hand without a system. . . Listen, it's taking shape. It could happen. But we need cameras and equipment. We need sets and make-up and lights and taxi fare. We need money. We need sponsors . . . Move on to the next phase.

Getting Backers

Live Aid was the miracle child of ideas. In ten weeks Bob Geldof's idea went from having no backers and no budget to being a world phenomenon. This was the result of endless hours of telephoning. Executive producers Mike Mitchell and Hal Uplinger talked to performers, to record companies, to satellite companies, to hundreds of television and radio stations, to broadcasters in hundreds of countries, to potential sponsors—begging, borrowing, coaxing, cajoling, convincing, inspiring: "No, I'm not kidding. It's a sixteen-hour show. . . . No, I'm not kidding." If someone said no, they called someone else, until someone said yes.

But *Live Aid* was the exception. Normally, proposed production elements are broken down and costs are estimated, with charts and diagrams. When the proposed ideas exceed a realistic budget, they are changed to fit the budget.

Most independent producers consider themselves lucky if 20 percent of the programs they conceive and propose become realities. Most ideas die before going into production, even on a pilot basis. The customary process is to make a presentation that details the content and intent of the program as well as the budget that would support it. Two versions of a presentation for *Christmas Around the World* in Figure 6–1 offer an example of different content for a multinational coproduction.

Once backing is secured, the business manager (unit manager) works with the producer(s) on every phase of production. For the 1990 *Goodwill Games* I created a series of overviews that detailed the production plans minute by minute; they were translated into technical manuals and bar graphs to measure the budget on a show-by-show basis. The comparison between budget and production is ongoing. There can also be levels of investment. With most pilots presented for network consideration, the potential backer (the network) may put up part of the seed money to produce the pilot. The producer puts up the balance. As I will discuss in some detail later, the global nature of production and financing has made coproduction almost a preordained part of both idea development and production, and the budget is basic to both. As soon as a program takes enough shape to be considered a practical possibility, the budget becomes a critical factor. It never disappears.

On *Live Aid*, budget and production worked hand-in-hand in many ways, not only in planning facilities but in helping to raise funds for famine relief. For example, the record companies representing many (not all) of the performers agreed to pay $20,000 to the Live Aid Fund each time we inserted a shot of the record label during a star's introduction. One producer's sole job for the entire sixteen hours was to watch three monitors carrying the various feeds. He logged each act as it played on each show, noting the

"CHRISTMAS AROUND THE WORLD"

-A Coproduction of (Sponsor)
and Tony Verna Productions

"Christmas Around the World" is a two-hour musical extravaganza celebrating the holiday with the unique customs and cultures of eleven countries around the globe. We'll visit castles and palaces and churches, snow-covered mountains and sun-drenched beaches, each location being so visually unique that if the sound went off, you'd still know where we were. Traditions as diverse as Rudolph the Red-nosed Reindeer and St. Nicholas will help us celebrate the feeling of one-world that Christmas symbolizes.

At each of our eleven locations, a performing host will sing and dance with local performers who will reflect the traditions from which our melting-pot America has fashioned our holiday celebration. Each location will feature a medley of songs and will run approximately eight minutes. Each song will showcase our American star, linking our U.S.A. holiday with the rest of the world.

The message of the program will be the music and, through that music, the multi-national traditions our locations reflect, the unity of our world on Christmas. There will be no religious message beyond the spiritual feelings inherent in some of the customs and songs. We'll start at the North Pole with the biggest snowman in the world and that "Frosty Snowman" will lead us through two hours of musical magic, an international greeting card set to song.

"CHRISTMAS AROUND THE WORLD"

Script Points for the Host

OPENING: Welcome to (Sponsor's) holiday celebration, "Christmas Around the World." This is one time of the year when there seem to be no language barriers or cultural barriers between nations. No single day unites the world as does Christmas day, and during the next two hours we'll be showing you that on this day the world can be a little different, a little better, as a dozen of your favorite stars entertain you and make the distance that separates us from the rest of the world a little smaller.

In a world that sometimes seems dark and at odds with itself, we hope that the colorful songs and celebrations we'll share with you will show you how bright and magical our one-world can be. So put aside whatever day-to-day problems may be on your mind and let today's electronic miracles carry you across the oceans, back to the countries and traditions our ancestors knew. Our electronic cameras and satellites may be new, but the spirit of Christmas is ancient--and timeless. So come with us now and meet our galaxy of stars around the world that, for today at least, looks to peace on earth and goodwill to all. "Frosty the Snowman" will lead us, and you'll see him in a lot of different guises as he, too, reflects the customs and cultures of our different heritages. Relax, then, and let's share "Christmas Around the World."

CLOSING: That's our (Sponsor's) trip to bring you "Christmas Around the World." We hope we've been able to brighten your holiday, to share the spirit of the season, the common warmth and harmony that this day symbolizes. As we look once again at our earth from outer

Figure 6–1 Program presentations to potential sponsors are like job objectives on resumes: Each must be tailored to the specific client. The first of these two presentations was designed for a live globalcast. The shorter live-on-tape presentation reflects a similarly impressive but less costly *Christmas Around the World*.

space, perhaps we can sense the divine order that makes it easier to imagine a place where no boundaries or languages separate us, where Christmas is truly a reminder that we share the same world.

We hope we've helped to put your world in perspective, with perhaps one other thought, that while it's nice to travel around the world, it's even nicer to be home, to share Christmas. Happy Holidays!

"CHRISTMAS AROUND THE WORLD"

--The Format--

COLD OPEN: U.S.A., Night

A satellite view of our earth will be followed by a shot at the North Pole of "The World's Largest Snowman," to the strains of "Frosty the Snowman." The up-tempo notes of "Frosty" and the Snowman theme will orient the audience through a montage of eleven other countries. In each country--avoiding the need for supers--the "Snowman" will wear a hat identifying the country and, across his chest, the flag of that country. In each location the "Snowman" will be topical, made of ice, papier mache, sand, feathers, etc.
Featured at all locations will be the stars--all performers--revealing up-front the entertainment extravaganza that lies ahead. Each location will provide a mini-musical of approximately eight minutes in length, featuring the location host singing/dancing with local performers. Possible hosts for the program include: ....(List of potential hosts) ....The host will provide welcome, set the scene and provide transitions between each of the program segments.

SEGMENT ONE: The location: Tokyo, Japan (Day): with an Ice Snowman

Music from: (Star) or (Star) or (Star) or (Star)
Events during the musical medley:
  1) Ice-sculpting
  2) Life-size puppets dancing
  3) The Parade of the Costumed Children

SEGMENT TWO: The location: Montreal, Canada (Night): with a
Logging Snowman

Music from: (List four possible stars)
Events during the musical medley:
  1) The Canadian Mounties ride and sing among Christmas bonfires
  2) A logging camp celebrates Christmas
  3) The lights of Niagara Falls

"Christmas Around the World": The Format (Continued)

SEGMENT THREE: The location: Rome, Italy (Day) with a musical
Snowman

Music from: (List four possible hosts)

Events during the musical medley:
  1) The Piazza Navone celebrates Christmas with jugglers, mimes, etc.
  2) A Christmas Parade with the Carbinerri Band
  3) Our singer with the Vatican Choir

Figure 6—1 *(continued)*

SEGMENT FOUR: The location: Port au Spain, Trinidad (Night) with
                                  a Birdman as Snowman

                 Music from: (List four possible hosts)

                 Events during the musical medley:
                                 1) A sugar plantation celebration makes
                                    toy dolls from sugar stalks
                                 2) Birds of Paradise
                                 3) A Flotilla of lighted boats
                                    accompanied with fireworks

SEGMENT FIVE: The location: Vienna, Austria (Day) with a toy
                                  Snowman

                 Music from: (List four possible hosts)

                 Events during the musical medley:
                                 1) The Vienna Boys Choir helps celebrate
                                    Christmas at a castle
                                 2) Making unique toys and ornaments
                                 3) A parade of Christmas costumes as
                                    young people help decorate the city

SEGMENT SIX: The location: Rio de Janeiro, Brazil (Night) with a
                                 Papier Mache Snowman

                Music from: (List four possible hosts)

                Events during the musical medley:
                                1) A Sun Festival on the beach
                                2) A Papier Mache Carnival Parade
                                3) A Christmas water show

             "Christmas Around the World": The Format (Continued)

SEGMENT SEVEN: The location: Geneva, Switzerland (Day) with a
                                   Skier-Snowman

                  Music from: (List four possible hosts)

                  Events during the musical medley:
                                  1) Cable cars on a snow-capped mountain
                                  2) Ice skaters visit a unique hand
                                     bell-ringer ceremony
                                  3) A sleigh ride around Lake Geneva

SEGMENT EIGHT: The location: Mexico City, Mexico (Night) with a
                                   Cowboy Snowman

                  Music from: (List four possible hosts)

                  Events during the musical medley:
                                  1) A Christmas Rodeo (Charreada)
                                  2) The Folk-loric Troupe
                                  3) The Parade of the Cowboys (Charros)

SEGMENT NINE: The location: Stockholm, Sweden (Day) with a Sailing
                                  Snowman

Figure 6–1 (continued)

Music from: (List four possible hosts)

Events during the musical medley:
      1) The Children's Christmas Flag
         Procession and Dance
      2) Holiday ice sailing
      3) The live Nativity Scene

SEGMENT TEN: The location: Melbourne, Australia (Day) with a
Barbecuing Snowman

Music from: (List four possible hosts)

Events during the musical medley:
      1) A songfest at a massive outdoor
         barbecue
      2) Baking unique honey cakes and pies
      3) A hundred square dancers and one
         kangaroo

"Christmas Around the World": The Format (Continued)

SEGMENT ELEVEN: The location: Beijing, China (Day) with a Panda
Snowman

Music from: (List four possible hosts)

Events during the musical medley:
      1) Performing Pandas
      2) Bicyling through the streets of
         Beijing
      3) A Pied-Piper Procession of children
         to a songfest on The Great Wall

CLOSE: U.S.A. (Night)

Behind our host, skiers light up the mountain snow with flaming torches as we begin a round-robin of the world singing "Silent Night" in their various languages, ending with the host's farewell and wish for the warmth of our "Christmas Around the World."

To: (SPONSOR)

From: Tony Verna

Re: Christmas Around the World

"Christmas Around the World" is 100 minutes of Christmas magic, all live-on-tape, but designed to capture the feeling of a live two-hour musical extravaganza celebrating the holiday season in this country by bringing to life the unique customs and cultures of 8 countries from which our national heritage derives.

Each of the three base locations--Paris, Melbourne and New York--will feature one of our top-name stars performing traditional Christmas songs of this country while we view traditional local celebrations, all set to music and dance. We'll visit castles and palaces, snow-covered mountains and sun-drenched beaches. Each mini-musical will link our American holiday with the rest of the world.

Figure 6—1 *(continued)*

A sample segment might feature (Suggested host) in Melbourne, Australia. To songs like "We Wish You a Merry Christmas," "It's a Holly, Jolly Christmas," and "Christmas Day in the Morning," we might see several hundred Australians--and one kangaroo--dancing and singing around a vast open area the size of several football fields, featuring dozens of barbecues and a series of huge bonfires--a vivid contrast to the snow and torchlight skiing we might see at St. Moritz in Switzerland.

The eight suggested countries, all of which have been contacted and are eager to participate, are: Australia, France, The Soviet Union, Spain, Switzerland, Brazil (or Venezuela), Senegal (Or Kenya) and--of course--the United States.

To maintain the feeling of a live telecast, segments from around the world will be integrated into our three star-based locations: New York City, USA; Paris, France; and Melbourne (or Sydney), Australia. The main setting in New York will be Rockefeller Center; in Paris it will be the Eiffel Tower; in Melbourne it will be the skyline of the city behind the field display of barbecues and bonfires.

The three base locations and their time pattern would be:

          New York City, USA..................23 minutes
          Paris, France......................23 minutes
          Melbourne (or Sydney), Australia....10 minutes

                    Total base time.....56 minutes

The five "remote" locations and their time pattern would be:

          St. Moritz, Switzerland..................10 minutes
          Barcelona, Spain........................10 minutes
          Moscow, USSR............................10 minutes
          Rio de Janeiro (or Caracas, Venezuela)...10 minutes
          Dakar, Senegal (or Nairobi, Kenya)....... 4 minutes

                    Total "remote" time........44 minutes

These "remote" locations will be integrated into the base portions as our hosts MC the holiday celebrations by international stars highlighting their local Christmas ceremonies. In order to make such worldwide production possible, it is necessary for foreign rights to be bartered. The Production Company (Sponsor and Tony Verna Productions), however, will retain full rights to two runs within the United States and Canada.

"Christmas Around the World" will marry the centuries-old traditions of the holiday with today's latest television techniques, brought together with a unique feeling of live globalcasting made possible through the tested skills of the production team that pioneered live international TV with programs like " Live Aid," "Sport Aid," and "Prayer for World Peace."

"Christmas Around the World": a musical whose meaning grows out of the very music that celebrates the occasion. No preaching. Just joy and celebration, memories mixed with holiday magic, all star-studded and driven by the international language of music, the songs we all cherish, regardless of our background or beliefs. Truly a tribute to the season and the one-world meaning of the season: "Christmas Around the World."

Figure 6–1 *(continued)*

approximate time and length of each performance. This log served as a record of each of the telecasts—since programming was seldom the same—and a list of which album covers appeared where. Using this, we were able to bill the record companies quickly, rather than having to screen separate tapes of all of the shows. The production and business staff devised the system together. That's the way it usually works.

Getting Distribution

Philosophers sometimes ask: "If a tree falls in the forest and there is no one there to hear it, is there a sound?" The answer, if it's a globalcast, is that it would probably be an endangered rain forest, and there'd better be someone watching *and* listening.

Some globalcasts, such as *The Academy Awards,* are regularly scheduled network shows. Most of the international efforts, however, are independent productions that have to create their own ad hoc network of stations. There are syndication companies that specialize in placing programs on regional networks and local stations. The producers of a show like *Live Aid* usually contract with one of these companies to handle distribution. Some globalcasts, like many regularly scheduled syndicated shows, are bartered—given to stations without charge in exchange for a specified number of commercial positions during the show. The station gets the bulk of the commercial positions to sell. The production company retains enough to sell to cover their production costs.

Earth 90 had income from event sponsors, who got signage—their company or product name—on the stage sets and on the open and close audio/video billboards, and sponsors who just bought spots. A syndication company handled placement of the show on an ad hoc network. *Prayer For World Peace,* with no commercials during the program, received program grants from two national sponsors who got mention at the top and bottom of the show. My own company, Global Media, controlled *Prayer For World Peace.*

However it's done, a globalcast has to get a network of stations to carry the show. Without an audience, there is no show. Without a budget there's no show. So before we turn to the actual creation of a production, let's deal with the mechanics of budgeting.

Getting a Budget

Many shows that begin production don't make it to air because the money runs out. All television costs, but especially globalcasting costs, have become so high that coproduction is almost a necessity (see chapter 19).

Let me give one example from my personal experience. The show was called *A Worldwide Christmas Celebration,* a very fine and inspirational program featuring the singing of Sandi Patti, Ray Charles, Johnny Cash, Placido Domingo, and others plus inspirational messages from Billy Graham, Mother Teresa, and Terry Waite (before he was taken hostage). Sponsored by the Church of the Nations, the program was envisioned as a live, two-hour Christmas Eve production. All of the normal budgeting was put in motion. Sites were chosen, scenic designers hired, lighting designers hired, and so on. There were to be live injects from the Boys' Choir in Vienna and from the Great Wall of China, among others. On paper the project looked magnificent. But when a business manager analyzed the budget, the figures just didn't add up. As laudatory as the goals were, as ready as the talent was, the available funds simply did not match the dream. By applying some very skillful financial analysis, and with the reluctant but realistic acquiescence of the sponsor, we were able to scale down the production to match the budget. The show was done not live, but in tape segments, edited. The host segments linking the performances—providing the continuity—were taped in New York, at Rockefeller Center. Billy Graham was taped at his home in North Carolina. Sandi Patti and others were taped in Florida. The show was composited on tape and distributed globally by shipping copies of the tape, rather than by satellite. It was a highly successful program, accomplishing virtually all of the creator's goals. It just wasn't the program the sponsor originally envisioned, and the reason is that, always, the reality of the budget supersedes the desire, creativity, and inspiration of the sponsors, talent, and crew. The budget is basic to the production.

Globalcasting's Specialized Costs

It's relatively easy to break down above-the-line and below-the-line costs on a studio production. Whether you're an independent producer raising financing, getting sponsors, and securing air time or an employed producer who is assigned cost limitations, sponsors, and air time, you have to balance the budget. With an in-studio production or even an annual special like the Kentucky Derby, the cost categories remain relatively constant. Above the line is production: talent, music and musicians, scripts and writers, nontechnical staff, office and space rental, office support and supplies, and so on. Below the line is technical: engineering staff, including stage managers and teleprompter, engineering equipment, satellite/fiber optic costs, scenery and props, transportation, housing, makeup, costumes, and so on.

But when you enter the world of globalcasting, there are new categories and levels of cost to consider. Below the line, satellites, equipment rental, and inject producers and directors become significant factors. Because you're usually dealing with multiple time zones, communication costs become a major expense, and travel costs and satellite "rehearsal" costs are high.

One large expenditure often underestimated is the cost of volunteer talent. Gigantic shows like *Live Aid* would not be possible if famous singers and actors did not donate their services. But related costs borne by the production accrue at significant rates: house bands, transportation, accommodations, personnel to coordinate the movements of the talent and their instruments, clothes, and costumes.

Computers: A Basis for Budgeting

Computers can be used to simplify budgeting by providing flexible budget forms. Starting with a basic checklist, and checking for variances from the list, the budget format can be shortened or expanded, depending on the complexity of the production.

The first step is to assemble information. What is the scope of the production? How many injects? How many uplinks? Downlinks? What kind of equipment has to be rented? What are the communication patterns? Two categories require special attention for globalcasts. One of these is contingencies. Because of the nature of globalcasts, there are more possibilities for the unexpected to occur, so the amount allocated to contingencies must be greater than usual.

The other category, "artistic license," is really a subdivision of contingencies, but with sharper focus. It's a hedge against the unforeseen but predictable. It takes a dream to create *Live Aid* or *Prayer For World Pease,* and realizing these dreams is potentially expensive. The artistic license category pads the first budget in consideration of this.

Starting, then, with the basic computer breakdown, you begin to match the desired goal with the available budget. The method, inevitably, is to have meetings. The first meetings are conceptual, setting the parameters of the subject matter, how it can best be communicated. Specifics are negotiated. It's possible that a change of location or air time can save money without sacrificing the artistic goals of the program. In countries where communications are government-controlled, you may negotiate with a Ministry of Post, Telephone and Telegraph (PTT). Then there are the inevitable meetings with engineering and contractors. Lighting, stage, and sets have to be set up or created. Once the basic parameters are set, meetings have to be arranged with the unions, both to outline the needs of the show and to determine the union costs involved.

To keep everyone on track, the production manager and budget managers submit ongoing status reports detailing goals, current attainments, current and predicted costs. The time frame within which all of these events must occur is also a critical factor in budgeting. In creating *Live Aid* we had only ten weeks. That's not much time to negotiate, let alone create!

"What If?" plays a big part in budgeting as well as in production. But there is another question to ask in budgeting: "Does it feel right?" Often a project sounds good philosophically and looks good on a budget sheet. But it just doesn't feel right; it doesn't supply the right conceptual tone for the program. The answer is intuitive, and you ignore it at your peril. If if looks right but doesn't feel right, go over it again until the feeling matches the facts.

Dealing with cost versus concept on a globalcast is a balancing act, and not infrequently you juggle costs from one area to another. Conflicts are inevitable, but production decisions have to be based on available funds. "It can't be done" will be said more than once. And never stop asking "What If?" Never stop asking "Does it feel right?"

Tricks of the Trade

Reshaping The 1990 Goodwill Games For *The 1990 Goodwill Games,* I broke down the predicted action minute by minute and used this as a basis for outlining the production requirements: cameras, cables, monitors, satellites, trucks, and people. The combination of these breakdowns provided the basis for budgeting.

We had a limited budget but unlimited dreams, and we had to make the two meet in the middle. In the first draft, we were more than 100 percent over budget. The goal was USA $20 million, and the budget came in at USA $49 million. Managing Director Michael McLees and I worked with Turner Broadcasting officials: Paul Beckham, then Senior Vice President of Finance & Administration and TBS head of "The Goodwill Games"; Rex Lardner, then Senior V.P. & General Manager of TBS; and Barry O'Donnell, then V.P. of Operations/Administration. Together we cut the budget to $25 million. Here are some of the ways we did that.

The two main areas are personnel and hardware. We tackled the personnel costs by drawing up a series of alternative plans. With each version, we were able to consolidate crews by revising transportation plans—which crew would cover which event and how they would get from one event to another. By coordinating this plan with the location of each crew's housing, we were able to cut costs in all of these areas.

We analyzed the hardware in several ways, and we found alternative ways of attaining the desired goal. We didn't diminish the goal but rather tackled the approaches to the goal. In many cases we discovered that the "easy way" was the final solution. Large facilities were scheduled to be constructed—facilities that would have been convenient but were not, in fact, needed.

Another way we cut hardware costs was to focus on the physical equipment. This was a one-shot show. Two thousand people would descend on Seattle. They would need cameras, computers, cables, and copiers—and then those thousands of pieces of equipment would not be needed again. We worked backward, just as you do when you time a show: we measured the salvage potential of each major piece of equipment. And that's how we shopped, aiming to satisfy the massive technical needs of the production while at the same time maximizing the resale value after the show.

The budgetary analyses were difficult and time-consuming, but we cut the proposed budget in half.

Tape It If a script calls for a segment to be filmed, usually exterior scenes that will be repeated, consider the use of tape instead of film. With access to the proper video equipment, this could save thousands of dollars.

Transmission Costs Hourly rates for fiber optic cables are around USA $400, versus $200 to $800 for satellite transponder time, depending on the time of day.

There are some specific choices that can affect satellite costs. On a globalcast you will inevitably be working with multiple television formats—NTSC, PAL, SECAM, and, shortly, HDTV. Before these video signals can be processed at master control, they have to be converted to a single standard. Depending on the injects involved, the conversion can be accomplished either at the inject location before being uplinked or at the downlink site. (Video formats cannot be converted on a satellite transponder, but satellites are capable of changing band, as from C-Band to Ku-Band.) It is usually cheaper to transmit in the original format and convert to a common signal at the downlink. Occasionally, however, it's better to convert before uplink. For example, SECAM and PAL M seem to transmit better after conversion. Nonetheless, the budget may dictate converting before uplinking.

The good news on satellite costs is that they have diminished through the years. For many years Intelsat was the sole satellite source. Since then multiple commercial satellites have been launched, and costs are now about one-fourth of what they were in the early days. Even so, transponder costs are still a major factor in budgeting a globalcast. In the section on news you'll see that one-third of CNN's USA $15 million expenditure for the five months of Desert Shield coverage went to buy satellite time.

Occasionally the nature of the show can affect the budget positively, as was the case with *Earth 90*. In support of the environmental theme of the show and in support of UNEP and UNICEF, GTE SpaceNet donated two

Ku-Band transponders for the USA broadcast, saving us several thousand dollars. And *Live Aid,* of course, received many such contributions-in-kind that helped make that program possible.

Summary

Although there has to be a balanced budget, I recommend running a two-track race. You have to keep budget restriction in the back of your mind, but if you limit your imagination with a bottom-line mentality you will stifle creativity.

It's important to maintain the adversarial relationship between bookkeeping and production. Production personnel should always aim for the moon. Bookkeeping personnel should always keep the rubber band attached to the moon so it bounces back within the budget's parameters. Begin with the questions: What's the best way in the world to communicate this? What is the best date? What are the best locations? What is the best number of cameras to make this one soar? What does the audience really want to see and how do we show it to them?

Find the essence of the production and then and only then turn to the practical limitations of budget.

Chapter 7

How to Hire the Staff

Once the concept, budget, and format of the show are set, you have to create a production team and set up the physical operations. Let me reemphasize that most of these functions occur simultaneously. For example, once a syndication contract is signed, sales of spot commercials continue until all the available commercial positions have been sold or until air time, whichever comes first. Usually staff are hired, talent is booked, the script is developed, and the technical operations are set up at the same time. They are separated here only for the purpose of analysis.

Defining Job Titles

What jobs are we talking about in globalcasting? You've seen the range of material, from sports marathons to musical spectaculars to solemn religious ceremonies. Even with a short life span, globalcasts vary as much as any other telecasts, and the jobs parallel those for any production, with a few additions.

Not everyone comes on board at the same time. You need the engineer in charge, public relations staff, set designers, business manager, writers and researchers, and security staff early. As plans become concrete, more staff is added until the production crew is up to full strength.

Because freelancers come from different backgrounds, defining titles and responsibilities is critical. Giving the agent who represents the star of a series the "honorary" title (and salary) of executive producer doesn't carry over into globalcasting. Virtually without exception, people who have titles on globalcasts do the work.

The *executive producer* could be the supervisor of one or more producers, but in globalcasting it's more likely that the executive producer will be the initiator of the project, and possibly also the entrepreneur who put the package together.

The *employed producer,* who is subject to the authority of an employer or group of employers, usually shares the credit with the executive director. This person is usually called the line producer, referring to the fact that the job's responsibilities encompass below-the-line as well as above-the-line elements.

The *executive director/producer* controls all aspects of the globalcast—creative, technical, and crewing—while the *coproducer* stays on top of finances and administration, from inception to completion, including the selection, coordination, supervision, and control of all worldwide talent and craft unions, subject to all international broadcasters and their rights with provisions of all collective bargaining agreements and personal service contracts.

There are also potentially dozens of other producers on a globalcast: supervising producers, senior producers, segment producers, PSA producers, coordinating producers, and so on. The responsibilities of each of these will vary from show to show and have to be defined carefully and clearly for all concerned. Figure 10–25 is an example of a table of organization, which breaks down all of these categories.

Directors are generally chosen for their area of expertise. When I need directors skilled in shooting musical groups, I turn to experienced pros like Lou Horvitz and Sandi Fullerton. Most of the directors and producers for *The 1990 Goodwill Games* had had experience working on Olympic Games or other major sports events. On a globalcast there is no time to train anyone in a key position; everyone has to be up to speed before you start.

The same level of experience is vital for the stage managers for musical shows. Devising, timing, and supervising on-stage equipment changes is critical in formatting and controlling a show. Experience and competence—and diplomacy in dealing with the talent and agents—are musts.

Similar competence levels are always sought in the production assistant staff. The job of my personal production assistant has always been critical because the timing in these multinational events is so complex. But I have always made room in my productions for college interns. The best place for the next generation of globalcasters to learn the rigors of the business is on the job. (This is one reason I had audio from the control room made available to some college sites.) The work may seem menial—copying and distributing rundowns, faxing, and phoning—but these jobs are necessary, and the exposure to the content and pace of reality is worth as much as any graduate course. Baptism under fire is still the fastest, surest way to learn.

There are often unusual jobs connected with globalcasts. A very important part of our trailer compound in Seattle for *The 1990 Goodwill Games* was a videotape library, which housed more than five thousand videotapes that could be accessed via a computerized system. *Librarian* may not seem to be a likely title on a globalcast, but it was a major factor in our success in Seattle.

Confronting the Energy Level, the Gee-Whiz Factor, the Unexpected

It's vital to deal with three factors in working with all staff members on a globalcast: the energy level, the gee-whiz factor, and plans for the unexpected. Frequently, on a remote location, just the geographical shift in time zones will sap the energy of a globalcasting crew. This potential, plus the fatigue that builds up after long hours of work—often at odd hours—has to be guarded against.

The gee-whiz factor develops when people are confronted, possibly for the first time, with satellite techniques, overseas injects, and so on. To ensure that no one is distracted on the air by the electronic razzle-dazzle, I discuss new techniques in meetings and make sure everyone experiences them in "rehearsal."

Being prepared for the unexpected is always a factor on a live show. The need is compounded on a globalcast. Setting the tone of a production helps overcome this challenge. In the example of Paul McCartney's microphone going dead during *Live Aid,* the directors, camera operators, and audio staff were all primed to react to any given crisis. Within seconds we met the challenge creatively, and an unexpected crisis became one of the memorable highlights of the sixteen-hour show.

Chapter *8*

How to Survey and Work Remote Locations

Two of the jobs specific to globalcasts are preshow surveyor and remote producer. By definition these positions function by themselves, away from home base. The people filling these positions have to have such a total comprehension of and immersion in the production that they can fit the distant remote segment meaningfully into a coherent whole. Although we used a videophone between our New York and NHK Tokyo control rooms on *Earth 90,* that kind of visual contact between master control and a remote is still rare. The nuances you can detect in face-to-face conversation are not apparent under these circumstances, and there is never enough time for master control to uncover and cover all contingencies in the field. The staff members who are dispatched "out there" have to be so competent and so well briefed that they can foresee all of the what-ifs an executive producer is expected to foresee, and be prepared to solve problems no one could foresee.

Surveying Remote Locations

One of the earliest steps in a globalcast is to survey the headquarters location and any inject points. Even if you have used the location before, you should do a survey. New roads may have been built, new traffic lights put up, old trees cut down, new trees planted, equipment availabilities altered. Camera locations that worked last time may not work this time. Cabling may have been changed. Always do a survey.

And on a program like *Prayer For World Peace,* where you are setting up complex technical systems in locations—like local churches—not outfitted for telecasting, the survey is even more rigorous. Time zones, languages, facilities, communications—all of these have to be dealt with logistically.

There are three keys to a successful preliminary survey. One is to hire a person who has sufficient background to know what has to be done. There is no time to train in an area as critical as surveying. The second key is to brief that person so thoroughly that he or she understands the total production completely and can evaluate the remote situation as a part of the whole. The third key is for the person doing the survey to locate the right liaisons at the remote location. The on-site contact has to know the environment, the area, the people, the television stations, the available support services (engineers, camera equipment, lighting

equipment, transportation, power sources, food services), and local restrictions and politics.

Globalcasts may encounter restrictions imposed by PTTs—Ministries of Post, Telephone and Telegraph. In some countries these government agencies enforce financial, union, or geographical restrictions or restrictions on coverage. Bonds of up to USA$100,000 may have to be posted before you can import broadcast equipment. Union restrictions may dictate the role of personnel, including the directors. A vivid example of government restrictions was the forced shutdown of live TV coverage from Tiananmen Square in 1989. Local contacts are critical.

It is absolutely essential that the local "point person" accompany the production company surveyor *all the time.* The first step after meeting in person is to brief the local point person on the total production and on *all* of the needs for/from that particular location. It is vital to communicate not only the needs of the production, but the schedule and timing. The second step is to detail the itinerary, both to inform the point person and to be sure nothing has been missed that the point person might catch. The last step, of course, is to maintain phone contact so that both the remote location and production headquarters are kept up to date.

I detail various methods of briefing the staff, including the preliminary survey team, later in the chapter. Here I want to focus on the job of the surveyor, which I will do by giving a brief account of one of the world's most experienced and skilled producers, Robin Barty-King, a British producer who traveled the globe on *Prayer For World Peace.* This account covers a two-and-a-half day trip he made to Dakar, Senegal, to survey one of the sites for the program.

Making an Original Survey

One of the first hurdles for a surveyor is simply arriving at the destination. As Barty-King says, "It is never quite the same weather as England." You arrive clutching a handful of telephone contacts you believe to be willing/ very willing/doubtful/disbelieving of your telephoned commitments and promises. A lot of telephone preparation has been done setting up the location, but it is often difficult to persuade someone on the end of a phone several thousand miles away that you are going to bring their church/ sports stadium/music venue into the world television arena with one or two billion viewers, especially when their previous record capacity has been five thousand. There's nothing like a production meeting face to face to bring the realities of the production home to them and, more important, to you.

Dakar, Senegal, is a small, busy capital on the West Coast of French-speaking Africa. Prior to *Prayer For World Peace,* it had never been on global TV, and television facilities there presented a technical challenge.

Barty-King's handful of telephone numbers in Dakar included Dak, the cameraman; his interpeter, Sam; the governing commissioner; the local television station; and the hotel. Once through the airport formalities, Barty-King met Dak, who spoke no English at all, and Sam, who managed a healthy version of "Franglais," a mixture of French and English that seemed to work.

Barty-King had made an appointment to see the governing commissioner but discovered on his arrival at the offices in Dakar that there had been some miscommunication between the local commission and the Vatican in Rome. They were not ready to meet with him. So he shifted his efforts to the local TV station personnel, since it was they from whom the show would be demanding so much technical expertise.

Whatever is promised on the phone, the remote surveyor gets a much more realistic sense of the location's ability to deliver once able to see the gear. Before leaving London, Barty-King sent a telex listing the basic requirements: a minimum of three cameras, lighting, and sound. What greeted him, as is so often the case, was "bags of enthusiasm plus equipment (albeit somewhat archaic) capable of delivering, but never before having been used for something like this."

At the TV station, Barty-King spent most of his time discussing the concept of the show. To get what we needed to create the necessary teamwork, he knew there would have to be a massive cooperative momentum created between the telecommunications department and the television department, both government-owned. As a result of this first meeting with the TV staff, Barty-King got an agreement that, subject to successful negotiations with the commission, they would have a production meeting at the church the next day.

From the TV station, Barty-King went back to the commissioner's office, where his French was "tested to its limits and beyond." But the meeting at the church was agreed to. Shifting back to the hotel Barty-King reported to Dak and Sam, checked for telexes from London and the USA, and, finally, got dinner. One day's work was done, but matters were still tentative.

The next day's meeting was attended by the commissioner's representative, the production heads and technical heads of television and telecommunications, and the church priests. Barty-King felt pressured because the entire meeting was conducted in French, so he was not exactly clear about what was being agreed to and what wasn't. He was also distracted by the presence of Dak, who, it turned out, was filming the proceedings for posterity.

In 1987, Dakar did not have limitless resources, but Barty-King found sufficient basic, simple, and effective equipment. Barty-King left Dakar feeling pleased with the progress, that enough had been accomplished, and that the basics would be in order when the remote producer returned before the day of the show to sort out the final details. When we went on the air on June 3, what finally appeared from Dakar raised a cheer in the transmission control room.

As Barty-King noted in his report, "Communications provide the most important part of these shows. To get a signal out of Dakar to London and back was hard enough, but to get clear, uninterrupted four-wire and telephone line for the four hours of the show was always going to be touch-and-go. But if we can't talk to each location directly, then the coordination of the show becomes impossible." On most globalcasts it is necessary to communicate with non-English-speaking contacts. Language skills are vital.

Respecting time zones can be equally critical. Nine-to-five days may be comfortable, but they bear no relation to working on a globalcast. When you are in London, New York, or Los Angeles, communicating with New Zealand, the Philippines, or Tokyo, you need batteries of clocks telling you the correct time in the various time zones. You need staff awake in the middle of the night, which is afternoon somewhere else. You need stamina to survive long days with frequently questionable meals. And everyone who has any ambition to work on a globalcast needs an up-to-date passport, for globalcasters may be in Dakar one day and Bombay the next.

Working at a Remote Location

The producers who do the remote surveys and those who return to supervise during the globalcast act as the eyes and ears of the executive producer/director and the crew at master control. Whatever the complications in each local production, it is the original surveyor who makes the initial contact, who sets the tone and establishes the relationship—who can make or break that portion of the globalcast.

Follow-up on those initial contacts and plans, however, becomes the job of the on-site producer. Phone contact is maintained with each remote location, but it is the on-site producer who has to make the wheels turn. On *Prayer For World Peace* all of the remote producers met in London less than a week before the telecast. After they were briefed at a series of meetings, they flew to their destinations to put in motion the production elements necessary to air the show.

After reestablishing the contacts made by the original surveyor, the on-site producer has to establish a series of new contacts, with the crew members who will run the video, audio, and phones and nonprofessionals who, in this example, included church officials, political officials, and parishioners. Again, language is a factor.

Establishing and maintaining communication with the master control site is essential. On behalf of the master control crew, the on-site producer has to line up and work with the event promoter, local video company or station, engineer in charge, camera crews, electricians, telecommunications company, local director, security officials, transportation crews, and representative controlling funds involved in the local production.

Prayer For World Peace had seventeen natural locations; other shows may have set designers, costume designers, musicians, and so on, each offering unique challenges to the remote producer. The remote producer in Caacupe, Paraguay, was confronted with a special pilgrimage that was being climaxed by our globalcast, with 100 thousand parishioners. The historical shrine in Washington, D.C., faced an overflow crowd, with many parishioners unable to enter the church. In Czestochowa, Poland, Mother Teresa attended to share the prayer with the Pope.

The most unsettling report came from Gil Stose, in Zaragoza, Spain. Shortly before air time, vandals smashed the three monitors that had been set up at the church so the parishioners could see the Pope. There was just time for Stose and his crew to rig up substitute video so the Pope would be seen in the church. Apart from the technical problem of getting substitute monitors, Stose faced the challenge of combatting the fear engendered by the act of vandalism. There was nothing anyone in London could do to assist him in Spain. The situation was his problem, his responsibility. This highlights the need for reliance and decision-making ability to perform the job of remote producer on a globalcast.

Chapter 9

How to Communicate Within the Staff

There are three distinct forms of communication within the staff: communication during the production phases, before going on the air; communication during the air show; and communication after the show. All are critical.

Before discussing each of these, let me discuss the philosophy behind the techniques, both developed out of experience. Underlying all of my techniques is a philosophy essential to a globalcast: There is no room for secrets. Network staffs sometimes develop cliques, keeping facts secret from other individuals or departments. There is no room for this behavior on a globalcast. The tone has to be set at the top and carried down through every job function: everyone shares all information with everyone else. If it's not an informed team, it's not a team. And if there aren't face-to-face meetings, it isn't a show.

However thorough a minute-by-minute grid may be, by themselves grids are never sufficient to communicate all that has to be communicated. There have to be many in-person meetings, so many that people who "catch on" early will think they are a waste of time. They aren't. The purpose of the meetings is to establish personal rapport; to establish group rapport; to turn generalities into specifics; to communicate the goals and techniques of the show; and, perhaps most important, to get *feedback* and establish channels for continued feedback.

An executive director has to be sort of a benevolent dictator. Without minimizing where the power of final decision lies he or she has to establish a climate that makes all suggestions, comments, and questions welcome. Many, perhaps most, suggestions will have to be rejected, but the rejection of any idea or suggestion is couched in terms that will not shut off future feedback. There *must* be constant feedback, free and open. It has to be a team.

Meetings also help overcome the star-gazing and gee-whiz factors that can potentially distract staff members. Reviewing the show time and again makes everyone so familiar with it that its parts become second nature. Also—with everyone admonished always to ask "What If?"—problems not foreseen at the outset may become evident the second or third time around. On shows as complex as globalcasts there will always be another "What If?" This seemingly tedious ground has to be covered, because once the show starts there will *always* be some unforeseeable crisis, and when you're on the air, you don't have time to deal with both the unforeseen and the unforeseeable.

These meetings also force the executive director to think and rethink the what-ifs. It's a little like sifting sand until all the fine stuff falls in place and the clinkers are caught and removed.

One final point about meetings. Business communication textbooks tell you that the reason many people resent meetings is that they see no results from them. The books say follow-up is the most important factor, so that everyone involved knows that there was meaning to the meeting and that actions resulted. The textbooks are right. Someone *always* takes notes at my meetings. There is *always* specific follow-up with those notes, to be sure the points covered in the meeting have been taken care of. Television is communication. Communication is a process, a two-way process. Suffer the meetings. They'll lead you to success.

Communication Before the Show

Grids

The method of in-house communication is as unique to each show as are the format and technical structure. The format and timing have to be documented on hard copy so each staff member can read and reread them and carry the message that forms the basis for the show.

Figure 1–9 shows the grid I developed to communicate the complex production patterns for *Live Aid*. Because the nature of the shows was similar, I adapted this grid for use on *Our Common Future* and *Earth 90: Children and the Environment*. Neither of these shows was devised to raise money, as *Live Aid* had been. Both *Our Common Future* and *Earth 90* were created to alert viewers to the dangers to our environment and to offer small but practical steps individuals can take to help establish and maintain a balance between nature and humans. *Our Common Future* was an outgrowth of a study conducted for the United Nation's Commission on Environment and Development, and the show was scheduled for June 3, 1989, to precede World Environment Day on June 5.

A five-hour globalcast, the show aired from 2:00 to 7:00 p.m. Eastern Daylight Time. A collaborative effort of the Centre for Our Common Future in Switzerland and Uplinger Enterprises in the USA, the program was presented in cooperation with and in support of the United Nations Environmental Programme and UNICEF. Musical stars on the show included Elton John, Stevie Wonder, R.E.M., Melissa Manchester, Tom Jones, Kenny Loggins, Maureen McGovern, John Denver (solo and in a duet with the USSR's Alexander Gradsky), Phoebe Snow, Lenny Kravitz, Sting, Diana Ross, eight-year-old Korean violinist Sara Chang, and Vladimir Pozner with the Moscow Symphony Orchestra. Some of the hosts were Christopher Reeve, Lisa Bonet, Sigourney Weaver, Bob Geldof, and Angelica Huston.

The show was syndicated in the USA, with commercials, and was carried on a number of radio stations in the USA and worldwide. NHK televised it in HDTV, with their own hostess, and fed the signal by satellite to Japan and to two public monitors set up in the USA. The French cable channel Canal Plus unscrambled its signal on June 3 so the globalcast would be available to all viewers. For both of these shows, music was the medium that carried the message.

Our Common Future   On *Our Common Future* we aired a series of PSAs produced specifically for the show with introductory messages from heads of state, including Canadian Prime Minister Brian Mulroney, China's

Prime Minister Li Peng, USA President George Bush, and UN Secretary General Javier Perez de Cuellar. And the hosts contributed their personal feelings. Unlike *Earth 90,* however, we supplemented the messages in these PSAs and host comments with live interviews, on the balcony of Lincoln Center's Avery Fisher Hall in New York City. Norway's Prime Minister Gro Harlem Brundtland, who had chaired the committee preparing the report for the UN, interviewed live guests several times during the globalcast. New York news reporter Denise Richardson was with Brundtland to aid with the interviews, to control timing, and to make the transitions to and from the stage.

Technically this was a relatively simple show. It consisted of a single concert in New York City with live injects from the UK, the USSR, and others. Because we were dealing with a single concert, I was able to intercut incoming feeds as well as Brundtland within the same control truck. The staff used a normal rundown format, combining the world feed and the domestic feed, adding only a teleprompter/chyron column for added information.

Figures 9–1 and 9–2 show two pages from the final rundown from *Our Common Future.* The Domestic column indicates where and how the cutaways were made to the commercially syndicated feed. Rolland Smith anchored the World Feed and Nina Blackwood anchored the Domestic Feed. Figure 9–3 is a video print from the tape of the show. It shows Nina Blackwood, who stood just outside doors leading to the balcony at Avery Fisher Hall. When the doors were opened,

Figure 9–1 "Spirit of the Forest" was a 5:02 song that had premiered two days before *Our Common Future* at UN headquarters in New York City. It featured Donna Summer, Brian Wilson, Ringo Starr, Olivia Newton-John, Mick Fleetwood, and other stars in a song dedicated to the theme of the show, so it combined entertainment with message. Note that as soon as the Public Service Video (PSV) ended at 2:24:47, the Domestic Feed pulled away from the World Feed using a specially produced sixteen-second tag and an eight-second bumper to lead to a commercial pod of 2:30. Nina Blackwood, who hosted the domestic syndicated feed, then introduced Melissa Manchester's song, which we joined in progress.

```
OUR COMMON FUTURE                                                    Page Two
1ST HOUR  2:00:00PM (EDT)
```

	VTR/LIVE	SEGMENT	SEG TIME	RT	DOMESTIC	AUDIO	PROMPT/ CHYRON
11.	LIVE-NY	Smith O/C throws to:	(:30)	2:21:55		Sotto music	
12.	LIVE-NY	Brundtland & Denise O/C	(1:00)	2:22:55		ATR	
13.	VTR	PSV: OCF LAUNCH	(1:52)	2:24:47			
					Tag (:16)		
14.	LIVE-NY	Smith/#1 Lisa Bonet segue, applause & Lead Melissa Manchester	(:40)	2:25:27	Bumper (:08) COML #2 (LOCAL) (2:30) Nina V/O: ...Intros Melissa (JIP)		
*15.	LIVE-NY	MELISSA MANCHESTER (2 songs) "Sometimes I feel so sorry for God"(4:02) "Over the Rainbow" (to track) (LTT)	(3:40)	2:29:29 2:33:09	JIP		
16.	LIVE-NY	Rolland O/C	(:10)	2:33:19			
17.	LIVE-NY	Brundtland/Denise/Khosla	(1:30)	2:34:19		Sotto music	
18.	VTR	PSA: PM Gandhi/ PSV: India	(2:48)	2:37:37		ATR	
19.	VTR	OCF Geneva Announcement	(:10)	2:37:47	Tag (:16) Bumper Local (:08) COML #3 (LOCAL) (2:30) Nina O/C		
*20.	LIVE-NY	Smith throws to: Video	(:20)	2:38:07			
*21.	VTR	"Spirit of the Forest"	(5:02)	2:43:09			

OUR COMMON FUTURE   Page Three
1ST HOUR 2:00:00PM (EDT)

	VTR/LIVE	SEGMENT	SEG TIME	RT	DOMESTIC	AUDIO	PROMPT/
*22.	LIVE-NY	#2 Lisa Bonet: Intros Sara Chang	(:30)	2:43:39	(poss. L/3 Lisa Bonet domestic may not have seen yet.)		
*23.	LIVE-NY	SARA CHANG 1st Paganini Concerto Excerpt 2nd Paganini Concerto Excerpt	(3:00) (3:00)	2:46:39 2:49:39	Bumper National (:08) COML #4 (NATL) (2:30) Nina poss. fill to Lisa Bonet		
*24.	LIVE-NY	#3 Lisa Bonet: Thanks Chang, Makes Statement & Intros M.Dibango	(1:10)	2:50:49			Prompter
25.	LIVE-NY	MANU DIBANGO "Soul Makossa" & House band	(4:00)	2:54:49			
*25A.	LIVE-NY VTR	Smith intros L. Armstrong Video "What a Wonderful World"	(:20) (3:30)	2:54:09 2:58:39			
		TOTAL	58:39	2:58:39	EPPS cue in @ 59:20 EPPS cue out @ 59:50 Station Identification (:10)		

---END 1ST HOUR---

Figure 9—2 Item 22 has Lisa Bonet introducing eight-year old Korean violinist Sara Chang. The note in the Domestic column indicates that if Lisa Bonet had not been introduced and seen previously on the Domestic Feed, that crew would have inserted a lower-third identifying super. Note too that, because the commercial pod in item 23 ran only 2:30, host Nina Blackwood might have to fill until the second Paganini Concerto ended on the World Feed. Also, note that a teleprompter was to be used for Lisa Bonet's statement.

the camera could see the stage behind Blackwood. While she was performing her function on the Domestic Feed, the screen in the background was carrying the video of the World Feed, which went out without commercial breaks. The Domestic Feed, of course, had its own producer/director team in a separate truck to create that separate program-within-a-program.

Earth 90 *Earth 90* was much more complex, but I still found that the easiest and best way to communicate was a combined grid-rundown. After its successful HDTV coverage of *Our Common Future,* NHK Enterprises USA coproduced *Earth 90* and announced it as the first of a series of globalcasts planned through the start of the twenty-first century, aimed at focusing attention on the environment. *Earth 90* was a three-hour globalcast positioned on June 2, 1990 (USA time; June 3 in Europe and Asia), the day designated by the United Nations Environmental Programme as Worldwide Clean-up Day.

The show was set in three locations, each with two cohosts. In Yoyogi Stadium in Tokyo, John Denver and Yu Hayami cohosted a live concert with ten thousand people in the audience, two thousand of them children who sang with Bakufu Slump. The cohosts in France, on tape and sent in via satellite, were Olivia Newton-John and Herbert Leonard. The cohosts

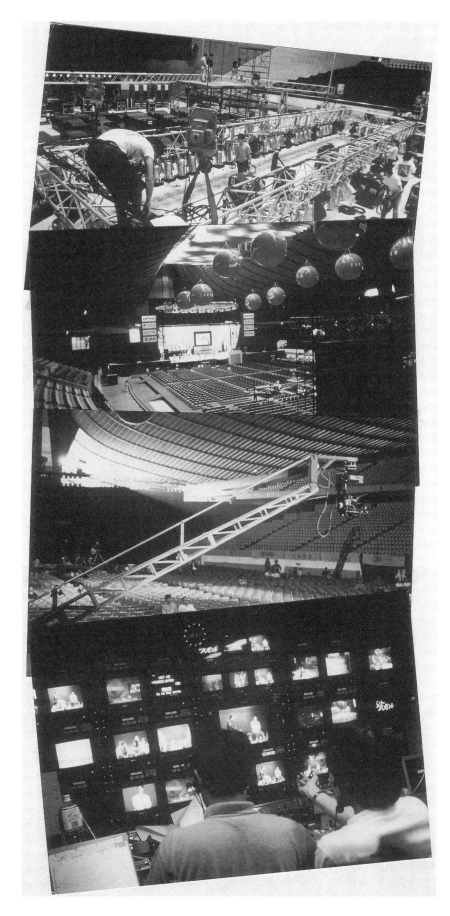

Figure 9–3 This video print makes two points. First, the large screen behind syndicated host Nina Blackwood provided sustained entertainment for the Lincoln Center audience by showing overseas performances. Second, Nina's position made the entire New York stage her background, without in any way interfering with the flow of the World Feed.

Figure 9–4 *Earth 90:* These *Earth 90* photos show the massive size of Yoyogi Stadium in Tokyo, plus the technical preparations behind the scenes, including Tokyo Director Masamitsu Obara's control room.

in the USA, live at the Brooklyn Academy of Music, were Debbie Gibson and Rolland Smith.

In addition, we had Gilberto Gil live from Rio de Janeiro, Brazil, with a studio audience; Alabama live from Fayetteville, North Carolina; Julio Iglesias on tape from a concurrent live performance at Jones Beach; Kitaro on tape from a previous live performance; Crosby, Stills and Nash on tape from a live performance the previous day, flown in; Falco, the Vienna Symphony Orchestra and the Children's Choir on tape from Austria; and a long list of taped performances for use on the World Feed. From that sketch you can see the scope of communication necessary to coordinate all of the parts.

Earth 90 was a very beautiful show. Figures 9–4 through 9–6 offer a glimpse of one of the settings, Yoyogi Stadium in Tokyo. The globes strung up on stage and around the stadium matched in style, but not in color, the globe on stage at the Brooklyn Academy of Music shown in Figure 9–7. We mounted two huge screens over the stage, which we used for multiple purposes: to let the audience in New York see and enjoy performances from overseas injects, to tie together performers on the New York stage with performers on stage in Tokyo and elsewhere, to provide

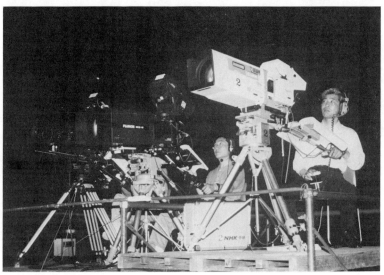

Figure 9–5 These *Earth 90* shots show NHK's HDTV equipment in action with a side order of cabling.

Figure 9–6 The twenty-five-screen monitor wall between Tokyo cohost John Denver and Russian star Alexander Gradsky added powerful visual images to their duet. John Denver wore an interrupted feedback (IFB) line so I could maintain instant control in Tokyo as well as in New York. The second photo clearly shows sponsor signage beneath the *Earth 90* logo.

Figure 9–7 This video print of *Earth 90* shows Debbie Gibson and the rest of the New York cast on stage for the finale at the Brooklyn Academy of Music. The two screens suspended on either side of the *Earth 90* globe logo carry a live picture of the show's cohosts in Japan, John Denver and Yu Hayami. The on-stage screens offer multiple uses: showing close-ups of a performer or of his/her hands or fingers on an instrument; repeating the *Earth 90* globe, restating the theme of the program and shifting the effect of the color scheme; showing the studio audience performances at the other global locations while the stage is being reset for the next performer; carrying the PSAs being aired via tape; or strengthening the global nature of the telecast by carrying video from another continent (pictured). The latter technique also allows for intercutting conversations or musical duets. Contrast this technique with the split-screen technique used on newscasts.

closeups of performers, and to provide video to supplement the on-stage performances. The screens could be raised or lowered, used or not used. By placing them on stage with the performers, we were able to establish a feeling more personal than the remote feel of the screen splits often used for two-way conversations on news programs.

Figures 9–9 through 9–11 show three pages of the rundown from *Earth 90*. The content of Figure 9–9, the May 27 rundown, is comparable to that of May 31 (Figure 9–10), but note how the expanded rundown/format has combined the World Feed and Domestic Feed under the item column and added a dozen new columns, allowing this one document to be used by everyone working on the show. Each person could insert his or her own cue notes under the appropriate column.

The two expanded sheets in Figures 9–10 and 9–11 cover the start of the third hour of the show. Notice the use of the double screens to reestablish the relationship between New York and Tokyo. Note, too, that we included applause in our timings. If the five or ten or twenty seconds of

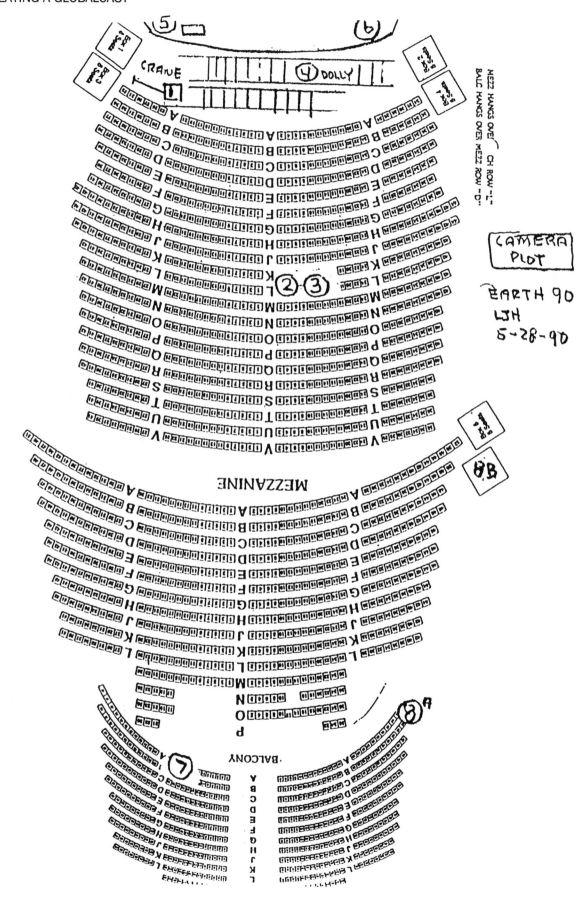

Figure 9–8 This floor plan of the Brooklyn Academy of Music shows the camera positions I used to cover the New York performances of *Earth 90*.

applause is not accounted for in the timing, before long a show of this length is running over so badly it becomes uncontrollable.

As an aside, let me reinforce the effectiveness of round-robin applause on globalcasts. It's a strong visual device to establish and maintain the worldwide feel of a show, but its use requires planning and split-second timing. Say your act is performing before a live audience in Brazil, with live audiences watching on DiamondVision or similar screens in Japan and the USA. By presetting shots of each of the three audiences—in such a way that the location is instantly identifiable—you can cut from the close of the performance to the applauding audience in Brazil, to the applause in Japan, to the applause in the USA—and perhaps back to a wide shot of the performer and applause in Brazil. The time pressures of the show dictate how long the sequence can hold. If you have to, you can show three different audiences in five or six seconds. It's better if you have ten or twelve seconds. Whatever the length, it's an effective way of reestablishing the global nature of the show. And it demonstrates again the interdependent nature of a globalcast. Everybody at each of the locations, including the talent, has to be briefed and ready for the split-second timing involved in airing the shots. Everybody has to be communicating.

Returning to the rundown, the scribbling on the second expanded page in Figure 9–11 reveals that even with timing as tight as it usually is on a talent-packed show like this, you have to be ready to cover any eventuality, including running short. While we were on the air, contrary to expectations it looked as if we might have time left at the end. We had to plan a procedure to cover. We penciled in a standby tape, a possible message, a

Figure 9–9 Compare this small *Earth 90* rundown with the expanded version in Figure 9–10. Note how the timing was refined between May 27 and May 31.

CONFIDENTIAL & TENTATIVE WORKING SHEET May 27, 1990 Hour 3
5:00 PM
EARTH '90
CHILDREN AND THE ENVIRONMENT

Program Format
HOUR 3

Page 7

		COMMERCIAL FEED (Orbis)	WORLD FEED	COPY	TIME	SEGMENT TIME
Live/USA/ Japan	39)	Re-establish World Hosts USA/Japan	same	Hosts re-cap split screen End at Debbie	11:00:00	:40
Live/USA Live/USA	40)	ACT #13: USA-BAM Debbie Gibson Intro Jeff Healey Perf #18 Jeff Healey "I Think I Love You Too Much" Applause	same same same	Debbie Intro Healey	11:00:40 11:00:55 11:05:25	:15 4:30 :20
VTR	41)	Animation/Celebrity Message #5 (Cousteau/Water Pollution in Zimbabwe)	same		11:05:45	2:00
VTR/France VTR/Europe	42)	Olivia Newton John Intro Richard Marx Perf. #19 Richard Marx "Help" Applause	same same same	Olivia Intro	11:07:45 11:08:05 11:12:25	:20 4:20 :20
VTR	43)	BUMPER	same		11:12:45	:15
(VTR)	44)	COMMERCIAL BREAK #9	World Act #9	(Jeanne Mass)	11:13:00	2:30
LIVE/Japan VTR/USA	45)	ACT #14: JAPAN/USA /EUROPE John Denver Intro Julio Iglesias Perf #20 Julio Iglesias "Everytime We Fall In Love" Applause	same same same	Denver Intro Iglesias	11:15:30 11:15:45 11:19:45	:15 4:00 :10

CONFIDENTIAL WORKING FORMAT May 31, 1990 6PM

ITEM	WORLD AUDIO	CHYRON	BAM STAGE	BAM PROMPT	BAM/BU ANNC	SCREEN	APPL	JAPAN STAGE	JAPAN PROMPT	JAPAN FAX CUES	VERNA	SEG	TIME
HOUR 3													11:00:00
39) RE-ESTABLISH WORLD HOSTS USA/JAPAN—Live —Double Screen												1:00	11:01:00
40) ACT #13 USA-BAM—Live												:15	11:01:15
BAM A) Debbie Gibson Intro Jeff Healey												3:10	11:04:25
B) Perf #18 Jeff Healey "I Think I Love You Too Much"													
Possible Japan applause												:20	11:04:45
41) ANIMATION CELEBRITY												2:00	11:06:45
MSG #5 VT (D) or (F) (Cousteau/Water Pollution) *2:03:05*													
42) ACT #13 FRANCE-VT(D) or (E)												:20	11:07:05
A) Olivia N. John Intro Richard Marx													
Perf #19 Richard Marx "Help"												5:10	11:12:15
Applause												:10	11:12:25
43) BUMPER VT (D) or (F) *Un # Sour # de # Pluie*												:15	11:12:40
44) WORLD ACT #9 Blues Troittoir US COMMERCIAL BREAK #9												2:30	11:15:10
Applause												:05	11:15:15

Figure 9–10 Note that both this expanded rundown and the shorter version in Figure 9–9 include time for applause. Applause, especially on a globalcast where it can reestablish the show's worldwide impact, is often overlooked as a time-consuming element. It's as much a part of the performance as the music.

possible reprise of the "Earth 90," song which Debbie Gibson had performed earlier. We didn't have occasion to use any of these planned standby items, but they were there if we needed them.

Communication was not, however, as simple as an expanded rundown. We were communicating among three continents in at least three languages. We had to develop special modes of communication to meet our needs.

Communicating Across Continents Figures 9–12 and 9–13 dramatize just a few of the complexities that arise when different languages and different cultures join to accomplish a single goal. Even with the best of intentions, getting-it-right is not always easy. Figure 9–12 is a page from the Japanese script for *Earth 90*. Although the show was broadcast in English, the staff in Tokyo prepared the script in Japanese also so the Japanese performers would be at ease in understanding the dialogue.

Figure 9–13 shows one of the interim versions of the opening of the show, which we faxed back and forth to each other until we reached agreement on all the ingredients. We started planning the opening verbally, by phone. Then we went to the written script. Finally, to be absolutely sure we were clear on both sides, we combined the "pictures" with the video/ audio breakdown. The sketches will never win a place in the Louvre, but they got the point across.

Because there are differences in style among European, USA, and Japanese directors, I initiated talks among us so we could keep our individual styles and still have an *Earth 90* look to the show.

Our methods of communication differed drastically, however. Contact between our New York control room and that in Japan was on multiple levels. The Japanese director, Masamitsu Obara, wanted as much personal contact as possible, so NHK rigged up a camera in each of our control

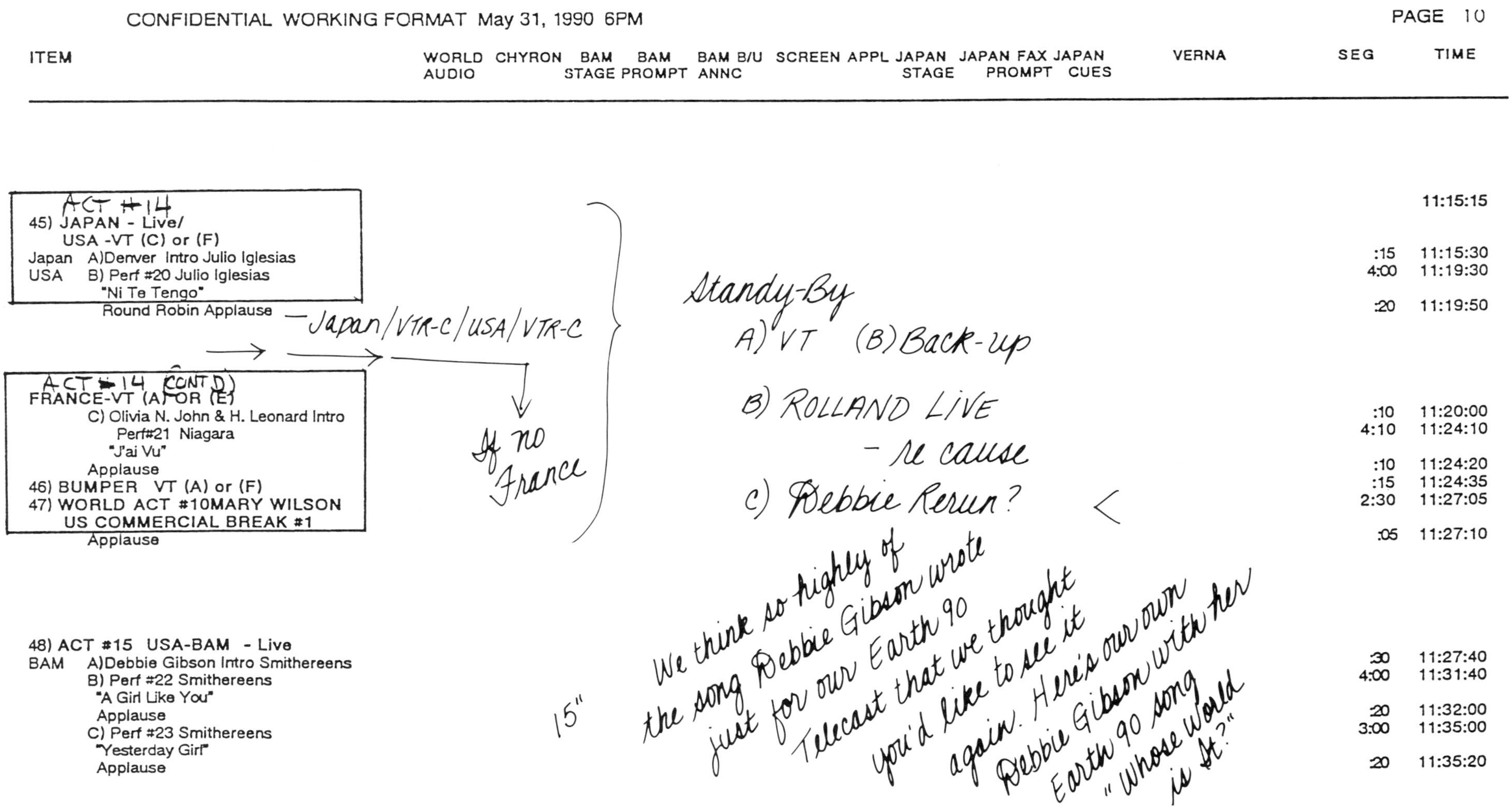

CONFIDENTIAL WORKING FORMAT May 31, 1990 6PM — PAGE 10

ITEM	WORLD AUDIO	CHYRON	BAM STAGE	BAM PROMPT	BAM B/U ANNC	SCREEN	APPL	JAPAN STAGE	JAPAN FAX PROMPT	JAPAN CUES	VERNA	SEG	TIME
ACT #14													11:15:15
45) JAPAN - Live/ USA -VT (C) or (F)													
Japan A)Denver Intro Julio Iglesias												:15	11:15:30
USA B) Perf #20 Julio Iglesias "Ni Te Tengo"												4:00	11:19:30
Round Robin Applause — Japan/VTR-C/USA/VTR-C												:20	11:19:50
ACT #14 (CONT'D) FRANCE-VT (A) OR (E)													
C) Olivia N. John & H. Leonard Intro												:10	11:20:00
Perf#21 Niagara "J'ai Vu"												4:10	11:24:10
Applause												:10	11:24:20
46) BUMPER VT (A) or (F)												:15	11:24:35
47) WORLD ACT #10MARY WILSON US COMMERCIAL BREAK #1												2:30	11:27:05
Applause												:05	11:27:10
48) ACT #15 USA-BAM - Live													
BAM A)Debbie Gibson Intro Smithereens												:30	11:27:40
B) Perf #22 Smithereens "A Girl Like You"												4:00	11:31:40
Applause												:20	11:32:00
C) Perf #23 Smithereens "Yesterday Girl"												3:00	11:35:00
Applause												:20	11:35:20

If no France
Standy-By
A) VT (B) Back-up
B) ROLLAND LIVE
- Re cause
c) Debbie Rerun?
15"
We think so highly of the song Debbie Gibson wrote just for our Earth 90 Telecast that we thought you'd like to see it again. Here's our own Debbie Gibson with her Earth 90 song "Whose World is It?"

Figure 9–11 The scribbling on this page shows that we were planning standby material even during the globalcast. On a live telecast you can't stop asking "What If?" until the show is off the air.

106 CREATING A GLOBALCAST

ＡＣＴ　＃９　　Following BAKUFU-SLUMP, TEKE TEKE SONG (Denver, Hayami & Nakano)
（サンプラザ中野のコメント）

デンバー：中野さん、あなた方のグループは環境問題を扱った曲が多いですね。皆さんの本拠地であるこの東京から世界に向けてメッセージを何かお願いします。

中野：日本では、地球の温暖化が問題になっていますけど、そのことについて考えるというよりは、ひとりひとりがまず行動を起こすことが大切だと思います。御存じのとおり温暖化の原因は二酸化炭素なんで、燃費のいい車に乗るとか、木をもっと植えて二酸化炭素を減らすことなんかも出来ると思うんだけど・・・。環境問題について話したり、コンサートで扱ったりするのは重要なことだと思うけど、個人が行動を起こさない限り、何の変化もありえないんじゃないかな。

Figure 9–12 This script page from Tokyo reflects the multicultural nature of *Earth 90*. Although the program aired entirely in English, the Tokyo script was prepared—as this page indicates—in Japanese as well, to put the Tokyo performers at ease with the material.

rooms that sent a picture back and forth every ten seconds. We could communicate visually as well as via the phones and headsets. On the air, of course, I had interrupted feedback (IFB) contact with John Denver. I needed that for all of Denver's cues, but especially for the intercontinental portions. We had cross conversations on the air between the USA and Tokyo and also between Tokyo and Brazil. That four-second delay that cancelled the Mick Jagger-David Bowie live duet also was a factor in these intercontinental conversations.

I also had a video variation on the pre-hear I always insist on for audio. I had them set up a fax machine that let me preview the teleprompter in Japan, so I could be sure the upcoming copy was correct before John Denver and Yu Hayami read it. So I had both pre-hear and pre-see on this show.

My relationship with the European director, Gerard Pullicino, was different. Pullicino and I had only phone conversations to communicate about the cutting style. The French crew, headed by world producer Linda Wendell, had to make the all-day drive in their trucks from Paris to Berlin, to tape Richard Marx in a live concert there and to get B-roll footage of the then-collapsing Berlin Wall to use with Marx's powerful rendition of the Beatles' song "Help." The gas they bought in East Berlin gummed up the fuel system in one of the trucks. That slowed them down on their all-night seventeen-hour drive back to Metz, France, near Luxembourg. In Metz, at the Smurf Park—after working the previous day and driving all night—they had to set up for taping the following day. They taped the performances of Olivia Newton-John, Herbert Leonard, Niagara, Jeanne Mas, and Blues Trottoir, using a helicopter to get the aerial shots we needed for

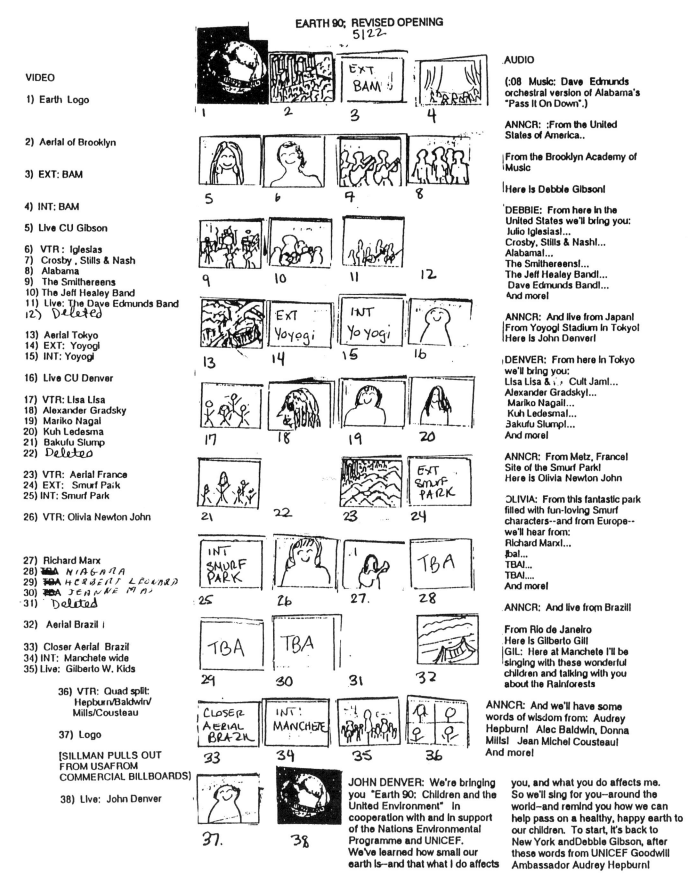

EARTH 90; REVISED OPENING
5|22

VIDEO

1) Earth Logo

2) Aerial of Brooklyn

3) EXT: BAM

4) INT: BAM

5) Live CU Gibson

6) VTR: Iglesias
7) Crosby, Stills & Nash
8) Alabama
9) The Smithereens
10) The Jeff Healey Band
11) Live: The Dave Edmunds Band
12) Deleted

13) Aerial Tokyo
14) EXT: Yoyogi
15) INT: Yoyogi

16) Live CU Denver

17) VTR: Lisa Lisa
18) Alexander Gradsky
19) Mariko Nagal
20) Kuh Ledesma
21) Bakufu Slump
22) Deleted

23) VTR: Aerial France
24) EXT: Smurf Park
25) INT: Smurf Park

26) VTR: Olivia Newton John

27) Richard Marx
28) ~~TBA~~ NIAGARA
29) ~~TBA~~ HERBERT LEONARD
30) ~~TBA~~ JEANNE MAS
31) Deleted

32) Aerial Brazil

33) Closer Aerial Brazil
34) INT: Manchete wide
35) Live: Gilberto W. Kids

   36) VTR: Quad split:
      Hepburn/Baldwin/
      Mills/Cousteau

   37) Logo

[SILLMAN PULLS OUT
FROM USAFROM
COMMERCIAL BILLBOARDS]

   38) Live: John Denver

AUDIO

(:08 Music: Dave Edmunds
orchestral version of Alabama's
"Pass It On Down".)

ANNCR: :From the United
States of America..

From the Brooklyn Academy of
Music

Here is Debbie Gibson!

DEBBIE: From here in the
United States we'll bring you:
Julio Iglesias!...
Crosby, Stills & Nash!...
Alabama!...
The Smithereens!...
The Jeff Healey Band!...
Dave Edmunds Band!...
And more!

ANNCR: And live from Japan!
From Yoyogi Stadium in Tokyo!
Here is John Denver!

DENVER: From here in Tokyo
we'll bring you:
Lisa Lisa & Cult Jam!...
Alexander Gradsky!...
Mariko Nagal!...
Kuh Ledesma!...
Bakufu Slump!...
And more!

ANNCR: From Metz, France!
Site of the Smurf Park!
Here is Olivia Newton John

OLIVIA: From this fantastic park
filled with fun-loving Smurf
characters--and from Europe--
we'll hear from:
Richard Marx!...
bal...
TBA!...
TBA!....
And more!

ANNCR: And live from Brazil!

From Rio de Janeiro
Here is Gilberto Gill
GIL: Here at Manchete I'll be
singing with these wonderful
children and talking with you
about the Rainforests

ANNCR: And we'll have some
words of wisdom from: Audrey
Hepburn! Alec Baldwin, Donna
Mills! Jean Michel Cousteau!
And more!

JOHN DENVER: We're bringing
you "Earth 90: Children and the
United Environment" in
cooperation with and in support
of the Nations Environmental
Programme and UNICEF.
We've learned how small our
earth is--and that what I do affects
you, and what you do affects me.
So we'll sing for you--around the
world--and remind you how we can
help pass on a healthy, happy earth to
our children. To start, it's back to
New York andDebbie Gibson, after
these words from UNICEF Goodwill
Ambassador Audrey Hepburn!

Figure 9–13 This early version of the opening of *Earth 90* shows how we overcame language barriers by supplementing audio with video. The homemade sketches were duplicated for later versions.

the opening of the show. As soon as they had finished taping and editing, they made a mad dash to get to Luxembourg, where RTL satellited the material to us via BrightStar—one hour before our air time. In New York, all we could do was wait, have faith, and be ready with our standby material in case trucks, tape machines, or satellites let us down. In the end, they all came through for us, and the European footage all aired, including some fabulous fireworks Smurf Park arranged for Olivia Newton-John's final number. But our ultimate success did not diminish the amount of standby planning and pre-show communication. In fact, it was because of all of the pre-show planning and communication that we could go on the air knowing we would have a successful, clean and effective show no matter what happened.

It is sometimes easy to forget the role of radio in globalcasts, but the fact remains that many people do not have access to television sets. As a result, millions of people experience globalcasts via audio only. For *Earth 90: Children and the Environment,* radio played two significant roles, one in carrying the show, one in our pre-show communication.

The major role radio played was in carrying the broadcast. The radio coordinator, Tara Gross, arranged for most of ABC Radio's 3,000 affiliates and most of the BC (British Columbia) News Limited's 250 affiliates to carry the program live. Voice of America (a potential 130 million listeners), Worldwide English Services (21 million), and England to Africa Services (27 million) carried the broadcast on a delayed basis.

The second role radio played was in helping swell our audience in New York and in preparing that audience to understand and enjoy the workings of a globalcast. New York radio station WSNR promoted both the show and the cause, airing environmental PSAs and running call-in contests for tickets the week before the globalcast. The day of the show, WSNR's afternoon DJ, Jim Douglas, acted as master of ceremonies and warmed up the audience prior to the show. I met with Douglas several days before June 2 and briefed him on the overall operation so he could explain it knowledgeably to the audience. Our writer briefed him again just before the show, to bring him up to date on changes. This type of pre-show communication was needed to prepare the audience for a globalcast.

Tapes

Communication for *Prayer For World Peace* differed dramatically from the musical shows, both in psychic distance and in form. In terms of conceiving the globalcast, the subject matter was totally predetermined. This was a regularly scheduled appearance of the Pope, and the Rosary is an historic, unalterable ceremony. This preordained format dictated two major requirements for the program. First was the psychic distance range—the approach. Second was the introduction to the ceremony, so that non-Catholics and Catholics alike would understand and appreciate what was about to happen.

Establishing the attitude posed no major problem. The narrator was Archbishop John Foley, who had a total grasp of everything that would take place and helped convey to everyone connected with the globalcast that we were presenting a ceremony, not an "event." No reporter would quiz the parishioners for crowd reaction, and no one would declare, "This is giving me goosebumps; I've never seen anything like this." The psychic distance had to be remote, with true reporting of the ceremony and no interpretive comments, so that each viewer could experience a pure and personal relationship with the ceremony.

The second major goal was the introduction to the recitation of the Rosary. We selected from the collection of art at the Vatican some of the most famous religious paintings appropriate to a recitation of the Rosary and showed these in connection with the reading of the Mysteries. We started the show with a live, digital GOES satellite picture of the earth, with music and voice-over, then went to a live shot of St. Mary Major in Rome with Archbishop Foley's voice-over (Figure 2–23), which was simultaneously translated into thirty-five different languages for transmission via Vatican Radio. The Archbishop explained the Rosary and listed the languages that would be used.

And so the stage was set for *Prayer For World Peace.* Next those sequences and meanings had to be communicated to the staff so they could convey them through global television.

On paper it sounds simple. The Mystery will be read. Cameras will show appropriate art work with appropriate music in the background. The Pope and his parishioners will recite one Our Father, ten Hail Marys, one Glory to God. They'll do that five times, and then the Pope will leave. Simple. Just set up the cameras and microphones and let the Pope and the people do their thing.

It may sound easy. And it looked easy on the air. But the *London Times* (June 3, 1987, p. 38) had it right when it called the program "the most complicated in the history of television." Fortunately, it happened just the way it was supposed to happen. But the simplicity of the ceremony depended on all of the one thousand-plus members of the technical and production team doing precisely the right job at precisely the right time. While Vatican Radio was broadcasting the narration in thirty-five different languages, I was giving camera commands I'd memorized in five different languages: English, French, Italian, German, and Spanish. Figure 2–17 is a reproduction of an original sheet from London, defining camera commands. It reflects the constant stress on the need for clean communication. Because we were working with so many countries speaking so many languages, it was critical that everyone know exactly what each command meant. Understanding the difference between *take* and *switch* could make the difference between being right and being wrong on the air. This printed sheet supplemented verbal communication.

Putting the Grid on Tape   Another indication of the uniqueness of each production is that we did not use a grid as the basic mode of communication on *Prayer For World Peace.* We used a lot of satellites and transponders and had to solve a lot of technical problems, but the structure of the program was not fluid, as it had been with *Live Aid,* and *Sport Aid,* where there had been hundreds of unpredictable elements. With *Prayer For World Peace* I was able to detail the program structure in time and place, which is what I did. We used on-paper rundowns during the show, but for pre-show briefings, I made a tape.

With stand-ins for all of the principle participants, I made a videotape of the ceremony as it would be performed in St. Mary Major, even duplicating the artwork in the opening. But if the participants and pieces of the program were substitutes, the structure was not. At each briefing session, all of the inject producers, directors, production assistants—everyone—saw the videotape. This visual aid was especially helpful to the inject producers who were to be off on remote, struggling like Robin Barty-King with languages and customs with which they were not familiar. Before they left London, they had a fundamental grasp of the elements the program would contain and what it would look like. They could see me on videotape, showing the path the Pope would follow in St. Mary Major. They could see his relationship to the audience in the church. The combination of the magnificence of the church, the artwork, and the music offered a sense of the grandeur the hour would communicate to viewers of the real thing. They could see how their locations would be integrated into the whole, and how the split screen would put multiple locations on screen at the same time.

The tape was helpful to the technicians planning the coverage, too. They could see precisely what areas had to be miked and cabled, without in any away disturbing the ceremony. And the production assistants (PAs) could see the overall structure of the ceremony of which their inject locations would be part.

Using the Dummy Tape for "Rehearsal"   The Friday before the Saturday globalcast was the first time we had all of the program PAs together. Figure 2–18 shows how they were set up in a large room directly adjacent to master control. We had visual communication through a large glass window on the wall between those two rooms and through another between master control and the audio room. We briefed the PAs in the morning, showing them the simulated tape and giving them a kit containing rundowns and other information describing the program.

After lunch we reassembled the group, one assigned to each remote location, in the PA room for a "rehearsal," which consisted of audio checks and a simulated run-through using the dummy tape.

Because of the large number of remotes (eighteen counting Rome as a remote from London) and because of the complexity of the mix-minus feeds, audio on this globalcast was a bigger challenge than video. We had no satellite time booked for Friday. It was strictly for audio checks.

As you can see in Figure 2–18, there was a producer standing by the window, in headset communication with me. He also had direct lines to the audio producer, standing beside and working with the audio man, and to the producer standing behind me, who was in constant contact on an open line with Rome. The PA producer had visual contact with me and the Rome-phone producer through the window in the wall.

In one section of the PAs' room were the three phones in touch with our subcontrol locations (Frankfurt, New York, and a production house in London). The rest of the room was taken up with two long tables seating the sixteen PAs, each with a phone to contact an inject producer. The chart lists the production assistant's name and phone number, the city he or she was responsible for, the name of the producer in that city, and the phone numbers of the inject location.

Figure 2–19 shows the two-page information sheet distributed to each of the PAs. This supplemented their audio briefing and provided a reference to which they could go and on which they could make notes.

Figure 2–20 shows the first two pages and the final page of the outline (rundown) from which the PAs worked. As would any rundown (the British call it running order), it gave an overview of the entire telecast *and* an indication of the points at which each production assistant's specific inject location was scheduled—always with the admonition that there might be a last-second audible to change what was preplanned. If you see mistakes and misspellings on these forms, remember that these are copies of the originals. Under the pressure of time, a lot of niceties go by the board. Spelling is often one of them.

Figure 2–21 shows the three-page time schedule we set up for the day of the globalcast, followed by the one-page checklist given to each of the PAs. The time schedule indicates the geographical sequence by which the technical crews planned to verify our signals, both audio and video. It indicates the language the technicians wanted the PAs to use in their checks with the inject producers, including the four-wires, the two-wires, and so on. The PAs' checklist starts with the critical mix-minus feeds.

I wish I could tell you it went as smoothly as the time schedule indicates. It didn't, and I didn't expect it to. Due to the complexity of dealing with thirty transponders on eighteen satellites, the technical crews were working until minutes before air simply to get acceptable contact with all of our inject locations. The fine tuning you see in the schedule, planned for a full hour, was accomplished in less than half that time, with the PAs learning a lot of technical language very quickly. I can assure you that every one of them will forever be able to explain exactly what mix-minus means—and probably how to say it in the language of his or her inject location.

Figure 2–23 moves one step deeper into the production—it shows the first two pages of the worksheet that contained an early draft of the words the narrator would read. You can see how the copy for the sixteen locations matches up with the sixteen wide shots called for on the PAs' rundown.

The first thing the PAs had to check during the Friday afternoon "rehearsal" was their phone-line connection with their remote producer. Believe it or not, there were some lines that did not work. There was simply no answer. That's why you "rehearse." That Friday we also had to check the Tech-Coord (technical coordination) lines, the Prod-Coord (production coordination) lines, and my headset line to all locations.

Once we had checked those audio lines, we simulated a run-through, using the dummy tape of the show and the audio lines. This "rehearsal" revealed the technical problems I referred to, some so intense they were not solved until after we went on the air.

It also revealed some production problems. On paper, the decades looked easy: ten repetitions of the Hail Mary. However, the languages threw some of the staff. They could not keep track of where we were. The first three repetitions were no problem. After that people could be heard asking, "Was that number four or five?" As a result, a new job category was opened. A PA was posted, standing in a position visible from the control room, the audio room, and the PAs' room and did nothing throughout the entire globalcast but keep count of the sequence in each decade. She held up large pieces of paper, numbered one through ten. Without a "rehearsal" to discover this need, we could have had unforeseen problems during the globalcast.

Overviews

The 1990 Goodwill Games is a good example of how patterns of communication have to be tailored to each show as carefully as are the format and technical cablings. *Live Aid,* despite its length, was a one-shot, as were *Sport Aid* and *Our Common Future.* And though *Earth 90* was conceived as the first of a series, it stood alone as a single three-hour program. (And the projected series was interrupted by history. *Earth 91,* scheduled for April 1991, was cancelled because of the Persian Gulf War crisis.) But *The 1990 Goodwill Games,* in Seattle and Washington state, was a series of twenty-three shows spread over seventeen days. Some two thousand people were scheduled to arrive in Seattle for those few weeks. All of them had to be brought up to speed before their arrival.

We used grids on *The 1990 Goodwill Games,* but we needed communication beyond that to keep the hundreds of technicians and production staff aware and updated as plans for the *Games* were made over the months and years of preplanning. I personally worked on the *Games* for two years, part-time for the first year and a half and full-time for the final six months.

The communication beyond the grids was in the form of *overviews.* With the help of the skeletal pre-show staff and the head writer, Brian Brown, I developed an overview for each of the twenty-three scheduled shows. Each contained anywhere from twenty-five to fifty pages and described the events that would take place, the athletes expected to participate, predictions of possible outcomes of the competition, special equipment we planned to use, special production pieces we were developing for each show, and a minute-by-minute grid of the show, breaking down the coverage, the commercials, everything. Figures 10–2 through 10–4 show different stages of a grid. Figure 10–2 is an example of the grid as it appeared in each overview. I'll discuss the other grids later.

This information could have been compiled in one gigantic book and sent out in that form. There were half a dozen reasons why we didn't do it that way.

1. It wouldn't have been read. I cannot conceive of anyone sitting down to read a book—however dedicated the individual, however fascinating the shows—covering twenty-three shows, each several hours long.

2. By spreading out the delivery of the overviews, we were able to create and sustain the interest of hundreds of people busy in other locations doing other jobs.

3. I needed feedback from these professionals who would be creating the coverage of the Games. I needed people to tell me that my visions were too expensive or could be improved or combined. As a result of the overviews, I got many suggestions, comments, and corrections that helped develop the final on-air product.

4. We had to budget the show. Developing a minute-by-minute breakdown of the coverage helped us to estimate the costs.

5. The technical staff had to devise, design, price, and plan the equipment necessary to put the plans into effect. As the overviews were published, the technical staff created and distributed a series of four operational manuals covering each show.

6. We had to detail for the Seattle Organizing Committee (SOC) our exact plans and needs so that they, in turn, could design traffic, housing, and transportation patterns to accommodate the influx of people who were to descend on the area.

Of course, there was one other reason. *I* needed to know. I had to translate Ted Turner's dream of a globalcast of an event that would use television to its maximum potential to share the sports competition and to unify the world in ways extending far beyond the reach of swimming or diving or running or racing. The overviews were the tool I used to bring this all together.

Communication During the Show

Establishing a Master Control Room

Now let us turn to the technical structure of a globalcast during the show. As soon as a program is formatted, the technical structure has to be designed to assure that the program is really possible. That structure, of course, depends on the format and is unique to each show.

On programs with an extremely large number of remotes, for example *Sport Aid* and *Prayer For World Peace,* the technical challenge becomes that much greater. Each time you add an inject from a different country, you add another set of transmission lines, another set of phone lines, another set of IFB lines, potentially another transmission standard, potentially another language, potentially another time zone. Each detail has to be pinned down. Schedules have to be established to test all of the detailed plans. Budgets have to be checked and rechecked. Everybody has to work together. Everybody has to communicate, and communicate thoroughly and well.

All of the audio and video channels on a show have to be funneled into and out of the central master control room. I directed *Prayer For World Peace* out of an existing control room at the Limehouse Studios in London, supplemented by a quad feed from a local production center, Molinare. *Sport Aid* came out of NBC's studio 8-H in New York and also included quad feeds. *Live Aid, Our Common Future, Earth 90,* and *The 1990 Goodwill Games* were all directed out of rented trucks. In short, you go where you have to go and you do what you have to do to get the job done.

The best option, of course, is to locate an existing control room with established transmission office and gateways to allow for easier conversion and reconfiguration. Assuming the master control meets those requirements, the first order of business is to search the area for public or private transmission carriers. This determines how many incoming or outgoing signals can be rented for that area of needed technology (earth stations, land lines, fiber optic cables, and so on).

The master control room switcher should be as large as possible so that there are enough inputs for the size of the event. As Figure 1–2 shows, to accommodate the switching and monitor layout required for *Live Aid,* we latched together two of the largest remote vans. For *The 1990 Goodwill Games* the crew custom-built a device specifically for my use.

The audio section of the control room has to be enhanced to accommodate the show's special needs, especially to tie in all of the intercom links. The private line phone (PL) should be configured to deal with all of the inbound and outbound communications as well as new communications channels, subchannels, and so on. In general, cameras, graphics, feeds, closed circuit monitors, and so on will overpower any normal control room and monitor panel.

As you can see in the list of keys for my audio box for *Sport Aid* (Figure 3–12), the overall interconnection of production audio from master control rooms to and from each remote inject is done within layers of the phone communication system, with five to eight duplex lines the norm for a globalcast.

Once the parameters and the number of injects have been decided, the next step is to match the audio/video signals between master control and the inject sites.

The surveys will have established whether you can use an established earth station gateway or have to bring in a portable station uplink. From each inject survey it will be determined how many satellites/transponders are needed to relay your signal back to master control. International venues traditionally utilize their domestic satellite, which then interfaces with one of the international satellites (Intelsat). Outbound links are determined by distribution needs—regional, quasi-regional, quasi-international, or fully international.

For financial and technical reasons, it is sometimes necessary to use compressed video patterns and assign transmission of that video to a sub-control, as we did on *Sport Aid* and *Prayer For World Peace*. This spider web type of transmission network can be very confusing to a new control room staff, so "rehearsals," including simulations, have to be arranged.

If you multiply all the feeds around the world with the number of communication lines that are needed, you can readily see why executive directors insist on coproducing shows—to protect themselves by having the proper equipment ordered and formatted.

Audio

The next step is audio communication during the actual show. The show is the fulfillment of all of the meetings, all of the phone calls, all of the rushed meals, all of the lost hours of sleep. This is what it's all about. Everyone is keyed up. The adrenalin is flowing. A million decisions have to be made. Each camera operator has to be alert to every move by every performer. Every audio operator has to follow the flow, from performer to performer, from microphone to microphone, from continent to continent, from live to tape to live to split screen. Doing the show is a physical and technical rush. But the show isn't over until it's over. There is a lot of carefully designed communication required while a globalcast is in progress.

To analyze audio channels, I will review *Sport Aid,* because it remains one of the largest, most complex globalcasts every produced. It's important to keep in mind that the technical complexity arose not out of a desire to create razzle dazzle, but out of need, out of the subject matter of the program.

Audio Channels *Sport Aid: The Race Against Time* was Bob Geldof's follow-up to *Live Aid.* Sponsored by Geldof's Band Aid Trust and the United Nation's Children's Fund (UNICEF), *Sport Aid* was a two-hour globalcast, starting at 10:00 a.m. EDT, Sunday, May 25, 1986, climaxing Sport Aid Week. The week before the telecast had been celebrated with sports events around the world: Grand Masters Tennis in Australia, National Physical Activities Week in Canada, a French Open Tennis exhibition near the Eiffel Tower in Paris, a badminton tournament in Hong Kong, a mountain climb in Ireland, a jump rope competition in Japan, football matches in Malawi, and on and on.

The globalcast covered millions of runners in more than two hundred cities, all starting at precisely the same moment in A Race Against Time. The start of the race was signalled when Sudanese world-class runner Omar Khalifa lit a torch in front of the United Nations in New York City. For the Sport Aid Week prior to Sunday, May 25, Khalifa made a one-man, worldwide trek from Khartoum, Sudan, to the UN. Figure 3–5 shows Khalifa and friends making the final run to light the torch at the UN. To explain the pattern of the audio channels, let me first explain the pattern of the show.

Our control room was in New York City, in NBC's studio 8-H. As the satellite schematic in Figure 3–3 indicates, this was the largest satellite broadcast to that date. We used twenty-four transponders on fourteen satellites, coordinated by BrightStar Communications and Visnews International (USA).

We had live injects from twelve cities: Athens, Greece; Auckland, New Zealand; Barcelona, Spain; Brisbane, Australia; Budapest, Hungary; Dublin, Ireland; London, England; New Delhi, India; Ouagadougou, Burkina Faso; Port-au-Spain, Trinidad; Rome, Italy; and Toronto, Canada.

As you can see on the satellite chart, to fit the inject feeds from seven countries onto the satellite signals from London, we had to use quad-feeds, four inject locations on the same signal. We accomplished this by using digital video effects (DVE), which convert the analog signal into digital information that can be manipulated to produce the computerlike effects TV viewers are familiar with. This signal processing is not to be confused with digital *compression,* which selects the most significant portions of the analog signal and compresses them so they fit the satellite transponder or cable system bandwidth. The DVE process does not remove any portion of the signal or compress it to fit into a smaller bandwidth; it makes possible manipulation of the existing signal. In the case of *Sport Aid,* DVE, in effect, squished the existing signals from the four inject locations and made four small pictures into one full-sized picture.

You can see that one quad-feed carried video from Greece, India, Ireland, and the UK. The second quad carried Spain, Hungary, and Italy. This system gave us access to all of those countries, but the limitation of a quad is that—unless you air the entire quad or call for a "hot switch"—you can air only one image from the quad at a time. Let me explain. With separate feeds coming in from Africa and Australia, the technical director (TD) in New York could switch, at will, from one country to another. All I had to do was steady the camera in question and call for the cut. The TD hit the button and the video changed. With Greece and India on a quad-feed, the TD in New York could not switch between Greece and India on the air. It was possible for me to call for a hot switch, meaning that the sub-control in London, which was controlling the quad-feed, would make the switch from one country to another. Figure 3–4 shows part of our New York control room, with the two quad-feeds at the bottom of the screen. This is an off-the-tape video print, taken off the air. I wanted the audience watching our globalcast to share the technology that was bringing it to them.

To make the two hours of *Sport Aid* easy for the viewers to follow, we broke the show down into three basic units: the setup, the race itself, and the postrace interviews. Interspersed, of course, came PSAs carrying messages about the situation in Africa.

The Setup   We began live at the UN, then did a quick round-robin of the twelve inject locations. This accomplished two goals: it established the vast scope of this undertaking and it introduced the audience to the locations from which we would be following the race. Our job was to use these twelve cities as symbols of the 237 cities that held races that day.

Next we set up the meaning and significance of Omar Khalifa's symbolic run from Africa to the UN, flashing back from our live cameras at the UN to a tape of Omar's journey across Europe to the USA. We then picked up Omar Khalifa as he ran the last mile, through the streets of New York City to the UN to light the torch, the signal to start the race. Dozens of celebrities ran with Omar Khalifa on his symbolic last mile. While they ran, we flashed back again, on tape, to highlights of the worldwide events that had filled the week prior to the date of the race. Finally, we round-robined the world again, taking time to establish local commentators at the locations and to build up the excitement leading to the start of the race. As we got close to the start, we intercut the race through the streets of New York with preparation around the world on a minute-by-minute countdown to the lighting of the torch.

The Race   The second phase of the show began the second hour of the show. It started with Khalifa completing his race, his journey to the United Nations, and lighting the torch.

Coverage of the race itself was amazing. If the buildup had been paced, the race itself was fast and exciting. Dozens of sports, screen, television, and music stars ran the race that day, and their presence and their numbers were impressive. But as impressive as the stars were, they were overshadowed by the overwhelming force of the millions of people who ran The Race Against Time. It was the biggest single sports event in the history of the world, devised to tell the United Nations representatives how important people considered the issue of starvation in Africa. The next day the United Nations General Assembly *did* vote aid to Africa.

The video of the race was a joy to behold and to call (Figures 3–6 to 3–9). The tricky part was the audio. We had local commentators in some locations, but not all. Because it was impossible to predict ahead of time to which location we would be going, I had to have absolute coordination with Rolland Smith, the host of the show. This was through the IFB in his ear. He was not on camera during much of the race.

The arrangement with Smith was very simple—and very challenging. He had to be ready to talk about any of twelve cities with no more than a second or two notice from me in the control room. If I'd scheduled a local commentator, I would tell Smith where we were going and to introduce the commentator and then not talk until the next cue.

The local commentators with whom I was in touch via IFB phone lines were all bilingual, English-speaking announcers. I could control their commentary by giving verbal commands. But we also had indigenous, non-English-speaking announcers at various points along each race route. I had no audio connection to them. The bilingual commentators, to whom I could talk, had to cue and control these on-site reporters. This combination of commentators and positions allowed us to air multilingual accounts of each race's progress and to conduct interviews with the winners in their native tongue, the distancing coming closer to the conceptual tone I was looking for. If Smith felt an English translation was necessary, he had the ability to call for one, in any location.

If there was no local commentator, Smith had to be ready to ad-lib. He was chosen for the job, of course, because his years as a New York news anchor had honed his on-camera skills. But he also had to be briefed and updated about each of twelve locations while we were on the air. We accomplished that goal by setting up twelve desk-stands, the kind typists use, one for each location. The name of the city was written in large letters at the top of each stand, which contained sheets of information about that city and country. My writer and several assistant directors heard my calls to Smith on their headsets, and whoever was closest to the called location would hand Smith the sheets for that city so he would have statistics at his fingertips—the length of the race at that location, the number of people running, the names of celebrities involved, and background on the city.

The assistant directors were one step ahead of Smith because they wore two headsets, one to hear the IFB calls to the talent, one to hear me making the calls to the crew. They could hear me give a camera-ready call immediately before my instructions to Smith.

The most complicated part of directing a globalcast with so many injects is to keep your commands to everyone clear and in sequence. You develop a strengthened peripheral vision when you direct shows with multiple video sources. No one can watch twelve monitors at one time, and yet you have to. And you have to be able to watch and decide quickly enough, if you select one of the cities on a quad-feed, to let the crew know which of the four available lines you want to put on preview before airing it. Within a matter of seconds you have to see, decide, indicate the camera, steady the camera, cue the talent, and take—and get ready for the next one.

It is not a relaxing procedure, and it doesn't take long for a crew to understand why I put out the edict that no one talks to me unless it's an emergency. The audio lines *have* to be kept clear because so many people have to hear and understand what comes next. The chart in Figure 3–11 shows the audio channels we used on *Sport Aid.* It demonstrates how complex listening can be for someone on camera, on a switcher, or with audio or tape. *Listening* is as important as *doing* for every member of the crew, and everyone has to respect the need of everyone else for clear communication channels.

The complexity of the internal communication system on a globalcast is demonstrated in sketches relating to the audio box I used on the show. This box is not a standard piece of equipment. One is built each time by the crew, designed by my production head, Michael McLees, to meet the specifics of that particular production. Figure 3–12 shows the geographical destination of each of the thirty keys on the audio box. Figure 3–13 lists the geographical destination plus the method of transmitting the audio signal (land line, satellite, microwave). Figure 3–14 shows the line leading from the keys on the audio box, along the signal carrier, to the specific destination (to the commentator on site, to the control room, and so on). Figure 3–11 is a schematic of the whole system, showing how the lines fanned out from the control room in New York to the twelve inject locations around the world.

The thirty keys on the audio box are in two rows of fifteen. The side-by-side keys go to the same destination; that is, key 1 and key 16, beside it, went from the New York Control Room to another destination in New York, one to the commentator, one to the control room. Key 15 and key 30 both went to Budapest, Hungary, to similar destinations. The second set of numbers 1 through 30 indicate the business phones that went to the same destinations, but to other people there, not necessarily the same place my lines went to.

My squawk-box lines were all one-way. The person at the other end could not talk to me. You can imagine the chaos that would erupt if thirty people decided to talk to me at one time. The other lines were duplex systems, opened both ways. Generally speaking, there are five to eight such duplex lines used to serve the needs from central master control to each inject. In this case there were five circuits: one provided the "hot" mike—my one-way squawk-box line to each site; one accommodated the assistant director's position; one allowed engineering interglobal communications; one allowed me to talk directly to the stage managers around the world, to coordinate such things as performer's entrances and time cues; and one allowed the transmission people to have conferences.

A few things are readily apparent as you study these charts. First, the person who designs and constructs the system and box has to be extremely knowledgeable about the globalcast. Second, the executive director needs a vast array of communication channels to keep control of all of the variables and has to be decidedly dexterous to manipulate the keys that operate that array of communication channels. Finally, you have to plan what to do if certain keys don't work.

If there are times when a globalcast director seems autocratic and curt in giving commands, you have to remember that there are hundreds of people at the other end, each in need of sharp and instantaneous communication. There is no time cushion when a decision has to be relayed to a control room, a camera operator, or the on-air talent.

Figures 3–15 and 3–16 show the production grid that we used on *Sport Aid*. Figure 3–15 shows the blank grid with eleven job categories and one comment column across the top and the numbered items from the rundown running down the left-hand column. Figure 3–16 shows the second page of what was an eleven-page grid. (The initials R.S. in the audio and copy columns stand for Rolland Smith.) Note the warning that the grid could change more than once a day. Those who work on regularly scheduled shows may assume that a rundown is "the" rundown, but that is never true on a globalcast. Whatever your job, you have to stay in constant communication with the rest of the staff, and one of the most important ways to do that is to keep in touch with the staff member issuing updated grids.

Post-Race Interviews   Figure 3–9 shows a runner at one of the twelve remote injects crossing the finish line. Because the various races covered different terrain, took place at different times of the day and night, and involved thousands of runners, it was impossible to predict when each race would finish. We had to capture the winner of each race and still pace the interviews with winners by local commentators (see Figure 3–10). (And to show you how unpredictable live globalcasts can be, the winners at two of the locations crossed the finish line at precisely the same moment. I had to show one on taped replay later. Couldn't happen in a million years, but it did.)

The executive producer had planned the first hour of *Sport Aid* to move slowly and build up to the next hour. The second hour, featuring the race and the post-race interviews, was a real challenge. As I noted before, *Sport Aid* attained the goal set by Bob Geldof and UNICEF and went into the books as a success.

Video Channels

Prayer For World Peace was successful in large part as a result of careful planning. Although the audio was a greater challenge than the video, the video was no snap. We were dealing with thirty transponders on eighteen satellites plus a plentiful supply of land lines. Also, the large number of inject locations required using four quads and, to combine them all, a super quad. The satellite schematic in Figure 2–7 details the video lines.

Central master control was in the Limehouse Studios London rather than Rome because of complexities in union rules in Italy and the threat of a strike during the time of the telecast. There were three submaster controls to handle the overload, one in New York City, one in Frankfurt, Germany, and one in the Molinare production house in London.

There were seventeen injects. Four quads accounted for sixteen locations, and the seventeenth had to be hot-switched within its quad. The four quads were built as follows:

Quad 1 (New York)	**Quad 2 (Frankfurt)**
Washington, D.C., USA	Lourdes, France
Quebec, Canada	Czestochowa, Poland
Manila, Philippines	Mariazell, Austria
Guadalupe, Mexico	Frankfurt, Germany
Quad 3 (Molinare)	**Quad 4 (Limehouse)**
Knock, Ireland	Rome, Italy
Bombay, India	Fatima, Portugal
Rio de Janerio, Brazil	Zaragoza, Spain
(or Caacupe, Paraguay)	Dakar, Senegal
Lujan, Argentina	

Figures 2–14 through 2–16 are reproductions of the charts we used in London, showing how this was set up on the master control switcher, the location of the countries on each of the quads, and the numerical arrangement of the Super Quad. Figure 2–16 shows a breakdown by country of the Super Quad but does not indicate the hot-switch capability of changing the video from Rio de Janeiro in Brazil to Caacupe in Paraguay in Quad 3. These two locations could not appear together. On my command, however, the sub-control could hot-switch between them, so that on-air we could change from Rio to Caacupe or vice versa.

The Molinare production house, where Quad 3 was formed, came to our rescue technically. We had so many incoming feeds that Limehouse could not accommodate them all. The cost for microwaving the signals would have been astronomical, and we selected Molinare because they were capable of bringing in the four signals for a quad. They fed two lines to us at Limehouse. One line carried the four quad feeds, which could be used separately or together. The second line was a preview line where I could see either the full quad or any given city before I put it on the air. In Figure 2–25, a diagram of the monitor wall, you can see the quad monitors on the bottom row. Immediately above each quad is the monitor on which I could call out, full-screen, any one of the locations on the quad. Figure 2–12 shows Quad 4, combining Rome, Fatima, Zaragoza, and Dakar. Figure 2–11 shows Fatima, one of the locations from Quad 4, full-screen. These are all video prints directly from the tape of the show.

The Super Quad (Figures 2–13 and 2–16) was not set up in numerical sequence. Because I always wanted the Rome video to appear first on the screen, the Super Quad was set up with Quad 4 in the upper left corner (see Figure 2–24).

Again the Super Quad was created not to demonstrate the technical skill of the crew, but out of necessity, to bring the world together. When the formal ceremony had ended and the Pope was leaving St. Mary Major, we cut to all of the inject locations, where each parish rang bells. Hundreds of different-toned bells rang on my cue as the parishioners waved white handkerchiefs, a tradition to celebrate the Pope's presence. Using our electronic skills to the maximum, we put those four quads on the air at once—live images of all sixteen parishes. It was a powerful moment. Even then—especially then—our narration kept the psychic distance at its proper level, reporting, not reacting. For me it was a special moment, since I was able to create something that I had visualized.

To direct the program and to control all of these inputs, I sat with the monitor wall in front of me, the audio room to my right (with a glass window between us), the PA room behind me (also with a glass window). The the stool I sat on was raised so that I could see and be seen from all of those locations so we could use visual signals to supplement audio communication and also so I could see the bottom row of the quad monitors, which had been added to the base of the existing monitors.

Sending Signals Beyond the Video Channels In many countries, television sets are limited to large cities and telephone lines haven't reached everyone. But generally speaking, telephone signals reach more people than do television signals, and short-wave radio reaches most areas not yet served by telephones. We transmitted *Prayer For World Peace* through all of these media.

The television commentators narrating in non-English languages were in a studio next to the control room in Limehouse. I had no visual communication with them, but our communication was nonetheless instantaneous.

We had a PA and a technician, on headset, in the studio with them throughout the globalcast.

The commentators doing radio translations in thirty-five different languages were at the Vatican in Rome, therefore the distance between us was considerable. Any significant changes in format had to be relayed not just to the sixteen inject locations and five television commentators, but also to the thirty-five announcers, each speaking a different language. My communication with the Vatican was via telephone lines.

Prayer For World Peace was also carried by the United States Armed Forces Network, The Voice of America, and a number of independent radio stations in the USA. For people who could not receive the program by TV or radio, a 900-number phone line was set up for audio of the program. And, as I indicated earlier, I had separate phone lines made available to some college campuses so students could call in to hear the audio output from the control room during the program.

There is inevitably more communication going on within a globalcast than is going out over the air. On *Prayer For World Peace* we had dozens of producers, directors, and PAs, each of whom had to understand what was happening and how each job fit in. Everyone working on a globalcast has to understand all of the jobs and how they interconnect.

Converting the Standards   With the many incoming and outgoing signals, obviously the crew dealt with multiple transmission standards. This is so automatic for the crew that producers seldom think about these complexities, but obviously someone has to.

Here is just one example from *Prayer For World Peace*. The signal coming to the Frankfurt subcontrol (Quad 2) from Lourdes was in SECAM, the French standard in 1987. At Frankfurt it was converted from SECAM to PAL, the British standard in 1987, and sent to London in PAL. Our London control room was dealing strictly with the PAL standard, so the incoming signal from France simply blended in with all the other signals. That was the incoming conversion.

The process was, of course, reversed on the way back. The outgoing signal destined for France was converted from PAL back into SECAM and was transmitted in that standard to France. The signal going to the USA was sent out in NTSC. The signal to the UK stayed in PAL. There was a similiar pattern for each location. A lot of work goes unseen—and often unappreciated—on a major globalcast.

A Final Reflection on Communication   Any discussion of communication on *Prayer For World Peace,* I think, has to focus on Pope John Paul II. It was John Paul I who pioneered global communication by approaching Guglielmo Marconi, inventor of the wireless telegraph, to install radio facilities at the Vatican as a means of communicating with his worldwide flock. Pope John Paul II has followed in his predecessor's footsteps by letting the power of television carry the message to the faithful. It was John Paul II who approved the globalcasting of the ceremony and who approved the format I designed for it. The Popes of all ages have used the arts to communicate with worshippers. Great churches were built for that purpose. To inspire, many great artists painted and sculpted in an effort to convey their faith to others. In a sense, when electronics came of age, it was Marconi who replaced Michaelangelo. It is with these thoughts that one tries to use one's talent, to touch and carry on the inspiration that we inherited. *Prayer For World Peace* was listed as the single most-watched hour of television in the world, more than 1.5 billion people.

Communication After the Show

I always debrief the inject producers after the program to identify problems we can solve for next time and possible better ways to work that can be added to the system. I debrief the rest of the crew as well, because we learn by not repeating our mistakes, by sharing our experiences. A debriefing may seem like just one-more-meeting. But since all of the crew members have long ago learned that the meetings have meaning, no one has complained yet. They are exhausted, but they hold on long enough to remember.

Very often the first reaction after a major globalcast is "Never again!" But after a week of rest and eating at mealtimes, most of us are chomping at the bit, and we begin looking for "the next one."

Chapter 10

How to Produce a Series

As I indicated earlier, *The 1990 Goodwill Games* was different from other globalcasts we've discussed because it was a series instead of a one-day production. I devised the overviews as a method of sustaining interest in and gaining feedback from my production staff before we gathered in Seattle. By writing an overview for each of the twenty-three shows and distributing them over a period of a year, I was able to communicate information that would have been overwhelming had it been presented in a single book. The overviews were also the basis for plotting the technical and budgetary needs of the series. In this chapter, using excerpts from the overviews, I will detail the ways in which a global series differs from a one-time production.

The Scope of the Globalcast and the "Overviews"

The 1990 Goodwill Games took place over seventeen days and were also spread over the greater Seattle area and Washington state. Within Seattle, events took place at Hec Edmundson Pavilion, Husky Stadium, Seattle Center Coliseum, Seattle Center Arena, Seattle University, Mt. Baker/ Lake Washington, the Weyerhaeuser King County Aquatic Center and at Shilshole Bay Marina, in Puget Sound. Thirty miles to the south, in Tacoma, there were events at Tacoma Dome and Cheney Stadium. Thirty miles in another direction was the King County Fairgrounds, and a few hundred miles away were the Spokane Coliseum, near the Idaho border, and the Tri-Cities Arena, near the Oregon border.

There were three basic sources of programming available to the World Broadcasters: (1) an International Composite Satellite Feed, available to all countries that bought television rights; (2) five hundred hours of Host Broadcast coverage for rights-holding broadcasters to use to develop their own composite feeds, and (3) taped summary packages available through Televisa (Spanish), Nordic Group, TVNZ (New Zealand), and Host Broadcast. There was also a daily satellite news service provided to all USA news organizations for insertion in local news programming. There were an estimated three thousand accredited press writers/photographers and broadcasters present. While TBS was responsible for television coverage, the Seattle Organizing Committee (SOC) was responsible for coordinating everything connected to this massive operation, including planning for the huge influx of athletes, news representatives, television staffs, and equipment.

The 1990 Goodwill Games
The World's Best Athletes, Exclusively On Cable.

1990 Goodwill Games Schedule

SUN	MON	TUE	WED	THU	FRI	SAT
Jul 15 Key: M-Men W-Women SF-Semifinal F-Final	Jul 16	Jul 17	Jul 18	Jul 19	Jul 20 Gymnastics-M Swimming Volleyball-W	Jul 21 Welcoming Ceremony Gymnastics-M Marathon-M Rowing Swimming

Jul 22
 	
Athletics
Gymnastics-M
Marathon-W
Rowing
Swimming
Handball
Volleyball-W
	Jul 23
	
	Athletics
	Basketball-M
	Swimming
	Handball
	Volleyball-W
		Jul 24
		
		Athletics
		Basketball-M
		Swimming
		Handball
			Jul 25
			
			Athletics
			Basketball-M
			Volleyball
			-WSF
			Water Polo
				Jul 26
				
				Athletics
				Baseball
				Handball-SF
				Volleyball-WF
				Water Polo
					Jul 27
					
					Baseball
					Bsktball-MSF
					Gymnastics-W
					Ice Hockey
					Rhythmic
					  Gymnastics
					Handball-F
					Water Polo
					Wrestling
						Jul 28
						
						Baseball
						Boxing
						Cycling
						Gymnastics-W
						Ice Hockey
						Rhythmic
						  Gymnastics
						Water Polo-SF
						Wrestling

Jul 29
 	
Baseball
Basketball-MF
Boxing
Cycling
Gymnastics-W
Ice Hockey
Water Polo-F
Weightlifting
	Jul 30
	
	Baseball-SF
	Boxing
	Ice Hockey
	Modern
	  Pentathlon
	Synchronized
	  Swimming
	Volleyball-M
		Jul 31
		
		Baseball-F
		Basketball-W
		Boxing
		Diving
		Ice Hockey
		Modern
		  Pentathlon
		Volleyball-M
			Aug 1
			
			Basketball-W
			Boxing-SF
			Diving
			Ice Hockey
			Volleyball-M
			Yachting
				Aug 2
				
				Basketball-W
				Boxing-SF
				Diving
				Figure Skating
				Ice Hockey
				Yachting
					Aug 3
					
					Diving
					Figure
					  Skating-F
					Ice Hockey
					Judo
					Volleyball
					  -MSF
					Yachting
						Aug 4
						
						Basktball-WSF
						Boxing-F
						Diving
						Figure
						  Skating-F
						Ice Hockey-SF
						Judo
						Volleyball-MF
						Yachting

Aug 5 Basketball-WF Boxing-F Closing Ceremony Diving Figure Skating Ice Hockey-F	Coverage begins weekdays at 8:05 PM eastern. Coverage begins weekends at 3:05 PM and 8:05 PM eastern. On your TBS channel. **TBS** SuperStation					

+ Only the top 8 teams or individuals in 21 sports will be invited to compete
+ Over 50 countries represented
+ 2,500 participating athletes

+ 86 hours of live coverage
Don't miss the international event of the year, cablecast only on your TBS channel on participating systems.

SEATTLE '90

GOODWILL GAMES

Uniting The World's Best

Figure 10–1   This TBS ad for *The 1990 Goodwill Games* outlines the scope of the events and indicates the criteria for participation.

Courtesy Turner Broadcasting System.

Saturday, July 21, 1990   Tony Verna

AB	12:05	ADVERSARIAL BILLBOARD - VTR
	12:06	CONT'D (W/MENTION OF TOP 8 FORMAT)
CB	12:07	COMMERCIAL BILLBOARD #1
B	12:08	ANCHOR ... BLIMP #1 - LIVE - MOUNT BAKER - BLIMP #2 - AQUATICS CENTER
V	12:09	VTR/CROWDS/DAY ONE ... LIVE CROWDS TODAY (VOYAGER)
	12:10	VTR - THE WORLD WAS BROUGHT TOGETHER
I	12:11	LIVE - RUSSIA INJECT - IMPRESSIONS OF DAY ONE
	12:12	TODAY'S ANNOUNCERS - GREG LEWIS, BART CONNER, JOHN NABER, FRANK SHORTER
	12:13	VTR - RUNNERS LINING UP FOR START OF MEN'S MARATHON - V.O. FRANK SHORTER(LIV
NC	12:14	NATIONAL COMMERCIAL A&B NATIONAL 57&58
LC	12:15:10	LOCAL COMMERCIAL A&B LOCAL 41&42
MC	12:16:10	GOLD MEDAL COUNT #1
	12:17	GOLD MEDALS TO BE AWARDED: SWIMMING (6) ROWING (8) MARATHON (1) GYMNASTICS (
B	12:18	ANCHOR TO JOHN NABER - BLIMP - AQUATICS CENTER
	12:19	LIVE - SWIMMING - SET UP WOMEN'S 100M FREESTYLE
J	12:20	VTR - JOURNEY - SYLVIA POLL - COSTA RICA - (SYLVIA SPEAKS FLUENT ENGLISH)
J	12:21	CONT'D JOURNEY
I	12:22	LIVE INJECT - COSTA RICA (CHANNEL 7)
NC	12:23	NATIONAL COMMERCIAL A&B NATIONAL 59&60 (CUE-TONE) 10 SECOND BUMPER
LC	12:24:10	LOCAL COMMERCIAL A&B LOCAL 43&44
	12:25:10	EXT: AQUATICS CENTER (LIVE OR VTR) ... INT: LIVE LANE INTRO
	12:26	CONT'D
***	12:27	LIVE - START WOMEN'S 100M FREESTYLE
	12:28	END OF RACE
	12:29	SLO-MO - ANALYSIS
I	12:30	ANCHOR - POSSIBLE LIVE W/COSTA RICA INJECT (INTERVIEW)
	12:31	CONT'D
	12:32	ANCHOR - SET UP MARATHON W/FRANK SHORTER IN STUDIO W/MAP
	12:33	VTR - TOP 5 MARATHONERS POSITIONING FOR START (LIVE V.O.)
TP	12:34	TIME-PLATE - WOMEN'S 400M IM W/JANET EVANS (33 MINUTES)
NC	12:35	NATIONAL COMMERCIAL A&B NATIONAL 61&62 (CUE-TONE) 10 SECOND BUMPER
LC	12:36:10	LOCAL COMMERCIAL A&B LOCAL 45&46
PR	12:37:10	PROMO #1
***	12:38	VTR - START OF MARATHON (SOT TAPE - RECORDED CALL)
	12:39	VTR - MARATHON (RECORDED CALL) W/UNITUS MASCOT
	12:40	VTR - MARATHON (RECORDED CALL) W/EMERALD CITY MARATHONER
	12:41	VTR - MARATHON (RECORDED CALL)
	12:42	VTR - MARATHON (RECORDED CALL)
	12:43	VTR - WEIGH IN FOR WOMEN'S LIGHTWEIGHT SCULLS - LIVE STUDIO V.O.
	12:44	CONT'D W/LIVE BUMPER
NC	12:45	NATIONAL COMMERCIAL A&B NATIONAL 63&64 (CUE-TONE) 10 SECOND BUMPER
LC	12:46:10	LOCAL COMMERCIAL A&B LOCAL 47&48
	12:47:10	VTR - WOMEN'S LIGHTWEIGHT SCULLS W/LANE INTROS (RECORDED CALL)
***	12:48	VTR - WOMEN'S LIGHTWEIGHT SCULLS (RECORDED CALL)
	12:49	VTR - WOMEN'S LIGHTWEIGHT SCULLS (RECORDED CALL)
	12:50	VTR - WOMEN'S LIGHTWEIGHT SCULLS (RECORDED CALL)
	12:51	VTR - WOMEN'S LIGHTWEIGHT SCULLS (RECORDED CALL)
	12:52	VTR - WOMEN'S LIGHTWEIGHT SCULLS (RECORDED CALL)
	12:53	VTR - WOMEN'S LIGHTWEIGHT SCULLS (RECORDED CALL)
	12:54	VTR - WOMEN'S LIGHTWEIGHT SCULLS (RECORDED CALL) ... (MORE TO COME)
NC	12:55	NATIONAL COMMERCIAL A&B NATIONAL 65&66 (CUE-TONE) 10 SECOND BUMPER
LC	12:56:10	LOCAL COMMERCIAL A&B LOCAL 49&50
	12:57:10	VTR - CONT'D - WOMEN'S LIGHTWEIGHT SCULLS
	12:58	VTR - CONT'D - WOMEN'S LIGHTWEIGHT SCULLS (SET UP ELECTRONIC MAT - FINISH LI
	12:59	VTR - END OF RACE - SLO-MO - CLOSE FINISH
	1:00	LIVE ANCHOR W/SHORTER W/MARATHON MAP
	1:01	VTR - MARATHON UPDATE (LIVE V.O./CUE TRACK/INTERNATIONAL SOUND)
TP	1:02	TIME-PLATE: NEXT - JANET EVANS - WOMEN'S 400M IM (5 MINUTES)
NC	1:03	NATIONAL COMMERCIAL A&B NATIONAL 67&68
NC	1:04	NATIONAL COMMERCIAL A&B NATIONAL 69&79

Figure 10-2 The text explains this minute-by-minute grid in detail. Compare this pre-show estimated grid from my Show 2 overview with the grids in Figures 10-3 and 10-4.

GOODWILL GAMES
TONY VERNA

SHOW GRID – SHOW 2
SATURDAY, JULY 21

Code	Time	Description	Notes
A	12:05	(ACT 1)	
AB	12:05	VTR – ADVERSARIAL BILLBOARD	
AB	12:06	VTR – ADVERSARIAL BILLBOARD	
CB-1/B	12:07	LIVE – COMMERCIAL BILLBOARD #1; LIVE – BLIMP – SEATTLE	
I	12:08	LIVE – HANNAH, NICK INTRO/SET UP SHOW – (SWIM, MARATHON, ROW)	
	12:09	LIVE – SET UP AUS INJECT/HANNAH/NICK LEAD TO VTR FEATURE	INJECT – BRISBANE, AUSTRALIA
TF	12:10	VTR – THE WORLD WAS BROUGHT TOGETHER/RETROSPECTIVE SHOW	
	12:11	VTR – THE WORLD WAS BROUGHT TOGETHER/RETROSPECTIVE SHOW	
TP	12:12	LIVE – HANNAH/ NICK LEAD TO COMMERCIAL/TIMEPLATE, W/100M FREE	
NC	12:13	NC A&B NC 57 & 58 (CUE-TONE) 10 SECOND BUMPER	
LC	12:14:10	LC A&B LC 41 & 42	
	12:15:10	(ACT 2)	
LC	12:15:10	LIVE – NICK TO AQUATICS, NABER PREVIEWS SWIMMING	
	12:16	LIVE – NABER PREVIEWS SWIMMING, THROWS TO STUDIO	
	12:17	LIVE – STUDIO, HANNAH LEAD TO VOYAGER	
V	12:18	LIVE – VOYAGER/LARGENT	
MC/TP	12:19	LIVE – NICK MEDAL COUNT/TP–W/100M FREE; EVANS 400M IM; M/800M FREE	
PR-1	12:20	PROMO #1	
NC	12:20:30	NC A&B NC 59 & 60 (CUE-TONE) 10 SECOND BUMPER	
LC	12:21:40	LC A&B LC 43 & 44	
	12:22:40	(ACT 3)	
TF	12:22:40	LIVE – NICK TO FEATURE – USA VS. EAST GERMANY	
	12:23	VTR – USA VS. EAST GERMANY SWIMMING	
	12:24	VTR – USA VS. EAST GERMANY SWIMMING/NICK TO AQUATICS	
	12:25	LIVE – SET UP WOMEN'S 100M FREESTYLE – LANE INTRODUCTIONS	
	12:26	LIVE – SET UP WOMEN'S 100M FREESTYLE – LANE INTRODUCTIONS	
***	12:27	LIVE – START WOMEN'S 100M FREESTYLE	
TP	12:28	VTR – REPLAY / TIME-PLATE (SWIMMING; MARATHON)	
NC	12:29	NC A&B NC 61 & 62 (CUE-TONE) 10 SECOND BUMPER	
LC	12:30:10	LC A&B LC 45 & 46	
A	12:31:10	(ACT 4)	
	12:31:10	LIVE – INTERVIEW WINNER WOMEN'S 100M FREESTYLE TO STUDIO	
I	12:32	LIVE – STUDIO TO AUS INJECT	LIVE – INJECT – BRISBANE, AUSTRALIA
TE	12:33	LIVE – STUDIO W/INJECT TO SALNIKOV VTR TO AQUATICS – SET UP RACE	

Date and Time Printed: 7/21/90   1:33 AM

Page 1

Figure 10–3 This grid was prepared by David Raith in Seattle after consulting with the Seattle Organizing Committee. Note the date and time to identify the grid.

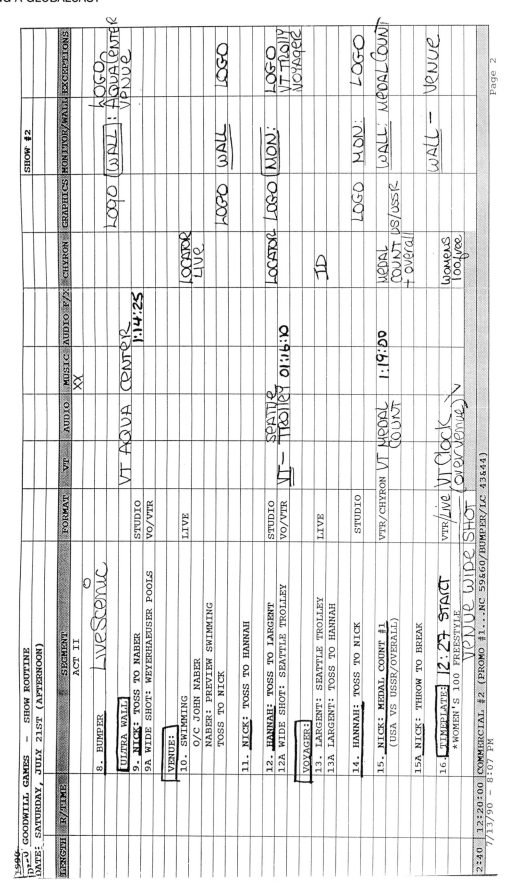

Figure 10–4 The estimated information in the grid in Figure 10–2 was updated to that in Figure 10–3 and ended the day of the telecast in this form, as prepared by the coordinating producer's staff. The penned-in times are segment times. The page's running time began with Nick Charles's toss to John Naber at 12:15, ended (as indicated) at 12:20.

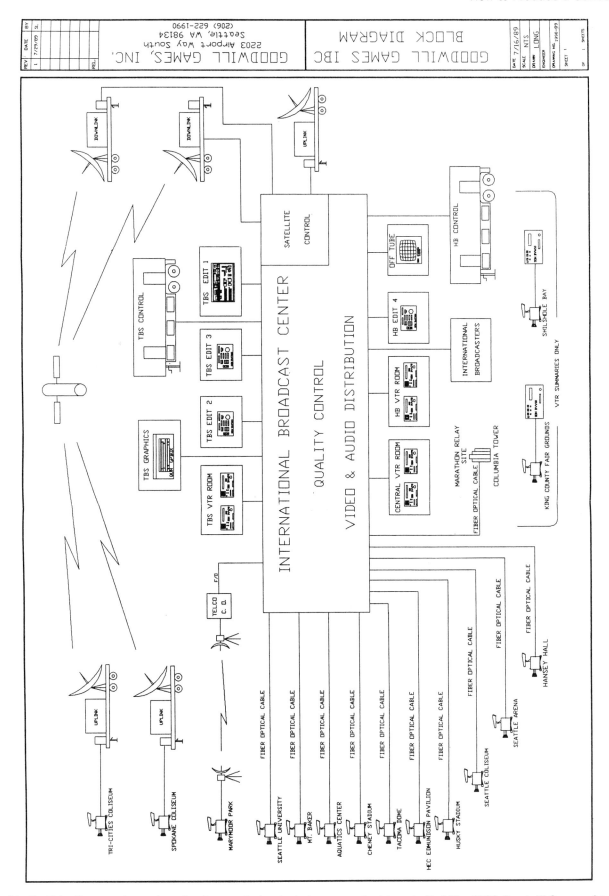

Figure 10–5 This diagram shows the mix of transmission techniques required to get all of *The 1900 Goodwill Games* signals from the multiple venues to the IBC. My control panel was in the IBC control truck in the upper right-hand portion of the diagram. See Figure 10–6 for details.

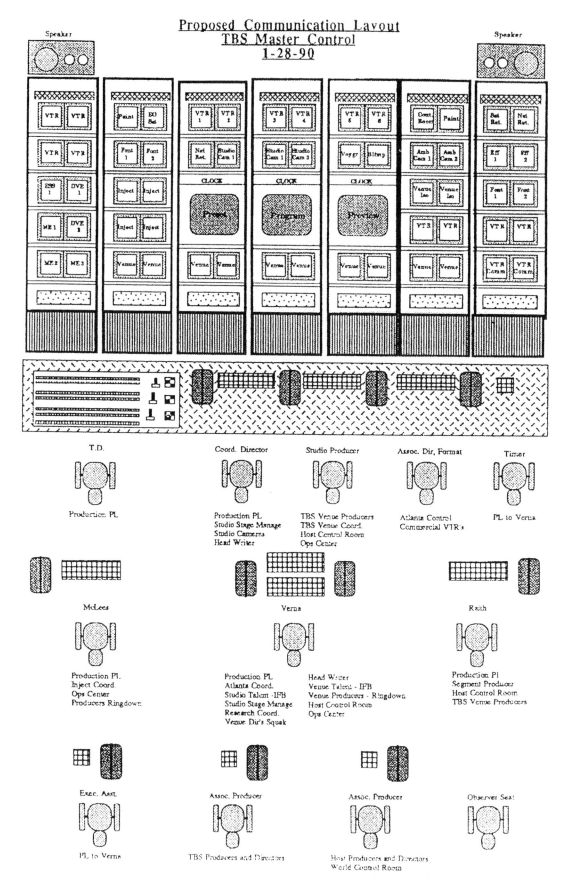

Figure 10–6 This diagram shows the positions of multiple inputs on the monitor screens and the seating in the master control truck. By positioning myself in the middle layer, I was able to communicate in person with the rows of personnel both in front of me and behind me. Note the details of my audio contacts listed behind my position.

VENUE	SPORT	PRODUCERS & DIRECTORS	TALENT	VENUE
AQUATICS CENTER	Swimming	Kim Whitelaw (HP)	John Naber (PP), Rowdy Gaines (A), Tracy Caulkins (A)	AQUATICS CENTER
	Water Polo	Andy Young (HD) ALL AQUATICS	John Naber (PP), Terry Schroeder (A)	
	Synch. Swimming	Mike Klatt (DP) AQUATICS	John Naber (PP), Tracie Ruiz-Conforto (A)	
	Diving	Doug Wren (DD)	John Naber (PP), Steve McFarland (A), Cynthia Potter (A)	
CHENEY STADIUM	Baseball	Glenn Diamond (HP), Doug Freeman (HD) / Ken Edmundson (DP), Tom Smith (DD)	Skip Caray (PP), Don Sutton (A)	CHENEY STADIUM
HEC EDMUNDSON PAVILION	Wrestling	Tom Huet (HP), Mike Beck (HD), Doug Brooker (HD) / John Barry (DP/D)	Craig Sager (PP), Russ Hellickson (A)	HEC EDMUNDSON PAVILION
	Men's Volleyball	Gary Milkis (HP), Ted Nathanson (HD) / Alan Winter (DP), Phil Martino (DD)	Chris Marlowe (PP), Karch Kiraly (A)	
HUSKY STADIUM	Welcoming Ceremonies	Lou Horowitz (HD)		HUSKY STADIUM
	Athletics	Akec Weeks (HP-T & F), Andy Sidaris (HD-T & F), Tim Rockwood (HP-Field), Garth Fowlie (HD-Field) / Craig Silver (DP), Henry Paella (DD)	Bob Neal(PP), Dwight Stones(A), Craig Masback(A), Dave Sims(R)	
KING CO. FAIRGRDS	Modern Pentathlon	Tom Huet (HP/D)		KING CO. FAIRGRDS
MARATHON COURSE	Men's Marathon & Women's Marathon	Doug Beeforth (HP), Martijn Lindenberg (HD), Kent Samul (HD), Doug Freeman (HD), Moira James (DP), John Delisa (DD)	Frank Shorter (PP), Craig Masback (A), Kathrine Switzer (A)	MARATHON COURSE
MARYMOOR PARK VELODROME	Cycling	Martijn Lindenberg (HP), Mark Warner (HD) / Kent Gordis (DP), Tom Courtney (DD)	Gary Gerould (PP), Brian Drebber (A)	MARYMOOR PARK VELODROME
MT. BAKER PARK	Rowing	Tom Huet (HP), Mike Beck (HD) / Jerry Adler (DP), James Burns (DD)	Ernie Johnson, Jr. (PP), Christopher Baillieu (A)	MT. BAKER PARK
SEATTLE CENTER ARENA	Team Handball	Thom Hastings(HP-THB/BOX), Gary Clem (HD-THB/BOX)	Ernie Johnson, Jr. (R)	SEATTLE CENTER ARENA
	Boxing	David Stern (DP-BOX), Frank Belmont (DD-BOX)	Steve Albert (PP), Mel Proctor (PP), Randy Gordon (A)	
	Women's Basketball	Ken Noland (HP), George Wasch (HD) / Kent Samul (DP/D)	Ron Thulin (PP), Ann Meyere (A)	
SEATTLE CENTER COLISEUM	Men's Basketball	Same as Women's Basketball above / Thom Hastings (HP), Gary Clem (HD)	Ron Thulin (PP), Rick Barry (A)	SEATTLE CENTER COLISEUM
	Boxing	David Stern (DP), Frank Belmont (DD)	Steve Albert (PP), Mel Proctor (PP), Randy Gordon – (A)	
SEATTLE UNIVERSITY	Judo	Tom Huet (HP), Rick Raiford (HD)	Ernie Johnson, Jr. (R)	SEATTLE UNIVERSITY
SHILSHOLE BAY	Yachting	Mike Beck (HP/D) / Jerry Adler (DP/D)	Gary Jobson (R)	SHILSHOLE BAY
SPOKANE COLISEUM	Women's Volleyball	Gary Milkis (HP), Ted Nathanson (HD) / Alan Winter (DP), Phil Martino (DD)	Chris Marlowe (PP), Tauna Vandeweghe (A)	SPOKANE COLISEUM
	Rhythmic Gymnastics	Jim Mellanby (HP), Rick Raiford (HD) / Jerry Adler (DP/D)	Susan Hutchison (PP), Wendy Hilliard (A)	
	Weightlifting	Same as Rhythmic Gymnastics	Doug Cooney (PP)	
TACOMA DOME	Gymnastics	Bob Rosburg (HP), Bob Lanning (HD) / Peter Bleckner (HD), George Finkel (HD) / Chris Carmody (DP), Ralph Abraham (DD)	Jim Simpson (PP), Tim Daggett (A – Men) / Kathy Johnson (R – Men, A – Women)	TACOMA DOME
	Figure Skating	Rohan Backfisch (HP), Ralph Abraham (HD) / Chris Carmody (DP), Bob Lanning (DD) / See Below	Jim Simpson (PP), Debi Thomas (A), Peter Carruthers (A) / Ken Wilson (PP), John Davidson (A)	
	Ice Hockey	Ralph Mellanby (HP), Henry Irizawa (HD) / Bryan Seip (DP/D)	Ken Wilson (PP), John Davidson (A)	
TRI-CITIES COLISEUM	Ice Hockey	Andy Sidaris (HP/D)	Ken Wilson (PP), John Davidson (A)	TRI-CITIES COLISEUM
WESTLAKE MALL	Closing Ceremonies	Kim Nye (SP)	Leandra Reilly (R)	WESTLAKE MALL
ATH. VILLAGE		Tim Garrigan (SP)	Steve Largent (R)	ATH. VILLAGE
VOYAGER			Larry King	VOYAGER
!BC		Brad Schreiber (HSP), Arlando Smith (HSD) / Jeff Blankman (DSP), Dave Caldwell (DSD) / Shari Bell (SP), Chris Glidden (SP), Vicki Metz (SP) / Ginny Sept (SP), Carolyn Skinner (SP), Tracy Verna (SP) / Dave Vik (SP), Linda Wendell (SP)	Hannah Storm, Nick Charles / Don Harrison	!BC

Figure 10–7 This personnel chart details the producers, directors, and talent at each of the venues. Note the need for duplicate production personnel to produce the domestic (commercial) feed and the host (world/noncommercial) feed.

Figure 10–5 shows how the signals came from each of the remote locations to the control room at the International Broadcast Center (IBC). The points close to the IBC were linked by fiber optic cable or microwave. Some of the distant points, such as Spokane and Tri-Cities, came in via satellite. Figure 10–6 is the plan for the control room at IBC, and Figure 10–7 details the venues, the sports events occurring at each, the production teams, and the talent. Note the breakdown of Host producer/director and Domestic producer/director. Both feeds had to be accomplished simultaneously; therefore, one team could not do both jobs.

The Equipment

The overviews detailed the use of equipment as envisioned in our globalcast. The following summary of the basic gear reflects the psychic distance range, the conceptual tone, which in this case was aimed more at world relationships than mere sports coverage.

Figure 10–8 shows the scope of the set. The fifty-by-sixty foot studio which was in a separate building in our trailer city (sixty-three trailers and a parking lot full of satellite uplink and downlink dishes). We used three hard cameras, two on wheeled pedestals, one on a boom crane. There was a photographic panorama of Seattle behind host Larry King, and cohosts Nick Charles and Hannah Storm were on either side of him. King sat behind a desk that featured a thirty-two-inch color monitor mounted on a hydraulically operated platform. When King was interviewing a guest on a remote inject or talking with one of the sports venues the monitor was raised. When not needed, it was lowered out of sight. When King was hosting, the cohosts stood at their podiums with a nine-monitor "Ultra Wall" beside them. Charles was positioned on a four-foot raised platform. The nine monitors in the Ultra Wall were separately controlled with a routing switcher and could be programmed to air nine parts of a single image, nine separate images, or combinations of those. (The upper portion of the set featured multi-nation flags to reflect the theme of world unity.)

Overall, we had more than 400 technicians using 137 hard cameras, 69 hand-held cameras, 8 radio-frequency (RF) cameras, 13 bruce cameras, and seven Matthews remote cameras (Figures 10–9 through 10–11).

Figure 10–8 This photo of our set in Seattle, Washington, shows (left to right): cohost Nick Charles at his podium in front of his monitor wall; anchor Larry King at his pivotal desk in the center, backed by a monitor available to show overseas injects, venue talent, and winners; cohost Hannah Storm at her podium and monitor wall. Note the multinational flags of participating nations. The photomural behind Larry King shows the skyline of Seattle.

Courtesy Turner Broadcasting System.

Figure 10—9 Nothing is so intrepid as a television camera operator. The hand-held operator in this shot appears about to be run down by a cyclist at the starting line. Note how the cameraman secured and controlled his camera cable.

Courtesy Turner Broadcasting System.

There were two super slow motion systems consisting of a camera with 300 feet of multicore camera cable, the camera control unit, and a one-inch videotape machine. Each Slo-Mo unit was combined with a self-contained utility van with three crew members (camera operator, videotape operator, and video operator/maintenance).

We had four telestrator 100 units, which we moved from venue to venue as needed, and thirty POV cameras, positioned to let viewers share experiences they had never had before. We mounted miniature POV cameras under the rider's seat on racing bicycles; on the heads of hockey referees; on the front and back of racing sculls; on basketball nets and over the courts; on horizontal bars, the vault, and the parallel bars in gymnastics; underwater near the flip turn position in the pool, and so on.

Our "Voyager" unit was a self-contained satellite news gathering (SNG) truck powered by a generator and capable of supplying a Ku-Band uplink throughout the state. This unit allowed viewers to experience the environment the athletes were experiencing in person. Seattle Seahawks former record-holding wide-receiver Steve Largent, our roving reporter, took the Voyager unit to sites like the Seattle Space Needle, Underground Seattle, a square dance in the company of athletes from Eastern Europe, a hydrofoil ferry ride across Puget Sound, and rides in the Goodyear Blimp and a fighter plane. Equipment on the unit included an RF camera, RF microphones, private telephone line, interrupted feedback system, tape machine, audio board, program audio circuit, a cellular phone, two-way radio, and return program video. To ensure that communication remained technically sound, we used a repeater system for the interrupted feedback system with a cellular phone back-up in all Voyager reports. With these, as with all of our injects from remote locations, we had pre-hear and pre-see just before air.

We had access to two blimps. One was the full-sized Goodyear Blimp with a Gyrocam 360 TV camera mounted below its gondola, remote controlled by a camera operator inside the blimp. Because the blimp's maximum speed was thirty-five miles per hour, travel time had to be calculated if it was to be used at more than one location during a single show. The other blimp was a mini-blimp ("Top Shot"), tethered first at Husky

Figure 10—10 *The 1990 Goodwill Games* interviews with event winners did not wait on formality, as this in-pool interview indicates. See Figure 10—11 for another perspective on pool coverage.

Courtesy Turner Broadcasting System.

Figure 10–11 Both camera equipment and camera operators have advanced well beyond the original "fourth wall" theatrical approach with which television started. As this photo shows, equipment and operators provide underwater POV shots as well as above-water coverage.

© 1990 Caryn Levy. Courtesy Turner Broadcasting System.

Stadium (literally a top shot) to cover the Welcoming Ceremonies during Show 3 and track and field events. After this the stationary mini-blimp was moved to Seattle Center.

There were two ambient cameras, one manned, one unmanned. The manned camera was positioned on top of the seventy-six-story Columbia Center Tower with a 360-degree view of the Seattle/Puget Sound area. The unmanned (bruce) camera was on Queen Anne Hill, which overlooks the Seattle skyline and Mt. Rainier in the distance.

We had five edit suites, which were used literally around the clock. The people slept a little, but the machines never slept. The largest of these suites contained a Grass Valley 200 Switcher, a Chryon 4100 EXB, an Abekas A-42 Still Store, an Abekas A-53 DVE, a 1/4″ audio tape play-back, three Beta SP or one-inch videotapes, and a Grass Valley VDE-141 editor. The four smaller edit suites contained a Grass Valley 100 switcher, a Chyron 4100 EXB, a basic AB roll system, a beta SP videotape, and a Sony BVE-900 editor.

The audio truck had a specially built console to pre-hear all remote audio. There were half a dozen times when the audio was fine on pre-hear but failed on the air, but most often the audio pre-hear worked as designed. Our resident musicologist, Artie Butler, created dozens of musical interludes to fit varying moods. The audio producer said we used an average of one hundred music carts per five-hour show and two hundred on the weekends, when we did two shows each day.

We had cue tones for commercials and replays. Before each day's coverage, the technicians built ten-second taped cue-tone packages for use between the national and local commercials in our commercial pods. Five of the six two-minute pods in each hour consisted of a one-minute national commercial followed by a one-minute local commercial. These national/local pods could not be separated because of an auditory transition device—the cue tone—enabling many of the cable broadcasters around the country to synchronize their commercial playback machines with the national spots as rolled from TBS in Atlanta. These playback machines were activated once the cue tone sounded and rolled tape, at speed, ten seconds after the tone. This system brought our typical commercial segment to 2:10—a one-minute national spot, the ten-second cue tone package, then the one-minute local commercial. Video during the ten-second cue tone was a TBS promotion.

We used a cue tone also to alert the world commentators using our feed at the IBC that a replay was about to be aired. We also set up visual cues for other effects. To signal a transition from live coverage to a slow-motion replay, we used a wipe from left to right on the screen. Returning to live, we reversed the wipe, going right to left.

Routing Signals from the IBC to the World

International Composite Satellite Feed was self-contained, and the Host Broadcast coverage could be shaped as individual broadcasters might choose. Following is an example of routing of video from a single event (Carl Lewis's final attempt in the long jump event). For ease of comprehension, I've limited the number of signals to five, although in reality there were many more;

From Husky Stadium we follow these five potential signals:

- Two Host Broadcast mobile units with equipment focused on the jump.
- Additional cameras in the TBS truck to add to the available host coverage.

- In the stands, an electronic news-gathering (ENG) camera operated by TVNZ to add a shot of their New Zealand commentator to supplement the Host Broadcast coverage.
- Canal Plus isolating a French competitor to Carl Lewis.
- On the field, at the World Broadcaster interview position, Gostelradio waiting to do a live interview with Carl Lewis immediately after the TBS interview.

These five discrete outbound transmissions were all sent to the IBC for routing, via fiber optic cable, terrestrial microwave, or a portable satellite uplink. At the IBC a *distribution router* directed each of the incoming signals to its proper destination. The router acts much like a telephone switchboard, routing a series of calls. This router transmits each signal to the proper independent broadcaster's production area, where any desired audio signals are added. Then the combined audio/video is returned to the distribution router, which sends the composite signal to *Transmission Control.*

Transmission Control was the final stop for each signal before it left the IBC en route to a satellite transponder. With different languages being added to the video, Transmission Control had to sort out and identify each signal and provide each with a means of transmission. While the circuits bringing signals to the IBC could be different (fiber optics, microwaves, uplink), all outbound transmissions from the IBC were sent via satellite.

Booking satellite time to provide coverage to each geographical area was the responsibility of each broadcaster. The IBC was responsible for uplinking the correct signal to each satellite transponder as ordered by the broadcasters.

The five signals we transmitted from Husky Stadium to the IBC were routed as follows.

The Host Broadcast signal was uplinked at IBC to the Galaxy 2 satellite; downlinked for turnaround purposes at Staten Island in New York; and uplinked from Staten Island to the Atlantic Ocean Region (AOR) satellite 332.5 for downlinking by official rights-holding broadcasters.

All TBS signals from IBC were backhauled to TBS in Atlanta, Georgia, for the addition of commercials, promos, and so on. Just as we had separate satellite feeds to back up the signal from Rome on *Prayer For World Peace,* we backhauled the signal from Seattle to Atlanta on two satellites. One uplink went from the IBC to the Westar 5 satellite, the other went via the G-Star 2 satellite. Both signals were B-MAC encoded (scrambled) and downlinked at TBS's "Satellite Farm" at the Techwood facility in Atlanta. The proper commercials, etc., were added and the final composite signal was uplinked to US cable operators via Galaxy 2 Transponder 14 (the *Games* was a cable exclusive). This TBS feed, minus the TBS announcer, was also sent out from Techwood to broadcasters who had purchased rights and wanted the TBS coverage rather than Host Broadcast coverage. Each of these countries could downlink the signal and add their own voice-over commentary.

The TVNZ signal: New Zealand did not own any satellite or transponder space in 1990 and so had to use the Intelsat system to get their signal from the USA to New Zealand. (Australia does own transponder space, frequently used by New Zealand, but it was not available at the time of the 1990 Games.) As a result, the TVNZ signal uplinked to a domestic satellite, which downlinked it to IDB, a transmission operations company in Los Angeles. TVNZ had contracted with IDB to turn around the signal. The signal from IDB was uplinked to the Pacific Ocean Region (POR) satellite Intelsat 180, which can be received in New Zealand.

The Canal Plus signal: This uplink followed the same pattern as TVNZ's signal. Canal Plus contracted with a common carrier to uplink their signal to a turnaround station at Lake Cowichan, Vancouver Island, in Canada. From there the signal was uplinked to AOR satellite Intelsat 307, which is received in France.

The Gostelradio signal: The USSR used its own set of satellites. With special permission from the FCC, Gostelradio uplinked their signal directly from the IBC to the Soviet-owned satellite Molniya, which was received directly in Moscow. From Moscow the signal was uplinked again via another USSR satellite, Intersputnik, to OIRT for distribution to Eastern Europe broadcasters.

These five signals were only a few of the feeds handled by the IBC router and Transmission Control. Multiply this by the number of countries involved and you will get a sense of the complexity of the IBC setup.

Communicating the Philosophy

The overviews were designed to communicate the philosophy behind the globalcast, the techniques to translate that philosophy into pictures and sound, and the physical equipment we would use to effect those techniques. The technical specifics—types of cameras, scheduling of edit suites, and so on—were left to the operational manuals developed from the overviews. In the overviews themselves I spoke in general terms because the readership was so diverse. Following is the broad vision I communicated in the first overview, explaining the material just covered here:

> This overview addresses TBS programming for the Goodwill Games, and though some of you will not be actively involved in the World Feed, you should be aware of our responsibilities to international broadcasters—and, for that matter, other broadcast activities that will be taking place simultaneously around you.
>
> In essence, TBS, the host broadcaster, will produce a basic program at each venue, a feed with generic coverage, and an international audio track. This program will be released to the international broadcasters either at the venue from which it was produced, or at the International Broadcasting Center (IBC).
>
> The IBC will act as a hub for all incoming and outgoing communications and transmissions. Inbound circuitry originating from all venues will ultimately be funneled to the IBC. Within the IBC, there will be a distribution router, a device which will permit TBS to distribute these incoming signals to various locations within the IBC. Whichever signals are requested by a specific international broadcaster will be routed to that location in the IBC.
>
> As host broadcaster, TBS will cover all sports live, with the exception of yachting and modern pentathlon, which will be taped for an edited replay.
>
> In addition, all international broadcasters will be able to access a composite World Feed, a generic package of coverage produced each day that will be switched by our International Coordinating Director, according to pre-instructions. Present arrangements call for this overseas mixed feed to be transmitted over the Atlantic Satellite, with an English commentary guide track provided by a team of six Seattle-based announcers pre-schooled on procedure.
>
> On Day 1, July 20, there will be a composite feed from 5:05 p.m. to 9:05 p.m. (Pacific Daylight Time) (TBS programs begin five minutes after the hour). The men's gymnastic team competition will be satellited live, followed by taped accounts of all eight swimming events.

The text then concentrated on the specifics of that first four-hour program. I used this technique in all twenty-three overviews, stating and reiterating the broad view of my philosophy for the Games and the broad technical sweep and then translating these generalities into the statistics and timing of events for each particular program.

I stressed in the overviews that one of the main goals in creating them was to clarify the concept of our coverage, that television production, by its nature, is a collaborative effort, a shared experience. I shared not only the program outlines but the process by which they evolved, including the processes that occurred before anyone received the overviews. I wanted total understanding, total participation, total input. I had to create a team before we met and the Games began. The following excerpt from the overviews shows how I expressed these thoughts.

General Motors makes cars. Boeing makes airplanes. Nabisco makes cookies. We're making a television program.

To set up our production cycle, I looked for systems that would allow us to get all we need to have done by the first day of the Games. What needed to be understood? How was information going to be circulated? How could we make sure it would be understood? That's why the concept of a production *cycle* appealed to me.

I knew we were not going to get from-here-to-there with a thousand phone calls taking the place of solid preparation. It was necessary to think *schematically* about this project. NBC's production budget for the 1988 Olympics was $120 million. We have $20 million. I thought we could do more with less, have a network-caliber program. But our margin of error surely would be smaller.

I needed to know how fast we could go. How long would it take us? How many people? What kind of load? How could we work faster? What were the best ways of doing this?

We are making a product from scratch. When you're doing that, first you find out how much time you have and you act accordingly. A new Jaguar model, I have read, takes 14 years from blackboard to assembly line. A Japanese car may take as little as two years—which is about the length of our production cycle.

Part I of our cycle is already completed. That was the time when ideas were sketched, when bold, unfinished strokes were slapped onto the canvas. We are in the midst of Part II—the distribution of overviews, the forecasting of the 23 shows, a way to see patterns in our work.

Part III has just begun, the first stages of pre-production, as the overviews are hardwired with operational manuals. The largely conceptual work of the overviews is being translated into marching orders. Part IV will follow the completion of the overviews and operational manuals. Part IV will be absorption, will be the time for exchanges between producers and directors, exchanges between this office and all our talent. This will be the time when we question until we are all satisfied.

Part V will begin June 4, as daily meetings in Seattle are held to finalize the grids for each of the twenty-three shows and we nail down the specifics of our world feed. This period should take us to mid-July. Some ten days out from the Games, we will begin Part VI, dry rehearsals—making final technical preparations in our studio, in the control room, at the venues. Our dress rehearsals, to take place July 18 and 19, constitute Part VII. Part VIII is the assembly line. The show begins. Reality.

In my years with CBS, you could almost count on a sixth sense with some production crews. You worked with some people for years, and many of the details were taken care of automatically. It wasn't necessary to write a detailed treatise for a Super Bowl or a Kentucky Derby. These were manageable events.

The Goodwill Games, however, are unique—as are the Olympics. There are few veterans in these endeavors. These multi-million dollar sports television extravaganzas are like crusades. The forces are assembled. A great effort is expended. The conflict is exhausting. Then everyone goes home. For many, one crusade—one Olympics, one Goodwill Games—is enough.

Before I signed up, this production had become a formless blizzard of memos. And a blizzard of memos can be just that—a storm in which no one can gain a point of reference, in which good ideas are lost among mediocre ideas, in which there is no central, final voice.

Because of our limitations on time and money, this needed to be a production based on the force of insights. But ideas and insights are fragile things if not preserved, if not made available to all. Hence the overviews.

From the beginning, *The 1990 Goodwill Games* were more than just sports. They were grounded on the thesis that sports could serve as a vehicle for an understanding of our common humanity. ABC had to suggest this notion to its Olympic viewers, but it was implicit in Ted Turner's vision of the Goodwill Games. But all of us know, also, that presenting a worthy cause does not guarantee a great work of television. Hype is no substitute for production.

To turn Ted Turner's vision into television reality, I had to come up with a practical way to translate philosophy into pictures. I began with some grandiose terms (philosophers *need* grandiose terms!) and then detailed them into specifics. For example, *global telepresence* was translated into *video specifics* and, eventually, *remote injects*. Here's how I developed that in the overviews.

In general, we will not stray from the fundamental principles of solid sports coverage. Since our budget is considerably lower than a network outlay for an event of this kind, I'm sure we won't suffer from the smell of excess. And that can be a blessing. Extravagant toys can only get you so far.

The notable difference I am seeking for our production is the emergence of a *style*, one that is a logical outgrowth of the use of global broadcasting technologies that already exist. Ours will be a truly global production—a production with a *global telepresence*. We will give the viewer a better seat, a fresh perspective, as some three decades ago the isolated camera and instant replay altered the nature of our viewing.

The question is: How can we make this global telepresence happen? . . . It can mean live visits to the corners of the globe so that we can gain a perspective on the variety of competitors who will come to Seattle for the 1990 Goodwill Games. . . . It can mean using a live satellite feed from Moscow to watch the mother of a Soviet track star as her son wins his race in Seattle. . . . It can mean a live shot of a scene at the center of a small Italian town as it celebrates the victory of a native son. . . . It can be as simple as a postrace interview with the tiny Hungarian teenager who has just won the 100-meter backstroke.

Our global telepresence could be a message as fundamental as the joy shared simultaneously in a mother watching her son triumph and the son knowing the mother is watching. Follow me a little longer, if you will, because we can take this a step further. We can establish a circle of interest. TBS will allow the son to communicate with the mother after the contest; at the same time, those present at the event, and the viewers at home, will be affected by the interaction between mother and son.

It is Friday, July 20, the first day of the Games. A phone call is made to Budapest, Hungary, where it is early Saturday morning, nearly 2:00 a.m. We tell the mother of Joszef Szabo that her son is about to compete in the 200-meter breaststroke. She has already given us permission to set up a satellite feed from her home.

Just before the race begins, the large-screen television at the Aquatics Center in Federal Way allows spectators to see Mrs. Szabo at her home in Budapest. Periodically, the TBS audience is given glimpses of the mother as she waits to hear our audio of the race.

During the race, we all watch as the mother applauds her son. And when it is over, we allow the son to speak with the mother. A few minutes after the race, swimming play-by-play commentator John Naber, accompanied by a Hungarian interpreter, interviews Joszef Szabo. Szabo is already aware that his mother has watched the race and now, with Naber acting as an intermediary, we let mother and son share their postrace feelings.

We have not sacrificed spontaneity. The drama is not scripted. But now technology permits the story line to be enhanced. Television, in this case, creates its own special kind of drama. It creates a continuum of shared interest, connecting viewer and spectator and athlete and the athlete's family. Hometowns are brought to life and into the formula of our program. Now the same satellite that is sending the event to the viewer is sending the viewer to the event.

Figure 10–12 shows one page of these remote interviews, which we called *injects*. Because they reflected the heart and core purpose of the Games, we aired about fifty of them during the twenty-three shows. They came from around the USA and from around the world. They came from East St. Louis, where we met Jackie Joyner-Kersee's high school track coach; from Budapest, where we talked with Kristina Egerszegi's parents; and from Brisbane, where we got background on swimmer Glen Housman. They gave us a chance to be personal and close in a world that isn't always personal and close.

One of our Injects gave us a chance to be fairly spectacular. During the first Goodwill Games in 1986 in Moscow, there had been a conversation with the two Cosmonauts in the USSR's MIR Space Ship. We planned to repeat that space talk, with Vladimir Pozner sitting with Larry King, acting as both guest and reporter. What we hadn't planned was a near tragedy with MIR, a technical failure that put the lives of the Cosmonauts in danger just days before our scheduled Inject. With our Inject, we were able to reassure the world that the Cosmonauts and the MIR Space Ship were safe and sound—and answering Larry King's questions.

One of the keys to this communication was the interpreters, who enabled our English-speaking reporters to ask questions of non-English-speaking athletes and allowed conversations with families, relatives, and friends in overseas countries during many of the injects. We had forty interpreters working with the production staff, in addition to several hundred interpreters helping SOC with overseas teams and delegations.

age No. 1
7/10/90

GOODWILL GAMES INJECTS (revised)   INJECTS | 7/10/90

DATE	SEATTLE TIME	COUNTRY	CITY	ATHLETE	SPORT	VENUE	STATUS
A 07/20/90	17:12-17:12	URS	Moscow	1990 GWG (Soviet view)	look back at '86 Games	IBC	faxd Gosteleradio 6/29
B 07/20/90	17:45-17:48	GDR	East Berlin	1990 GWG (East Bloc spin)	Eastern European sports	IBC	McLees confirm needed
C 07/20/90	18:17&18:29	USA	Fullerton, Calif.	Evans, Janet	swimming, womens 800M freestyle	AQU, IBC	BOOK SATELLITE TIME
D 07/20/90	18:53 live	HUN	Budapest	Egerszegi, Kristina	Swimming, women's 200M backstroke	AQU, IBC	McLees confirm needed
E 07/20/90	20:13-20:30	URS	Moscow	URS Volleyball/Gymanstics	USA VS URS Volleyball, Gymnastics	IBC, SPC	faxd Gosteleadio 6/29
A 07/21/90	12:34 live	AUS	Brisbane	Housman, Glen	Men's swimming,if in 800M freestyle	AQU, IBC	BOOK SATELLITE TIME
B 07/21/90	13:07 live	AUS	Brisbane	Clatworthy, Jodi	Womens Swimming, 400M IM	AQU, IBC	BOOK SATELLITE TIME
C 07/21/90	13:39 live	USA	Charlotte, N.C.	Stewart, Melvin Jr.	swimming, Mens 200M Butterfly	AQU, IBC	BOOK SATELLITE TIME
A 07/21/90	17:40-20:00	URS	Moscow	Opening Ceremony (3)	Welcoming Ceremony	IBC, HUS	faxd Gosteleradio 6/29
A 07/22/90	13:47 live	SUR	Paramaribo	Nesty, Anthony	Men's swimming, 100M butterfly	IBC, AQU	faxd 6/22 need costs
A 07/22/90	17:08-19:30	URS	Moscow	Soviet Gymnasts	gymnastics	TDM, IBC	faxd Gosteleradio 6/29
B 07/22/90	19:58&20:38	URS	Donestk, Ukraine	Bubka, Sergei	Track & field, Men's pole vault	HUS, IBC	faxd Gostelaradio 6/29
C 07/22/90	20:30 live	KEN	Nairobi	Kenyan runner	Men's track, 800M	HUS, IBC	telexed KBC 6/22
D 07/22/90	21:04	URS	Donestk, Ukraine	Bubka, Sergei	Men's track, pole vault	HUS, IBC	faxd Gosteleradio 6/29
A 07/23/90	17:00-18:55	PUR	Bayamon	PUR basketball team	Men's basketball, US vs. PR	COL, IBC	BOOK SATELLITE TIME
B 07/23/90	19:38 live	GBR	Newcastle	Cram, Steve	Men's athletics, 1500 M	HUS, AQU, IBC	call Paul re production
C 07/23/90	20:37 live	USA	East St. Louis	Joyner-Kersee, Jackie	Women's track, heptathlon	HUS, IBC	BOOK SATELLITE TIME

Figure 10-12 This page of planned out-of-town injects reflects the catholic nature of *The 1990 Goodwill Games* coverage. While most of our injects involved overseas/satellite coverage, even some domestic sites (Fullerton, California; East St. Louis, Missouri) called for satellite bookings. Note the critical Status column on the right.

Capturing the International Flavor

In addition to injects, we used several other devices to capture the international flavor of the globalcast.

Journeys

We preproduced video profiles of some of the outstanding athletes competing in the Games. The *Journeys* didn't show their professional profiles—we saw that as they competed. Like the Injects, the Journeys touched on the personal stories of how each athlete found his or her way to the Games. They aired just before the individual competed and were told through the eyes and in the words of each athlete. Some were emotionally powerful. One athlete told of the suicide of her coach. A runner told of the death of his father and how each time he runs, he runs for his father.

This was part of the mosaic I knew we had to create: compelling stories—sound bites, images—that would compel our viewers to follow the athletes through their competition. In the overviews I recalled noticing when I worked on the 1984 Olympics that many of the female swimmers had dolls or teddy bears. During NBC's broadcast of the Seoul Games, there was an unforgettable close-up of diver Greg Louganis's teddy bear. It was by the pool with Louganis, and the teddy bear had a bandage on his head, in the same place as Greg's bandage, which had been applied after he hit his head on the diving board. It's subtle glimpses like these that create the mosaic that works on you subliminally, that makes moments.

Television is moments. For many, remembering *The 1990 Goodwill Games* would mean remembering half a dozen stories, human stories. Rules, scores, political overtones—these would fade away. It would be the faces, the stories that would remain. We were a microcosm of the Global Village. We had an Athletes' Village filled with normal people who could do supra-normal things. We had to find their stories and share them with our viewers, turning the cold metal of cameras and satellites into the warm emotion of human striving.

Video Diaries

Another way we personalized the athletes was to give a number of them an 8-millimeter video camera to record whatever they wanted to record. They took the cameras, taped themselves, their friends, places, and things. We helped edit, added some music, and put their directorial debuts on the screen.

Athletes' Village

Not quite so personal, but based on the athletes' experience away from home was the Athletes' Village segment, which had its own producer and its own hostess. They roamed through the area where the athletes were quartered and covered more objectively what the video diaries covered for individuals.

The Voyager

The Voyager, discussed earlier, combined the unique sights and sounds of the greater Seattle area with the impressions it made on the athletes, and vice versa.

Translating Events into a Format

Just as the "Overviews" were charged with translating TBS' global philosophy into practical television techniques, so were they charged with translating these techniques into specifics for each program. We needed a format that would accommodate the sports events and the philosophical elements and get the commercial breaks in on time. Because we were dealing with live sports, we could never predict how exciting the action would be in a given game, only that there would be competition and a winner. That's one reason sports are so popular. Life is generally an unpredictable string of days. But like drama, a sports event has a beginning, a middle and an end—and a winner and a loser. It satisfies our longing for certainty. There are rule books. There are record books. Life is a mess. Sports are neat. The job of the producer, director, and talent of a live sports event is to find a story line that will intrigue the audience whether or not the action of the game is exciting. You have to give each viewer something to look for, a "horse to ride," a hero to follow, a record to be broken, a handicap to be overcome, etc. That's the inevitable pattern for a single sports event. The Goodwill Games are a lot of single sports events, many often occurring simultaneously. A storyline has to be developed for each event. That's a basic responsibility for each commentator and play-by-play announcer. (For details of how each individual sport should be covered, check my two previous books. *Live TV* is still in print. It recounts the day-by-day activities of producers and directors of sports events—among others—in their own words. They tell what they do, why they do it and how they do it. My earlier book, *Playback,* is out of print, so check your library. It details camera coverage of each of the major sports events.)

For the Goodwill Games a large part of the burden of story-telling fell, appropriately, on the individual commentators and play-by-play announcers. My job in the overviews and in the grids was to establish a framework to intrigue the viewer to watch each of a variety of sports events, each with its own story to tell.

Here I will describe some of the techniques put forth in the overviews.

The Adversarial Billboards

On the top of each show we put together as dramatic a piece as we could assemble, visualizing the two or three major competitive clashes that would occur during the next four hours. Figures 10–13 through 10–15 show how we combined action shots with sound bites and whatever would establish and dramatize the upcoming events. Personalities are what drive competition, so we highlighted the outstanding athletes in competition. By definition, the Goodwill Games are the "best against the best." There are very few heats or trials—virtually every event is for the gold medal. The Adversarial Billboards were designed to hook viewers into watching and staying with us, to give the reporters on these major events a jumping-off point for their individual efforts and—combined with the next technique, the closing response—a device to "bookend" each show. The Adversarial Billboards took hours to prepare and used video effects that required four tape machines. Because there were unpredictable shifts in event schedules—athletes not arriving, injuries, and so on—there were times that the team preparing the Adversarial Billboards worked right up to air time.

```
                              ADVERSARIAL BILLBOARD
                              SHOW 7 -- VERSION 3
         VIDEO                         AUDIO
         ---------                     ---------
                                       STORY 7
                                       BOOTH ANNOUNCER (V/O)
                                       ------------------
         BRAZILIAN FLAG GIVES          BRAZIL'S ROBSON DA SILVA
         WAY TO DA SILVA STILL ...


         US FLAG GIVES WAY             AMERICA'S MICHAEL JOHNSON ...
         TO TO MICHAEL JOHNSON
         STILL                         THE TWO HUNDRED METER DASH ...


                                       TONIGHT!
         ------------------------------------------------------------
         DA SILVA VIDEO                A BRONZE MEDALIST IN

                                       SEOUL ... DA SILVA FINISHED

                                       LAST SEASON RANKED FIRST

                                       IN THE WORLD ...

         ------------------------------------------------------------
         JOHNSON VIDEO                 BUT AMERICA'S MICHAEL

                                       JOHNSON HAS COME FROM NOWHERE

                                       TO STAKE A CLAIM AS THE

                                       WORLD'S BEST ... HE BEAT DA SILVA

                                       IN FRANCE LAST MONTH ... CAN

                                       HE DO IT AGAIN?

         ------------------------------------------------------------
         SINGLE SHOT, DA SILVA          "MICHAEL JOHNSON IS RUNNING
         DVE W/GRAPHIC
                                       VERY WELL. I WILL HAVE TO BE AT MY

                                       BEST TO WIN."

         ------------------------------------------------------------
         SINGLE SHOT, JOHNSON           "WE'RE VERY CLOSE. IT'LL BE
         DVE W/GRAPHIC
                                       A GREAT RACE"


         SEATTLE SKYLINE               FIRST IN THE WORLD LAST YEAR ...
```

Figure 10–13 This is the first of three pages showing video/audio from the Adversarial Billboard planned to open *The 1990 Goodwill Games* show 7. The video effects we used in these billboards required four VCRs functioning simultaneously.

```
NEWSPAPER W/HEADLINE

---------                            AGAINST FIRST IN THE WORLD
JOHNSON, DA SILVA
SPRINT FOR GOLD                      THIS YEAR ... CAN ROBSON DA

                                     SILVA HOLD OFF UPSTART

                                     MICHAEL JOHNSON?
                     ----------------------------------------
                     MUSICAL DOUGHNUT/FAST CUT/DRIBBLING
                     THROUGH "MEN'S BASKETBALL"
                     ----------------------------------------

                                     THE UNITED STATES
SOVIET FLAG,
U.S. FLAG                            THE SOVIET UNION

STILLS OF KEY                         ... MEN'S BASKETBALL ...
PLAYERS SPIN
OUT ...                              ONE OF THE GREAT RIVALRIES

                                     IN SPORTS ... CONTINUES ...
---------------------------------------------------------------------
SHOW FOOTAGE FROM 1988               TWO YEARS AGO IN SEOUL ... THE
OLYMPICS FINAL ...
SABONIS DRIVING                      SOVIETS DOMINATED ...
FOR A SCORE AS
U.S. DEFENDS ...                     RECREATED SOT: "AND THE USSR

CUT TO FINAL SECONDS                 ADVANCES TO THE FINAL WITH
OF GAME ...
WHISTLE BLOWS TO                     A CONVINCING VICTORY
END GAME
---------------------------------------------------------------------
SINGLE SHOT: ANDERSON                SOUND BITE, KENNY ANDERSON:
DVE WITH GRAPHIC
                                       "WE'RE A BRAND NEW TEAM. BUT

                                     WE WANT THEM BACK.''
---------------------------------------------------------------------
SINGLE SHOT: VOLKOV                  SOUND BITE, ALEXANDRE VOLKOV:
DVE WITH GRAPHIC
                                       "WE CONSIDER THE U.S. THE

                                     TEAM TO BEAT."
```

Figure 10—14   This continues the Adversarial Billboard planned to open *The 1990 Goodwill Games* show 7. The video effects we used in these billboards required four VCRs functioning simultaneously.

```
----------------------------------------------------------------------
SEATTLE SKYLINE                    AMERICA'S BEST
NEWSPAPER W/HEADLINE
------------------                 AGAINST THE
AMERICANS, SOVIETS
IN HOOP CLASH                      PUNISHING SOVIETS ... TONIGHT
```

```
                  -----------------------------
                  MUSICAL DOUGHNUT/FAST CUT
                  GENERIC SWIMMERS SWIM
                  THRU "100-FREESTYLE"
                  -----------------------------
STILLS OUT OF SEATTLE              THE FASTEST MEN ON WATER ...
SKYLINE ... BIONDI
                                   MATT BIONDI ...
JAGER
                                   TOM JAGER ... RACE AGAIN AT

                                   100 METERS ...
-----------------------------------------------------------------
CUT TO VTR:                        BIONDI'S BEST EVENT ...
BIONDI WINNING 100-FREE
                                   ONE OF HIS FIVE GOLD MEDALS
IN SEOUL
                                   IN SEOUL ...

                                   RECREATED SOT: "AND BIONDI

                                   WINS ..."
-----------------------------------------------------------------
SINGLE SHOT                        SOUND BITE, BIONDI:
OF BIONDI
DVE GRAPHIC                         "IF ANYONE CAN BEAT

                                   ME, IT'S TOM ..."
-----------------------------------------------------------------
SINGLE SHOT                        SOUND BITE, JAGER:
OF JAGER
DVE GRAPHIC                         "BIONDI'S THE BEST IN

                                   HISTORY. MY GOAL IS TO

                                   NARROW THE GAP ..."
-----------------------------------------------------------------
SEATTLE SKYLINE                    CAN BIONDI EXTEND HIS
NEWSPAPER W/HEADLINE
------------------                 DOMINANCE ... OR WILL
BIONDI, JAGER

      ONE MORE TIME                    JAGER PULL AN UPSET?

                                       STAY WITH US FOR

                                       A MEETING OF THE BEST ...

                                       AGAINST THE BEST
```

Figure 10–15   This is the final page from the Adversarial Billboard planned to open *The 1990 Goodwill Games* show 7. The video effects we used in these billboards required four VCRs functioning simultaneously.

Billboard Responses

At the end of each show Larry King would ask one of the cohosts to recall the competitions we had set up at the start of the show and wrap up the outcomes. This required the Still Store staff's grabbing video from events as they occurred so they could be pieced together for the response. (They were, of course, also grabbing video for Medal Counts, broken-records, etc.) The copy also had to be written as the events progressed. We had one or more of our writers in the studio throughout each show.

Time Plates

Because viewers are not equally interested in every event, we used *Time Plates* as bumpers leading to some commercial breaks. Each Time Plate consisted of a video or audio/video preview of an upcoming event, with the number of minutes until the start of that event. Sometimes we used video of the competitors. If it was available, we frequently used a soundbite with the video.

Medal Counts

As the statistics build, viewers are inevitably curious about which nation is ahead in gold, silver, and bronze medals. They also want to know when a world, or national, or personal record has been broken. If there were a significant number of events occurring that day, there might be more than one Medal Count on a single show. Each gave us an opportunity to recap personal as well as national accomplishments, and to add to the feeling of unity at the Games.

Detailing the Specifics

The overviews contained not just philosophy, techniques, and tools, but also predictions of the competitions we expected during each show. Here are excerpts from Show 6, the evening program for July 23, 1990:

In Show 6 we are in for a bit of white-knuckle programming as we attempt to juggle four events, some happening simultaneously—swimming, basketball, track and field, and women's volleyball. These truths will hit the fan:

- In Swimming, a world record is quite possible in the 100-meter breaststroke . . .
- The U.S. men's basketball team, smarting from the bronze-medal finish at the 1988 Olympics, makes its Goodwill debut . . .
- There are world-record holders expected in virtually every track and field event in Show 6 . . .
- And the U.S. women's volleyball team may be playing a match that could determine if it will advance to the medal round.

Complete with several solid story lines, this show has two key elements: In the first half of the program we'll air the Goodwill Games debut of the U.S. men's basketball team, which will be playing Puerto Rico. A good bit of the remainder of the program—hours three and four—will be devoted to events in which several of Great Britain's best and brightest should excel.

- In the men's 100-meter breaststroke (5:39 p.m.), the favorite is Britain's Adrian Moorhouse, the world record-holder. Moorhouse is the defending Olympic champion in the event.
- In the 110-meter hurdles (7:19 p.m.), Britain's Colin Jackson forms half of one of sport's best rivalries. The other, and slightly better, half is American Roger Kingdom, the world record-holder. Kingdom and Jackson were 1–2 respectively at the 1988 Olympics.
- Britain's Steve Cram, expected to compete in the 1500 meters (7:38 p.m.), has been one of the world's top middle-distance runners for the last decade, and he holds the world record in the mile. And country-man Peter Elliott was the silver medalist in this event at the Olympics.
- And at 8:17 p.m., Britain's Linford Christie is expected to give chase to a Carl Lewis-led squad of American sprinters in the 100 meters. Following the disqualification of Ben Johnson at the Seoul Games, Lewis and Christie became, respectively, the gold and silver medalists in the 100 meters.

The overviews attempted to convey not just facts but also feelings. Here is an excerpt from the very beginning of the overview for Show 6.

Monday morning, July 23, 1990. In Seattle, the darkness gives way to the sunrise, at 5:36 a.m. It is about 50 degrees, but the local weather service is promising a pleasant, sunny day, with a high of 77 or 78 degrees.

At the International Broadcast Center, the first cups of coffee are being poured, the first eggs are being fried. . . This is Day Three, Show Six. . .

The sense of urgency is evident early, long before Show Six is to air. The IBC is nearly at full bustle by 8:00 a.m. While coffee is sipped, donuts chewed, papers shuffled, there is chatter about Janet Evans dominating a weekend of swimming at the Aquatics Center. And discussion about Edwin Moses reclaiming his position as the world's best in the 400-meter hurdles. And an argument—a cameraman saying the Soviets will win the men's basketball tournament, starting today . . . a producer disagreeing, saying it will be the United States.

Two of TBS' five production trucks packed and moved Sunday night, one making the 30-mile journey from the Tacoma Dome (site of the just-concluded men's gymnastic competition) to the Seattle Center Arena, for team handball. The other truck needed only to move across town, a 20-minute ride from the Mount Baker Rowing Center to the Seattle Coliseum, site of men's basketball.

The U.S. women's gymnastics team, competing Friday, is to arrive at SeaTac airport this morning at 10:00 a.m. Inside the TBS compound, a van carrying some of our production personnel is about to leave the IBC to greet them . . .

At times, to convey the texture as well as the construction of a program, we created projected dialogue. This selection occurred later in Show 6:

At the start of the basketball coverage, play-by-play announcer Ron Thulin will tell the viewers to be prepared for two LIVE interruptions in the first half—one related to the game, the other unrelated.

- Unrelated to the game, occurring about midway through the first half (5:35:10–5:42 p.m.), we will have live coverage of the men's 100-meter breaststroke.

- Related to the game, we'll have a live inject from Puerto Rico's Olympic training center. When there's a break in LIVE basketball coverage, Thulin will send our TBS viewers to the IBC and Larry King, who will talk with athletes and others in residence at the training center.

Thulin: There is an injury on the floor, with the U.S. leading Puerto Rico, 20–18 . . . We've got 12 minutes left in the first half and as we wait here, let's send you back to our anchor position and Larry King . . .

King (turning to monitor by his left shoulder): Excuse me, Ron . . . We'll be back with you in a moment . . . As you saw a few minutes ago, I'm with some new friends at Puerto Rico's Olympic Training Center . . . With me now is Miguel Gonzalez, a 17-year-old boxer-to-be . . . Miguel, right now your basketball team is doing all right. Can they keep it up?

Miguel: Oh yes, Larry, I think so . . . We take basketball pretty seriously down here. A lot of my friends want to go to American colleges and play . . . If Ramon Rivas can score 30 or 40 points, I think we can win.

King: Aw, come on, Miguel . . .

Miguel: No, really, Larry. I believe it.

King (smiling): OK, stay with me, Miguel. They're ready again at the Seattle Coliseum. It's the U.S. by two. Let's go back to Ron Thulin . . .

With all of these predictions and conjectures came always the admonition that this was paper preparation, not concrete conclusion. Witness this quote:

I have been successful in producing effective television events by making educated guesses in looking ahead, suggesting situations likely to happen, picturing the workings of people in advance. This allows a progressive state of readiness, one that is not affected by who wins the race, or which team wins the game, or who is participating. More important is our predisposition and preparation as we arrive in Seattle, some four and a half months from now. But remember, these overviews deal with predisposition, not predestination.

The Road to the Air Show

At the end of each Overview came the first of three grids that led to our on-air production. Figure 10–2 shows part of the grid at the end of Show 2. This was distilled and broken down, minute-by-minute, from the generalized information in the overview.

This preliminary grid, used as a guide for planning, was later refined in Seattle by TBS Vice President, Sports, David Raith, who was in constant contact with the SOC, so that we were always updated on which athletes had arrived, would participate, and so on. See Figure 10–3 for the first page of this updated grid, moving us one step closer to actual production.

The second grid, updated each night after that day's show, was the basis from which the studio producer and his staff developed the production grid. Figure 10–4 shows this final form that carried us each day through the show.

You'll recognize this final grid as one comparable to those used on *Live Aid, Sport Aid,* etc. There are 12 columns covering Length, Running Time, The Program Segment, The Format (Live, VTR, Venue, etc.),

Videotape, Audio, Music, Audio F/X, Chryon, Graphics, The Monitor Walls, and finally "Exceptions." This was the end paper-product. We had progressed from the propositions and predictions of the overview to the nuts-and-bolts of which button gets pushed when and by whom. All that was left was to do it.

Rehearsal

Once a series like the *The Goodwill Games* starts, it rolls mercilessly on, crushing anyone who can't keep up the pace. On a normal weekday we had one four-hour show, from 5:05 to 9:05 p.m. PDT. The crews and equipment had to be in place for each day's events, and no day's events were the same as those of the previous day. The studio producer's staff met at 7:00 a.m. each day to translate the updated grid into the on-air grid, copies of which had to be prepared and distributed; at the same time the staff was preparing for future shows. We had a meeting right after each program to solve any problems that lay ahead, to avoid repeating any that had occurred. Most of the production staff managed to leave by midnight. And then came the weekends.

Each Saturday and Sunday we did two shows, one from 12:05 to 3:05 p.m. PDT, the second from 5:05 to 9:05 p.m. That meant twice as much work with less time to prepare. Most meals, as you can imagine, were eaten at a desk or on-the-move. Sleep was a sometime thing. Social life? Forget it. And, of course, with the Games opening on a Friday night, we were launched into a weekend.

I knew that if all we had prepared was Show 1, we'd never make it. We would get through that Show 1 with no problem, but going into a weekend unprepared would create chaos and disaster. I decided that if we made it through the first three shows—Friday evening and the two shows on Saturday—we'd be all right. We would have faced every situation we would have to face for the full seventeen days. So I sent out the order: show me three shows—script, graphics, Ultra Walls—the works. This meant faking it, of course. No one could know the outcome of events that hadn't happened. But we had to rehearse. Everyone had to taste the mix before any of us could digest it.

With this paper product in hand, my next concern was how to communicate it to the talent without overwhelming them. You can't absorb twenty-three shows in one sitting. But you can absorb the *patterns* of those shows. You can see the gears and how they mesh, so that you're convinced the unwieldly machine will move. How could we show the gears—clearly, understandably, simply, convincingly?

I knew that scripts alone would not communicate what I needed to communicate. The script for a four-hour show is bulky and unwieldly. You can't lay it out and look at it. I had a narrative summary of the first three scripts prepared. Figures 10–16 through 10–18 show three of the nine pages of the narrative for Show 1. Pages 1 and 2 show the style of the narrative, and page 8 shows how a number of sports elements were sequenced and detailed. When I saw these narratives, I knew we were closer, but I also knew we weren't there. Those nine pages were easier to absorb than a bulky script, but you couldn't really look at them and see patterns. I had the narrative broken down into outline form, showing how the three hosts interrelated. Figures 10–19 and 10–20 show the two-page outline that was developed for Show 1.

Show#1

ACT I

The show will start with an "Adversarial Billboard," setting up the major competitions coming up on this show (Biondi versus Jager in Men's Swimming, USA's Janet Evans versus E. German's Kristina Egerszegi, etc.) so the audience is warmed-up and oriented from the opening gate. Commercial Billboards identify our sponsors.

An aerial from the Goodyear Blimp shows the expanse of the Seattle area, and a wide shot of the TBS set shows the expanse of our TV operation. And then the cameras will find Larry King.

First it's Hello and reaction to a great city and a great event--seventeen days of the best athletes in the world competing with each other, the top sports figures from more than fifty countries, sharing their skills, their lifestyles--meeting the USA, meeting each other, bringing us all a global sharing that rides right over boundaries and politics--with satellites and cameras allowing us to bring these athletes' friends and families on screen, here to Seattle. We'll show all that in a little while. Now it's time to meet CNN Sportscasters Hannah Storm and Nick Charles, who'll be keeping track of all the Goodwill events for us.

Hannah, what can we look forward to tonight? Hannah says world-class competition in swimming, women's volleyball and men's gymnastics. Looking ahead, Hannah may point out that in total the 1990 Goodwill Games will feature 183 events, with 183 Gold Medals to be awarded. They'll take place at 15 different locations in the Seattle, Tacoma, Spokane, Tri-Cities area of Washington State. And all of these monitors and dials you see here will keep Nick and me in touch with all of the events taking place at all of the locations--all the time. We'll be able to update you everything going on in the Games.

Thanks to Hannah, then: Nick Charles, who are some of the athletes who'll be starring in these battles for gold medals? Nick tells us about Carl Lewis and Matt Biondi, the fastest man on land and the fastest man on water. In fact, tonight we'll have Matt Bioni pitted against Tom Jager. They've been going back and forth as world record-holders, so tonight's match-up should be as exciting as a swim match can get. Some of the other superstars expected include USA swimmer Janet Evans, Soviet gymnast Svetlana Boginskaya, the Olympians of the new generation--Canada's Kurt Browning and USA's Jill Trenary, the world figure skating champions, Cuban Boxer Felix Savon, and Kim Zmeskal, who may be the best gymnast the United States has even produced.

Thanks to Nick, then the monitor comes up as a physical representative of the world. The Games are taking place in Seattle, but the backgrounds and the emotions the athletes bring with them also bring the world here. And we, in turn, send back

Figure 10-16 This is the first page of the narrative description of *The 1990 Goodwill Games*, Show 1, prepared as a production tool—combined with the outline in Figures 10-19 and 10-20 and an audio tape—to familiarize Larry King, Nick Charles, and Hannah Storm with not just the flow of the program, but also how their contribution fit into the overall structure.

hopes and dreams and newer emotions. Television will let us share all that, and one example of that is Marina Dymova of Gostelradio, who is in Moscow for us at this moment. We see a wide shot, an aerial shot, of Moscow with some subtly identifying music under, then a wide shot of the area, then a CU of Marina appears on the monitor screen. With her is a Soviet policeman, _____
_____. (Questions to Moscow or Budapest, etc., will be interpreted simultaneously by the on-site interpreter. In some cases, as in East Germany, the people being interviewed will speak English. We'll know in advance.) Are people in Moscow excited about the Games? Etc.

Thanks to Marina and _____. Just 17 minutes to go to that great swimming match-up Nick Charles told us about--and lots more to show you and tell you about! The fastest men on earth in the water! Don't go anywhere!

(Timeplate showing the Biondi/Jager match-up, 17 minute countdown)

ACT II

We're back live in Seattle, looking forward to seeing the fastest swimmers on earth. We'll be meeting them at the Weyerhaueser Pools, one of the 15 locations for events here in the Greater Seattle area, but we'll also be meeting the athletes in their more casual moments--if there are any really casual moments for the world's top athletes about to compete with each other.

(Monitor screen will show the Athletes' Village)

Every day we'll be visiting the Athletes' Village, where the competitors from around the world live and get to know each other on a personal basis. Joining us now is another of our TBS team here in Seattle, Leandra Reilly, who'll be guiding us on our Village visits. Leandra, what do you have to show us today?

(Leandra intros Village)

Thanks to Leandra, then Monitor shifts to Puget Sound. Let me show you something. This is part of Puget Sound on our monitor here. There's lots more of it. Just take a look.

(Aerial from blimp or camera of Seattle/Sound)

There's lots to see in this beautiful part of the United States, and we have just the man to show it to you. A former NFL star who played here for the Seattle Seahawks, Steve Largent, is going to be our man-about-Washington, taking us on an electronic caravan we call our Voyager. And these aren't taped pieces. Steve's going to be all over this area, live, giving us the feel of the site of these Goodwill Games. Steve Largent sparked the imagination of fans in this area when he played ball, and we think his live Voyager reports will do the same now. Today Steve is on a ferry boat out in the Sound. Steve, exactly where are you?

Figure 10–17 This is the second page of the narrative description of *The 1990 Goodwill Games*, Show 1, prepared as a production tool—combined with the outline in Figures 10–19 and 10–20 and an audio tape—to familiarize Larry King, Nick Charles, and Hannah Storm with not just the flow of the program, but also how their contribution fit into the overall structure.

ACT XXI

Out of commercial, Larry's monitor has a Moscow Inject of Red Square. Interpreter Marina Dymova will be with Soviet citizens watching our feed. The Soviet athletes will probably be well ahead in men's gymnastics and Soviet citizens will probably be happy to respond to Lary's questions.

At the end of the Moscow Q&A, Larry tosses to Jim Simpson and Tim Daggett for more Men's Gymnastic action. They follow the action and toss to break from Tacoma Dome.

ACT XXII

Our of commercial, Larry's monitor contains a scene of Puget Sound with the ferry from which Steve Largent will present his Voyager. During Steve's tour of the boat, Larry can question the Captain of the ship.

At the end of the Voyager, Larry tosses to Simpson/Daggett at Tacoma Dome. They cover more men's gymnastics and toss directly to break.

ACT XXIII

Out of commercial, coverage goes directly to Tacoma Dome. The action ends with a toss back to Larry.

Larry may comment on the world-record swim—the first of the Goodwill Games—coming up in seven minutes, as he tosses to break.

ACT XXIV

Out of commercial, Larry welcomes swimming expert John Naber, who is at the desk with him. They talk about Naber's play-by-play role and about Mike Barrowman, leading to a Barrowman Journey.

After the Journey, Larry gives any reaction he may have, then goes to Naber, who leads to his own call, on tape, of the men's 200 meter breaststroke.

After Naber's report, he and lrry have a one-minute interview with the winner of the race (expected to be Barrowman). Then Larry tosses to Hannah for a report on Women's Volleyball (assuming we did not show the whole game). After her report, Hannah tosses back to Larry.

Larry turns to Nick for Medal Count #2. After that, Nick throws back to Larry.

Larry turns back to Hannah for the Miller Moment. Hannah does the Moment, then tosses back to Larry.

Larry may be able to ad-lib about the USA having a current lead in medal count, but with the expectation that the USSR could make up ground, so the two countries could be even by Sunday. In any event, he takes us to commercial.

Figure 10–18 This is the final page of the narrative description of *The 1990 Goodwill Games*, Show 1, prepared as a production tool—combined with the outline in Figures 10–19 and 10–20 and an audio tape—to familiarize Larry King, Nick Charles, and Hannah Storm with not just the flow of the program, but also how their contribution fit into the overall structure.

ACT	LARRY	HANNAH	NICK
1	1:00 P. 3: Welcome + Set up GWG :20 P.3A: Intro Hannah :20 P. 5: Intro Nick 2:00 P. 7: Global/Setup Moscow :15 P. 8: Throw to Break 3:55	1:00 P. 4: Venues/Equipment :15 P. 4A: Toss back to Larry 1:15	2:00 P. 6: Athletes :15 P.6A: Toss back to Larry 2:15
2	:30 P.11: Toss to Leandra :15 P.13: Toss to Largent 2:00 P.15: Interview R. Helmick :15 P.16: Throw to break 3:00	NO STORM COPY	NO CHARLES COPY (Voyager/Helmick)
3	:40 P.18A: Q&A Nick: Swimming :15 P.19: Toss to Naber/Gaines :55	NO STORM COPY	:40 P.18A: Q&A re Biondi/Jager
4	1:30 P.23: Q&A Winner Men 50m :15 P.24: Toss to Naber/Gaines 1:45	NO STORM COPY	NO CHARLES COPY (Wom 50m Freestyl)
5	3:00 P.26: E. Berlin Inject Q&A :15 P.27: Toss to Nick (Medal C) :15 P.29: Toss to Break 3:30	NO STORM COPY	1:00 P.28 : Medal Count(Explain) :15 P.28A: Toss back to Larry 1:15
6	:15 P.32: Toss to Hannah(Gym)	:30 P.33: Setup Men Gym/ Toss to Simpson/Daggett	NO CHARLES COPY (Setup Men Gym)
7	:15 P.39: Throw to Break	NO STORM COPY	NO CHARLES COPY (Men Gym)
8	:15 P.42: Toss to Nick (Journey) :45 P.44: Setup Evans Inject :30 P.45: Toss to Naber/Gaines :15 P.46A: Toss to Break 1:45	NO STORM COPY	:30 P.43: Lead to Evans Journey
9	2:00 P.49: Evans Friends Inject :15 P.50: Toss Hannah(Welcome) :15 P.52: Toss to Break 2:30	:30 P.51: Lead to Welcome Ceremony Preview	NO CHARLES COPY (Evans Inj/Welcom)
10	:15 P.55: Toss Hannah (Gymnast) :15 P.58: Toss to Break :30	:20 P.56: Toss to Simpson/Daggett for Men's Gymnastics	NO CHARLES COPY (Men Gymnast)
11	:15 P.60: Toss Nick (Swim setup) :30 P.63: Setup Injct(Egerszegi) :15 P.64: Toss to Naber/Gaines 2:30 P.65A: Q&A Inject Egerszegi 3:30 (Budapest)	NO STORN COPY (Wom Volleyball)	:30 P.61: Setup Swimming :30 P.62: Lead Egerszegi Journ 1:00
12	:30 P.68: Set up W. Volleyball :15 P.69: Intro Karch Kiraly :30 P.70: With Karch: Players :30 P.70A: Karch ISO players :15 P.71: Toss to Marlowe & 2:00 Vanderweghe for W. Vyball	NO STORM COPY (Set up Volleyball)	NO CHARLES COPY (Wom Volleyball)
13	NO KING COPY (W. Volleyball)	NO STORM COPY (Wom Volleyball)	NO CHARLES COPY (Wom Volleyball)
14	NO KING COPY (W. Volleyball)	NO STORM COPY (Wom Volleyball)	NO CHARLES COPY (Wom Volleyball)
15	:15 P.80: Throw to Break	NO STORM COPY (Wom Volleyball)	NO CHARLES COPY (Wom Volleyball)
16	:15 P.82A: Toss Hannah (Gymnast)	:45 P.82B: Update Men Gymnastics :15 P.83 : Toss toSimpson/Daggett 1:00	NO CHARLES COPY (Men Gym)
17	NO KING COPY (Men Gymnast)	NO STORM COPY (Men Gymnastics)	NO CHARLES COPY (Men Gym)
18	:15 P.87: Toss to Break	NO STORM COPY (Men Gymnastics)	NO CHARLES COPY (Men Gym)

Figure 10–20 Because a script was too bulky (and on teleprompter) I had the staff break down the content in Show 1 in outline form to give the cohosts a quick visual overview of what the program looked like. This outline parallels the narrative copy in Figures 10–16 through 10–18.

ACT	LARRY	HANNAH	NICK
19	:15 P. 90: Toss to Marlowe/Vande to update Wom Volley :15 P.90B: Toss to Nick for Toshiba Analysis :15 P.91B: Toss to Hannah for Welcome Preview :15 P.93: Toss to Nick for Swim Update <u>:15</u> P.95: Toss to Simpson and 1:15 Daggett for Men Gym	:30 P. 92: Weekend Preview <u>1:00</u> P.92B: Toss back to Larry 1:30	:45 P. 91: Toshiba Analysis :15 P.91A: Toss back to Larry :45 P. 94: Update Swimming <u>:15</u> P.94A: Toss to Larry 2:00
20	NO KING COPY (Men Gymnastics)	NO STORM COPY (Men Gymnastics)	NO CHARLES COPY (Men Gymnast)
21	2:00 P. 98: Inject Moscow: Man-on-the-Street <u>:15</u> P.98B: Toss to Simpson and 2:15 Daggett for Men Gym	NO STORM COPY (Men Gymnastics)	NO CHARLES COPY (Men Gymnast)
22	:15 P.101: Toss to Largent <u>:15</u> P.103: Toss to Simpson and :30 Daggett for Men Gym	NO STORM COPY (Men Gymnastics)	NO CHARLES COPY (Men Gym + Voyager)
23	:15 P.105: Throw to Break	NO STORM COPY (Men Gymnastics)	NO CHARLES COPY (Men Gymnast)
24	:15 P.108: Welcome John Naber :30 P.109B: React to Barrowman Journey, toss to Naber 1:00 P.111: W/Naber: Intr.Winner :15 P.112: Toss to Hannah for Report on W. Volley plus Medal Count :15 P.115: To Nick for Miller Moment <u>:15</u> P.117: Throw to Break 2:30	1:00 P.113: Women's Volleyball Final :45 P.114: Medal Count #2 <u>:15</u> P.114A: Toss to Larry 2:00	1:00 P.116: Miller Moment <u>:15</u> P.116A: Toss back to Larry 1:15
25	1:45 P.118: With Hannah & Nick: Wrap-up :20 P.119: Goodnight	1:45 P.118: With Larry & Nick: Wrap-up END SHOW #1	1:45 P.118: With Larry & Hannah: Wrap-up

Figure 10–20 *(continued)*

Now it was graspable. Now it was visually clear. The host and cohosts could look at those two sheets and know exactly who was doing what, how long each of their segments lasted, how they related to each other, how much time they had between appearances to check on results of events and get other updates. And they could see the *whole* show, not just their part of it.

But even this was not the whole story. Charles and Storm had to coordinate their reporting with the visuals we planned to put on the monitor walls. As Figure 10–8 showed, each had a separate Ultra Wall. When Larry King was hosting at the center desk, Charles and Storm were on opposite sides of the set, each working with his or her Ultra Wall. To communicate this relationship, the studio director drew up a series of sheets, totalling about eight pages for the first show, both showing and describing in words what we planned to have on each Ultra Wall with which they would be working.

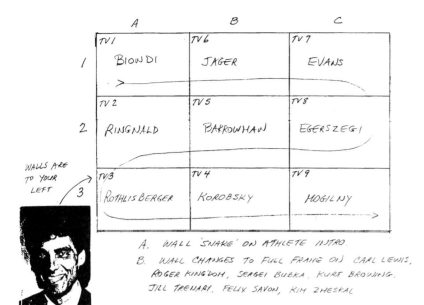

	A	B	C
1	TV 1 BIONDI	TV 6 JAGER	TV 7 EVANS
2	TV 2 RINGNALD	TV 5 BARROWMAN	TV 8 EGERSZEGI
3	TV 3 ROTHLISBERGER	TV 4 KOROBSKY	TV 9 MOGILNY

WALLS ARE TO YOUR LEFT

A. WALL 'SNAKE' ON ATHLETE INTRO.

B. WALL CHANGES TO FULL FRAME ON CARL LEWIS, ROGER KINGDOM, SERGEI BUBKA, KURT BROWNING, JILL TRENARY, FELIX SAVON, KIM ZMESKAL.

Figure 10–21   To orient Nick Charles and Hannah Storm to our plans for their monitor walls, the studio director prepared these sketches for Show 1. The "snake" in this diagram indicates that the video will appear one monitor at a time, replacing generic video with which the monitor wall would start. Item B indicates changes to occur after the video shown in the sketch.

Figures 10–21 and 10–22 show the first pages of Charles's and Storm's Ultra Wall "kits." Note the arrow indicating where the wall would be in relation to them on set. Because the monitor walls could be programmed with one picture or nine individual pictures, or any combination thereof—including sequencing ("snaking")—the arrow shows how the first set of images would be revealed. The notes at the bottom indicate how the images would be changed after the first nine had been revealed.

Thus we developed a visual aid that explained a process and, combined with the outline and narrative of the script, could show the patterns we would be pursuing. We went one step further. Because driving time can be used for study time, I had the narrative of the scripts dictated onto audio tape, which I gave to each of the hosts to play—at home or in their car.

So we were all ready for rehearsal, right? We'd faked our way through three scripts and outlines and audio tapes and monitor wall charts, and we were ready. The crews had checked out their equipment at the venues. The studio lights, audio, and effects had been tested. We were ready. Ah, but "What If?" Or, as poet Robert Burns wrote two centuries before a monitor wall was conceived, "The best laid schemes of mice and men gang aft a-gley." At *The 1990 Goodwill Games,* that means the monitor walls didn't work. They arrived. They were set up. They didn't work. Company technicians flew in. The company president flew in. The monitor walls would not work. Our Managing Director, Michael McLees, burned the phone lines calling around the country for another set of available monitor walls. He found one—in Chicago. A system there had just been used, could be torn down and flown to Seattle. He said, "Tear it down! Fly it in!"

Our replacement monitor walls—with a new technician, who had to be briefed—arrived at the airport shortly after midnight, seventeen hours before we were to go on the air. In those seventeen hours the system had to be transported to the studio, set up, tested, and programmed—while everything else that had to go on in the studio was happening.

We put a lot of pressure on the technical crew who had to assemble monitor walls at the last minute. We worked around their last-minute heroics, and we made it. The monitor walls on the first show did not work perfectly, but they worked. They were *there.* And all of the techniques we had developed to let the talent know how we planned to use them worked.

So the system of communication worked. The patterns we had hoped for were realized. Everybody but one survived the weekend: one graphics

Figure 10–22 Note the director's arrow in this diagram, orienting the talent to her physical relationship to the monitor wall. "Pictographs" were generic representations of individual sports used when specific video of athletes was not available.

	D	E	F
1	TV 12 BASKETBALL PICTO	TV 13 TRACK + FIELD PICTO	TV 18 SWIM - PICTO TO LIVE - WEYR HOUSE POOL
2	TV 11 BOXING PICTO	TV 14 VOLLEYBALL PICTO TO LIVE - SPOKANE COL.	TV 17 GYM - PICTO TO LIVE - TAC. DOME
3	TV 10 DIVING PICTO	TV 15 FIGURE SKATE PICTO	TV 16 ICE HOCKEY PICTO

A. PICTOGRAPHS - LIVE VENUE REVEALS (18, 17, 14)
B. WALL CHANGES TO FULL FRAME VENUE AERIALS
C. WALL CHANGES TO WAVING FLAG MONTAGE

WALLS ARE ← TO YOUR RIGHT

artist decided the pace was unbearable. We limped through the following Monday minus one graphic artist, but were able to lure a replacement from TBS in Atlanta.

Preshow Communication

I should add one postscript to this saga. While all of these scripts, outlines, and charts were being developed, I had to leave Seattle—not once, but twice. The first time was to attend a news conference in Los Angeles to brief reporters who would be covering the Games. Paul Beckham was there, the Turner Broadcasting System's Senior Vice President of Finance and Administration and the TBS executive in charge of *The Goodwill Games* (later promoted to become President of Turner Cable Network Sales and President of Turner Private Networks, Inc.). Rex Lardner was there, Senior Vice President and General Manager of the Games. There also were 1972 Olympic Gold Medal winner Olga Korbut and one of the most memorable participants in the 1976 Games in Montreal, Nadia Comaneci. I was called on to explain our concept of creating a global tele-presence, which I demonstrated by showing sample "Journeys" and "Adversarial Billboards." I explained our use of host Larry King as "Everyman," thrilling to the Games, asking Everyman's questions, but backed by sports experts Nick Charles and Hannah Storm, to satisfy Larry's curiosity and to keep official record of the outcome of the events. Then I turned Larry King loose on them, and "Everyman" took over. I also joked with some of the reporters later that we had arranged for a 70% eclipse of the sun to highlight our "Welcoming Ceremonies" Saturday, July 21. The eclipse was to occur from 6:52 to 9:12 p.m. PDT—right in the midst of our ceremony. No one can say we didn't plan these Games well.

My second call away from Seattle was to the White House. President Bush had freed time to tape a statement to include in the "Welcoming Ceremonies." which I got on one take, and then I boarded a plane back to Seattle.

The necessity to make these two steps away from the front lines emphasizes the need for a collaborative effort in globalcasting. *Every member of the team has to understand the production.* If one link is pulled from the chain of command, the chain has to heal and pull, instantly. On a globalcast

Figure 10–23 This photo translates Figure 10–22 into an on-air shot. The paper pictures become motion video. Note the use of multinational flags to keep the one-world theme of the Games ever present.

Courtesy Turner Broadcasting System.

Figure 10–24 The "war room" at *The 1990 Goodwill Games*. Operations and administration vice president Barry O'Donnell (standing) and coordinator of production Dan Baer (seated, facing camera, left) operated here before, during, and after each show, simultaneously meeting equipment and other crises, preparing for the "next one." Note the multiple monitors and charts needed to keep track of all the venues.

Courtesy Turner Broadcasting System.

you can't afford to miss a beat. Everyone has to pull his or her own weight and be ready, in an emergency, to pull someone else's weight, too. Communication is the key to creating the kind of team necessary to do the job.

Operational Manuals

This final note on the overviews deals with how the production patterns are translated into hardware and its use. The operational manuals that grew out of the overviews dealt not only with the technical equipment needed to accomplish the goals we set but also with the scheduling and use of that equipment. Again, this was not a one-shot globalcast, it was a series. Even as we were broadcasting live, we were preparing for the next day—and the day after that. We had to know, for example, that in Show 5 we had to tape Jackie Joyner-Kersee and her coach-husband for a piece that was scheduled to air in Show 6, to the tune "Every Breath You Take." We had to know that we needed sound bites from Day Two for use on Day Ten, etc.

The operational manuals accomplished this, in part, by detailing the structure of the organization. Using the overviews as a guide, we identified fifty areas of responsibility. To make this number manageable, we developed a chart that grouped the fifty areas into six major areas of responsibility: Production, Venue Operation, Facilities, Logistics/Finance, Engineering, and Information Services. In our trailer park we housed many of these services together in an operations trailer. Figure 10–24 shows some of the maze of monitors and wall charts that helped the supervisors of these six areas of responsibility to keep track of all of the venues and all of the events. The table of organization in Figure 10–25 details the structural breakdown of the overall operation.

The operational manuals, then, detailed the equipment, the uses of that equipment, and a schedule-of-use. Charts listed the known positions and schedules for equipment, for the edit suites, and so on. They translated production concepts into machinery and machinery into processes. They empowered us to identify ideas that were overdressed or under thought. They empowered us to know that in the frantic days of broadcasting we could open the throttle all the way without worrying that the pipes would burst. These manuals also, of course, provided us with a bar graph to keep constant control over the budget. They were a very basic part of our pre-show communication.

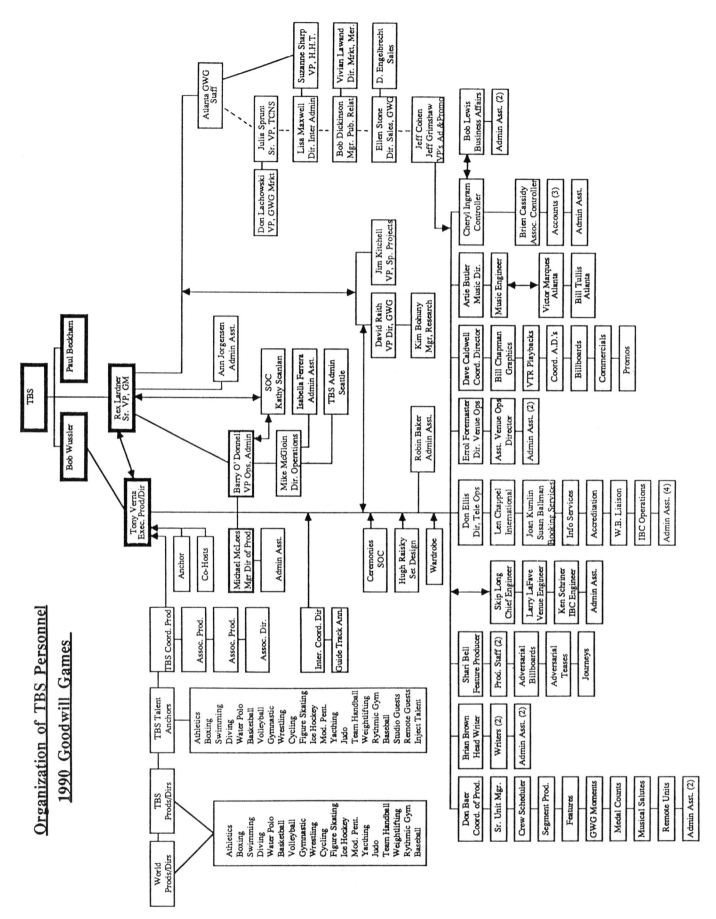

Figure 10–25 The table of organization drawn up for overview 1 of *The 1990 Goodwill Games*. Categories vary from program to program, but the basic chain of command—executive producer/executive director to producers to directors, and so on—remains constant.

Chapter 11

How to Book and Guide the Talent

When you're producing or directing a play or movie, if you cast the right actors chances are good the scenes will jell, the drama will soar. The sets will gain rave reviews, the costumes will be praised, the dialogue, music, pace, and subtleties will all convey a magic quality. Miscast the actors, and the sets, costumes, dialogue, music, pace, and subtleties will come off like tinsel on last year's Christmas tree.

Casting a globalcast may not be quite so critical, because the spotlight is usually shared. And the nature of the casting varies with the type of show. On *Live Aid,* we had a star-studded cast that promoters can only dream of. Although *Our Common Future* and *Earth 90* were considerably shorter, shows lasting three and five hours still require large casts and spread the glory.

For musical globalcasts, the talent booking is inevitably done by an agency or company specializing in that field. Contacts and contracts are much too complex for amateurs in this field. On the production staff, however, there are positions for talent/staff logistics specialists. Once booked, the talent usually requires special accommodations, transportation, and arrangements for equipment and costumes. Performances, interviews, photo sessions, make-up, rehearsals, sound checks—these and a host of other elements surrounding talent require coordinators on the production staff. It may appear glamorous, but it's work that requires stamina, tact, and the ability to forego sleep when necessary.

Because of time restrictions, there may be friction between performers, and staff members have to strike the proper balance between respect for the performers' audience appeal and the overall needs of the program. Diplomacy is not limited to international politics. Walking on eggs can be a productive exercise.

On a sports production like *The 1990 Goodwill Games,* the talent situation is different. The series nature of the production dictates vast changes in how the production staff works. The Games were spread out over seventeen days: my discussion will be limited to only eighty-six hours of domestic coverage, but the overall operation went far beyond that.

Figure 10–1 shows the athletic events covered in the USA during the 17 days of the Games. Beyond the domestic feed, however, there was vast global coverage of the sports competition. As Host Broadcaster, TBS provided approximately 500 hours of 183 events, including the welcoming and closing ceremonies bookending the sports events.

With a production of that scope, lodging, food, transportation, and laundry arrangements are overwhelming. The sports participants were determined by competition, but the hosts and announcers were selected carefully for their on-air roles.

Talent logistics are a major endeavor for a production of this size. The talent must be prepared to handle the unique nature of the series, not only the multiple sports events but also the political and philosophical implications beyond the events themselves. Using *The 1990 Goodwill Games* as an example, let me show you how setting the tone of a production determines the selection and guidance of the show's on-air talent.

Analyzing the Anchor Position

Heading into the 1990s, I felt obliged to question a process made stale by overuse. The anchor position for major TV sports events has always had an authoritarian flavor ever since Jim McKay did the Winter Games and Walter Cronkite—accompanied by his authoritarian anchor desk—hosted the Summer Games at the 1960 Rome Games from the CBS studios in New York City. This was still the stone age in sports television, years before Roone Arledge and the word Olympics became synonymous. In 1960 only the revered Cronkite was trusted to act as a unifying element for daily packages of videotaped action shipped back to him from my outpost at Rome's airport. This was the first attempt to televise a Summer Olympics in its entirety—or something approximating its entirety. There were no satellites, and videotape was just a recording device, not yet a production tool. The isolated camera and instant replay were still three years down the road, the satellites five years away.

In 1965, when the Early Bird satellite was scheduled to transmit the first sportcast—the English Derby—my announcers were once again drawn from CBS News, the veteran Charles Collingwood and the latest upstart, Dan Rather.

It seems that Cronkite, a newsman, set a precedent for having newsmen in the role of sports anchor. He would be followed at the Olympics by Jim McKay, who was a newspaper reporter before making the transition to television, and then by Bryant Gumbel, the prime mover in NBC's journalistic approach to the 1988 Seoul Olympics. These men were expected to offer the viewer a mix of pure sports elation (the simple, good news stories) with the intrigue of the behind-the-scenes maneuverings (the simple, bad news stories).

But in this authoritarian system, with a single figure completely responsible for the tone of the broadcast, the large picture can become obscured and bogged down in the inevitable drudgeries—the interviews, the staged transitions. This was especially true of the Goodwill Games, where our larger goal was to create global goodwill, not just to report on sports competition. The premise I developed for *The 1990 Goodwill Games,* therefore, stated that the anchor would serve the audience better if he or she were less perfunctory, providing a light wash of exposition, catching the romance of competition, identifying the hidden connections, tying in the world by hosting dozens of guests brought to us from around the world by satellite. Freed from the mental gymnastics and mechanical detail required by an event like this, the host could provide the viewers with a greater frame of reference, conveying a stronger sense of the human and global elements.

The electronic miracles at our command would allow our anchor to convey a sense of global ease—the facile way a Larry King can put papers aside, rest an elbow, and turn to chat with someone who has been effortlessly plunked down from across an ocean to sit next to him. This simplicity would become the symbol of our production—the wizardry of satellite communication with the straightforward manner of Larry King.

That's how we determined that we wanted a "Larry King" for our anchor post, and we were fortunate enough to get Larry King himself. We brought his *Larry King Live* executive producer, Tammy Haddad, to Seattle with Larry, to coordinate his appearances with us and with his talk show—and to help us graft the *Larry King Live* feel onto *The 1990 Goodwill Games.*

To free King to play the role of global goodwill ambassador and "Wow!" reactor to the excitement of the sports, we provided him with two cohosts, both sports experts: Nick Charles and Hannah Storm. These two were charged with keeping the show on-track in terms of sports coverage. King would be Mr. Everyman, asking questions. Charles and Storm would provide the technical answers, for Larry and our viewers. They shared the dual job of offering sharp, professional sports coverage and conveying the essence of the Games.

Analyzing Talent at the Venues

That thesis went for the talent at the venues, too. On a massive production like this, an executive producer can't just turn talent loose. You owe them guidance, just as you owe the producers and directors guidance. The producers and directors have to understand the content and pace and style of the show so they can work with the talent, otherwise the overall production becomes a mishmash of uncoordinated acts. So you establish guidelines. You set a tone, a goal everyone strives for.

Good reporting is a function of suspicion and intuition—knowing when to suspect the worth of information, when a small detail can lead to something bigger. And the way to ferret out the facts is to prepare. In guiding talent, I've always tried to use the two complementary methods of communication: tell them what *not* to do, then tell them what you expect.

To simplify *The 1990 Goodwill Games,* I wrote the series of overviews (see chapter 9). Each overview described the events that would take place, the athletes expected to participate, predictions of possible outcomes, and details of special equipment and production techniques being planned. These details included guidance for the talent. To the commentators I stressed the following.

> Prior to the beginning of your event, there will be press releases, phone calls, packets of team information, biographical information, the TBS research manuals, key information points you've studied on site, interviews with athletes, talks with coaches who told you what has worked and what hasn't worked.

> And prior to that, you'll have viewed tapes, with your co-announcers, producer, director, researcher and others. And you'll have started mentally to rehearse the type of comments you'll use on the telecast. You'll have rehearsed the type of replays to be used and the key points you'd like to make as the replays are shown.

> You'll have rehearsed the comments on the graphics that will be presented, so it won't seem to the viewer that it's the first time you've seen them. And with these rehearsals, you'll have talked about the types of camera shots, replays, and graphics that *won't* work.

So now here you are, after a week of getting ready. Your level of confidence—before and during the telecast—will be directly related to your level of understanding. That means being prepared, and that's where my advice comes in, because *increased preparation demands increased editing.*

Your most important material will be collected in the days and hours before your event because it is at this time that you'll get the latest information. Interpreters will be available so that you can canvas all the coaches and players and find out *why* one has done this, *why* another has done that.

It's critical that you capture the personalities of the players and coaches. In the days and hours before your event, you and your producer and director will talk about the tendencies of a team or player. You may talk about an addition to the graphics menu, or another possibility for a replay. Last-minute plans.

Then comes the telecast. Don't bury the audience in the opening minutes with too much background material. Viewers care about the NOW. They don't want an overload of information. Let that come parenthetically. Start by relating directly to the action on the screen. *To leave a lot of notes on your legal pad is not bad.* Use what fits. Leave the rest.

I then detailed the difference between the roles of the play-by-play announcer and the analyst. For the play-by-play announcer:

Don't get in the way of the game . . . Keep the style concise and sharp, not bursting to get back into the broadcast the minute the analyst finishes a comment . . . Play-by-play serves the broadcast best by setting up the analyst, but never leading the analyst in a certain direction unless it's pre-understood.

For the analyst:

The sin is talking too much and saying very little . . . Smiles and a soft nature don't work here . . . Good analysis is the product of preparation and experience. Facts aren't enough. You need inside information, the drama behind the event, the perspective of the coaches and athletes.

I quoted Vin Scully, who said in *Live TV* that a good analyst will probably use about 20 percent of the preparation. "The choice of *what* 20 percent is the key." The secret is to use what applies, and only what applies.

This was the basic philosophy I stressed in the overviews. And this last point was particularly applicable to a worldwide competition stressing not world competition but world cooperation. Our goal always was less adding up scores than relating scores to global concepts.

Of course, in other places I also detailed specific uses of practicalities like the telestrator and lane indication graphic for swimming coverage.

Integrating Talent and Equipment Techniques

Figure 11–1 shows why it was important for the talent to know about our technical plans, specifically about the monitors and screens at the venues and how that equipment tied into our letting both the studio and the worldwide audience share special coverage of the event. This figure shows diving star Gao Min speaking with her parents in Beijing, China. The photograph, from the *Seattle Times,* shows the equipment mounted on Min's back that let her hear her parents *and* Larry King *and* the interpreter *and* my IFB line. The monitor in front of her had to be a closed-

Figure 11–1 This photo from the *Seattle Times* shows three-meter springboard gold medalist Gao Min talking via satellite with her parents in Beijing, China. The photo shows—as our television cameras did not—the necessary wiring on Gao's back. Her headset let her hear her parents, the Goodwill Games interviewer, and me via an interrupted feedback (IFB) line. Read the text regarding special wiring for the monitors.

Courtesy Pedro Perez, *Seattle Times.*

circuit monitor. Had it been an air monitor she would have seen herself—in infinity—as well as her parents. So we installed the special closed circuit monitors for our Injects.

The large screen for the audience also was not an air monitor, it was a "switchable monitor." A switcher, as opposed to the technical director, controlled the video on that monitor, following guidelines set down in the overviews. Only gold medal winners went into close-up on the switchable monitors.

It was moments like these that put the greatest pressure on the talent and interpreters who—while interviewing an event winner—had to judge when to step in, when to lay back. No instruction book could help them at that moment, when victor or defeated athlete found emotional release through a satellite encounter. These satellite injects that tied the world together were a challenge to everyone—to the staff who set them up, both in terms of human contact and satellite booking and hardwiring; to the crew who had to line up and take the precisely right over-the-shoulder shot (Our TV picture did not include the equipment on Gao's back.); to the interviewer who had to set up the on-air meeting; to the interpreter who had to keep our audience informed without interfering or impinging on some very personal moments.

I might add that all of the injects were not so sensitive and personal. A few times we visited gyms and bars and family rec-rooms where the atmosphere was more like a Saturday matinee than a soap opera. Each inject, however, matched its meet and punctuated the sporting event of which it was a part. And although I assure you there was nothing in any of the overviews about the show's anchor singing, on one occasion I remember Larry King singing "The Sidewalks of New York" with a group of people with equally questionable voices. Even off-key moments can be great when electronics bring parts of the world together—and the talent is right. Getting things right with the talent is a major task of the production team of a globalcast.

Chapter **12**

How to Create Scripts/Schedules

A production schedule is just like a show schedule. You have to back-time from the end, and you have to stay flexible. I detailed the crisis with the monitor walls on *The 1990 Goodwill Games*. No one could have foreseen the need to schedule seventeen nonstop hours immediately prior to air for the technical crew to set up an entire replacement system for one that didn't work. You have to expect the unexpected, and the only way to do that is to have well-thought-out *non*crisis scheduling that allows for crises.

In terms of booking and scheduling talent for fully commercial multi-venue globalcasts, you aim for the stars who fit both your budget and the concept of the show. For shows seeking unpaid talent, the usual pattern is to aim for top stars who espouse the cause. If you can get a commitment from one major performer, that usually inspires willingness on the part of other stars to join the show. Booking and scheduling talent is a little like fitting together the pieces of a jigsaw puzzle. You place one piece—one star—that you know fits and anchors the picture. Then you scout around for matching pieces, hoping that no two are exact duplicates (both demanding top billing). You need enough stars to hook an audience on the opening, sustain their interest throughout, and hold them for a grand finale. Fortunately for producers, many singers and actors are sensitive to issues like the environment, stricken nations, and children. The telephone, fax machine, flexibility, patience, and an ability to get by with little sleep are the main factors in booking and scheduling a show.

Scripting is much the same. Whether it's intense research on performers or standby scripts, writing on a globalcast starts early and does not end until the show is over. There is writing up to air time and even during the show. These last-minute changes demand flexibility not only from writers, but from teleprompter operators, directors, the talent—everyone. Again, the focus is on the interdependence and interlocking of all of the roles on globalcasts.

I will use *Earth 90* as an example here because its structure required our devising a complicated series of standby situations that involved the script, the tape room, the teleprompter, and the talent throughout the show. More than once, crew members claimed I was driving them crazy, but there was logic behind every standby we planned. And we planned a lot of standby. Let me recap the format of the show and you'll see why.

The show was set up in three locations, each with a pair of hosts: John Denver and Yu Hayami live in Tokyo, Debbie Gibson and Rolland Smith live in New York, and Olivia Newton-John and Herbert Leonard on tape in

Europe. We had additional live injects from Brazil (Gilberto Gil) and North Carolina (the group Alabama) and additional tape injects: Crosby, Stills and Nash, taped the day before the show, and Julio Iglesias, taped at a concert running the night of our telecast. With master control in New York, the back-up considerations were as follows.

Tokyo was a live satellite feed. If we lost the satellite, how could we back up the Tokyo performances? Answer: Tokyo did a dress rehearsal the day before the globalcast and fed the rehearsal to New York to be taped. BUT. This "rehearsal" was performances only, *no* introductions. So we had a tape to cover performances but needed back-up copy introducing the Japanese performers. Copy was written for Rolland Smith to cover that eventuality. The standby tape reel had the performances, in sequence, ready to go.

Brazil was a live satellite feed. For a variety of reasons it was impossible to pretape Gilberto Gil's performances in Rio, so we had to create other taped standby performances, which were put on a standby reel. Standby copy was written for Rolland Smith.

North Carolina's feed of Alabama was live by satellite from the Cumberland Arena in Fayetteville. If that feed went down, back-up tape was needed. Standby copy was written for Rolland Smith.

Europe was all tape, but the only satellite time we could book—after their taping in multiple locations—was one hour before we went on the air. If we lost the satellite feed from France, a full 25 percent of the show would disappear. Standby performances had to be put on tape. Standby introductory copy was written for Smith.

Austin, Texas, was on tape, being flown in. If the Crosby, Stills and Nash tape did not arrive, standby was needed. Happily, they were able to supply us with a video version of their song to put on standby. Standby copy was written for Smith.

Julio Iglesias's concert began at 8:00 p.m. We went on the air at 9:00 p.m. A number of reasons prohibited our picking up the performance live, so it was taped shortly after 8:00 p.m. We made three tapes. One went on a motorcycle, another into a van. Both raced to get the tape to us. We scheduled this performance late in the show—11:15 p.m.—but there remained the possibility that the tape might not arrive. The third tape was made in our tape room, but there was always the possibility that it might not be usable. Again, a standby tape was selected and timed. Standby copy was written for Smith.

You have noticed, no doubt, that all of this standby scripting was directed at Rolland Smith. This was done for two reasons. First was his familiarity and ease with live TV and on-air changes. As a long-time news anchor Smith was fully accustomed to having changes made under pressure on the air. Second, putting the full burden on him left Debbie Gibson free to concentrate on what we fully expected would happen—a show on which all of the satellites would work and no standby would be needed.

Planning for all of these possibilities also meant that the tape room had to prepare tape reels in such a way that I could access anything needed at any time. In addition to all of the above contingencies, we prepared a generic Standby Reel in case cameras or cables went out at master control. And all of these standby conditions had to be added to the regularly scheduled tape inserts in the show. Fortunately, some of the tape crew specialists, like John Calabrese, had been with me on more than one show and worked as hard as I did plotting out these patterns.

Having all of these alternatives also meant that Rolland Smith had to be prepared to do any of half a dozen different shows. If we lost the satellite from Japan, he had to introduce all of those acts. If we lost the satellite from Europe, he had to be prepared to introduce alternative acts in the positions allocated to European acts. And so on down the list through all of the possible signal failures. He also, of course, had to be confident and jovial and go on with Debbie Gibson as if satellite failures were the farthest thing from his mind.

During the globalcast, Smith had an easel set up just off camera, on stage. Our writer/supervising producer stood by his side. Just before air the three of us sat and talked through all of the eventualities. On stage they had separate stacks of standby copy. The greatest relief came when we received the feed from France—and verified that the tape was there. Europe was in-house, so all of that standby copy could be discarded. I was on IFB to Smith and on headset to the writer/producer so both could be alerted simultaneously to any crisis, or any solution to a crisis. Throughout the three hours, this behind-the-scenes team was ready to go with any of the alternatives we had discussed. As each piece of the puzzle aired successfully, they dropped off those standby scripts and moved on to the next.

We had a couple of places where Smith and Debbie Gibson had to fill a short amount of time, but no major crises developed. Naturally, we had standby fill copy, which had been written in paragraph form so that any single paragraph or collection of paragraphs was a complete unit. The fill could run five seconds, ten seconds, fifteen seconds, twenty seconds, thirty seconds—and make sense however long it ran. Enough of this public service copy was put on teleprompter to cover an emergency. Additional copy was on cue cards and on index cards on the stage podium. The teleprompter operator was on headset, so he was kept aware of all decisions also. Our cohosts read the fill copy that was needed from teleprompter—our only standby to make it to the air.

Despite the crew's complaints, they knew we needed all of the fallback positions. It wasn't easy. It was necessary.

The script for most of these globalcasts, *Prayer For World Peace* being an exception, is what is called *continuity*, brief spurts of monologue or dialogue that tie together the multiple parts of the program. Much of this writing is necessarily mechanical, introducing the stars and their performances, including the research necessary to produce these mechanics. Frequently there are requirements. To get stars to appear, you may have to agree to plug their latest album or film, or mention a favorite charity or cause. Because of the tight timing, there is usually little or no time for ad-libbing by the stars. If they have a personal message to deliver, it is usually put on teleprompter (see Figure 9–2). But there are times when you have no choice. On *Our Common Future*, naturalist Jim Fowler loosed an eagle trained to return on call. However, the eagle liked a railing on the second balcony better than the stage and refused to return, despite call after call. The reluctant eagle created one of the audience-pleasing highlights of the show—all ad-libbed—and required our adjusting time later to compensate. But that's live television. That's why writers, like directors, don't go home until the show is over.

However many scripts we have to write or tapes we have to line up, however many alternatives we have to conjure up, so far as the viewer at home is concerned, there are no problems. There is only a well-planned globalcast that goes smoothly. That's what the production team has to aim for—and plan for—in creating every globalcast.

Part *Four*

Transmission Technology

Chapter 13

Transmission Via Satellite

A Caution About Convergence

I preface this report on direct broadcast by satellite (DBS)—and the rest of this book—with this caution: All that you have read so far has been practical data based on fact and experience. The facts have been researched and I have had the experience. I stand by both. What you read from this point forward, while based on fact as it is known in 1993, will require interpretation on your part. I will lay out the details and facts on DBS, pay-per-view (PPV), high-definition television (HDTV), cable TV, niche programming, and so on. I will call your attention to signposts that appear to indicate the direction in which global communication, and specifically global television, is traveling. But for the decade of the 1990s no one, including myself, can predict the future with certainty. Will Japan be fiber-wired by 2015? Will the USA accept that Japanese plan as a challenge and strive for the same goal? Will DBS be directed to every home with a TV set and, as a consequence, wipe out the cable TV industry? Will the digital compression that will expand channel capacity on both cable and DBS keep both methods of transmission active and competitive? Will the television viewing public accept the challenge of having to choose among 500 channels, or will it say no to both cable and DBS and continue to rent home videocassettes? Will production restrictions imposed by the European Community (EC) and individual nations create so much coproduction that national identities will be erased from production, or will those restrictions result in increased national identity of TV programming?

I don't know the answer to these questions, and neither does anyone else. Only time can determine the communication winners and losers of the future. This much I can assure you, and I will stress this point consistently through the rest of this book: the communication media are converging and will continue to converge. That convergence will affect the sets on which people watch television, the standards of the signals used to transmit what people watch on television, the content of the programs on television. That convergence will also affect everyone involved in the production of global television.

In the pages ahead, therefore, I will lay out the known facts, label my assumptions as such, and offer you the opportunity to fit the pieces of the future together armed with the best current information and background. Because you are reading this book, your reaction to these converging media may be one of the influences that will determine the direction of the future. Whatever lies ahead, it promises to be fascinating, fast-developing, and full of electronic fun.

DBS: Direct Broadcast by Satellite

Digital Radio

Audio signals broadcast directly by satellite are important to global television for four reasons:

1. Radio remains competition for television.

2. Radio/TV simulcasts can enhance programming, especially musical programming.

3. As the distribution patterns of most globalcasts have indicated, radio can carry the program content of global TV shows to hundreds of millions of people who otherwise would not have access to that communication.

4. DBS-Radio and DBS-TV use the same satellites.

Another tangential reason that satellite radio may influence the future of television is that many of the companies already involved in transmitting digital radio are TV cable companies. In the USA, some cable companies—with backing from organizations like Viacom International and Tele-Communications Inc.—are expanding the delivery of digital cable radio to subscribers around the country—for a fee, of course.

General Instruments, whose Jerrold Division builds and sells cable converters, also has a communications division that provides digital radio to cable households in Pennsylvania and Florida. Other USA companies reportedly starting up digital cable radio services are Digital Radio Labs, International Cablecasting Technologies, and Satellite CD Radio.

The lure for these radio subscribers, of course, is the digital transmission, the brilliantly clear sound that has seen the record industry go from vinyl disk to tape cassettes to compact discs (CDs) in a couple of winks of an electronic eye. The record industry has been wary about the possibility of home listeners recording the distortion-free digits, but the cable radio companies have vowed not to play albums in their entirety or to advertise a program schedule that would allow listeners to plan a home recording session.

Japan Does Digital by DBS

Digital Radio is transmitted directly by satellite in Japan also. Starting in December 1990, the Satellite Digital Audio Broadcasting Company began the world's first satellite digital radio station, Radio GIGA. Partly owned by Japan Satellite Broadcasting (JSB), which has also inaugurated DBS-TV service, Radio GIGA began scrambling its commercial-free broadcasts in April 1991, when it expanded its service from twelve hours daily to round-the-clock music. Subscribers had to invest in a satellite dish to receive the signal and a decoder to unscramble it. There is a small monthly fee (reportedly USA\$4.60). Remember that because Japan uses Ku-Band satellites, only much smaller antenna dishes are needed to receive satellite signals. As with USA cablecasts, subscribers who have digital audio tape recording devices can record the CD-quality music. The failure of one of the JSB satellites to orbit successfully slowed planned expansion, which would have provided additional radio stations, but Zipang Communication (involving Tokyo Broadcasting System, publisher Kadokawa Shoten, and others) started three new free channels in June 1992, with scrambling taking effect in December.

Competition from Digital Radio Broadcasting

Even though Digital Cable Radio and Digital Satellite Radio are in their infancy, they already face competition, from over-the-air broadcasting done digitally. Still in the testing stage, digital radio broadcasting might actually split radio audiences with cable/satellite services. Why? Because one of the great advantages of radio is that it easily follows anywhere people go—in cars, on boats, and so on. An estimated 40 percent of radio listening is done away from home. It's possible, then, that both digital cable/satellite services and digital broadcasting services might succeed. Theoretically, cable/satellite companies would satisfy home bodies and the broadcast stations would satisfy people on the move. Again, there is cost involved. Current automobile radio receivers are not capable of receiving digital signals. If digital radio broadcasting becomes a reality, redesigned radios would have to be installed in cars. Continued tests of digital broadcasting are reportedly taking place in the USA (NASA is involved in these tests), Canada, the EC, and Japan. (See RBDS in Glossary.)

Nondigital Radio: Single Channel per Carrier (SCPC) Format

In addition to digital radio DBS, *non*digital radiocasts are available via satellite. There are, at present, more than 150 channels of nondigital radio transmitted just from North America's group of communication satellites. These are generally feeds intended to be rebroadcast either in part or in their entirety by local radio stations, but with just a little added equipment a home dish owner can receive them—unless and until the signals are scrambled. Because these channels are not advertised, you have to read a publication like *Satellite TV Week* to discover and gain access to them.

These special radio channels are not like conventional audio subcarriers that can be tuned into by using any satellite antenna dish. They use a different transmission format, called single channel per carrier (SCPC). You can access this format only if you install a special SCPC receiver on your satellite dish.

There are two types of equipment that will connect a listener with the SCPC channels. One is a scanner, a special FM receiver priced from about USA$500 to just over $1,000. A dedicated SCPC tuner designed specifically to add to home satellite dishes is easier to use and priced around $500.

With 150 channels available, obviously the choice of subject matter is extremely wide. CNN, ABC, UPI, AP, National Public Radio, and Voice of America all use an SCPC channel. Canadian broadcasts, including French language broadcasts, can be accessed via Canada's Anik satellites. For Spanish broadcasts, you can tune in Mexico's Radio Felicidad and for Italian, Radio Italiano (RAI). Many networks have begun scrambling their feeds to affiliates, but as of this writing most radio feeds are still unscrambled.

DBS Radio Versus AM/FM, Shortwave

In addition to existing radio broadcasts via satellite, there is the promise of much more to come. Two of the most ambitious proposals for satellite radio were announced in 1990 by Satellite CD Radio and in 1991 by Radiosat International.

In *Satellite TV Week,* Mark Long reported that Satellite CD Radio Inc. asked the FCC for permission to build and operate two geostationary satellites to be launched in 1994–1995. The goal is to create a national satellite-delivered digital radio service. Three regional beams from the satellites will cover the eastern, central, and western portions of North America. Because solid objects can block satellite reception, these regional beams will deliver signals in two ways. In urban areas, where

Figure 13–1 The satellite's function is the same whether programming is delivered to the home viewer via cable (left) or directly to the viewer via DBS (right). The programming signal (usually scrambled) is uplinked from its source to a satellite transponder, where the frequency is shifted down to avoid signal interference on the return trip. In the case of a cable company, the television receive only (TVRO) dish is large and usually computer-operated, to bring in the sharpest satellite signal. The cable operator's TVRO antenna has a descrambler to translate the signal back into programming. The signal is then *re*scrambled by the cable operator to paying subscribers via cable, usually fiber to the curb and copper wire from curb to customer. The viewer's cable converter box descrambles the signal and enters the audio/video into the subscriber's television.

In the case of DBS, the pattern is exactly the same except no cable operator intervenes between the satellite and the viewer. Home dishes can range in size from ten-foot backyard dishes to six-inch windowsill antennas.

buildings may pose a problem, one set of signals will go to a network of terrestrial repeaters that will make the signals available to city receivers. The other set of signals will aim at areas outside the range of the earth-based repeaters. The CD radio receivers—estimated to cost around USA$200—will be designed to pick up whichever signal is stronger. They will also be designed to contain traditional AM/FM receivers in addition to the CD tuner. Home, portable, and mobile receivers are planned, plus CD adapters that can be added to existing home or car radio receivers. The National Association of Broadcasters is also reportedly exploring terrestrial methods of broadcasting digitally, in the hope of fending off satellite competition and allowing local broadcasters to keep control over their markets. A local station can buy and install a digital satellite earth station for less than $10,000.

An even more ambitious proposal is that of Radiosat International, whose long-term goal is to replace the existing shortwave radio service currently used by an estimated 500 million listeners around the world. The organization plans to launch and operate three satellites, starting in 1995, each of which will offer more than 200 channels to be leased to commercial and noncommercial organizations. The cost of receivers for this new service is estimated at USA$50, substantially less than the cost of the CD sets. Shortwave has historically been the main form of radio service in many third world countries, and there are substantial numbers in Europe. As a common carrier, Radiosat International will be available to all nations.

Aside from the immediate impact of digitally simulcast soundtracks of movies and music specials, it's difficult to predict the possible future impacts of DBS Radio on global TV. It is obvious, however, that there will be impacts.

Television: How DBS Works

DBS-TV will obviously have a greater impact on global television than will radio, but it's important to put DBS in perspective. Chapter 14 traces some of the forces that have altered television transmission techniques

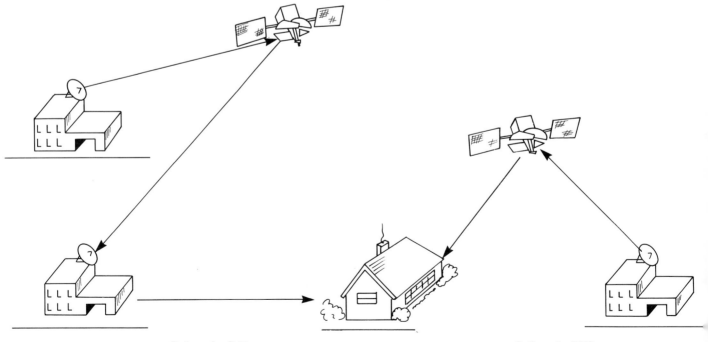

Delivery by Cable   Delivery by DBS

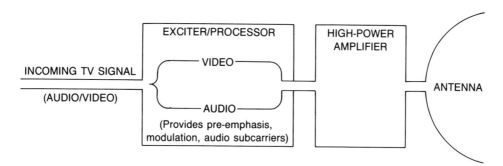

Figure 13–2 To correct a potentially bad signal-to-noise ratio of color and audio data, and to meet FCC requirements for FM broadcasting, satellite uplink stations provide preemphasis for both video and audio signals. This emphasis on high-frequency components is removed by matching deemphasis at the downlink station. The high-power amplifiers (HPA) increase the signal level sufficiently to drive the satellite transponder. Program audio and other audio channels (such as cue circuits) are put on separate subcarrier frequencies for transmission.

from simple broadcast to cable to home video to DBS. Here I review the basic mechanics of DBS and how they have influenced transmission patterns.

Figure 13–1 (left) shows how distribution works by cable:

- The cable programmer delivers programming to an uplink point on earth. The signal is encoded so it can be received only by people with decoders.
- The signal goes to a leased/owned transponder on a satellite in the geostationary orbit.
- The transponder shifts the frequency of the uplink signal, amplifies it, and sends it back to earth to a large earth receiving station (downlink).
- The local cable company receives the program and retransmits it—by cable—to subscribers. Distribution is obviously limited by the amount and kind of cable that has been laid and by the number of subscribers.

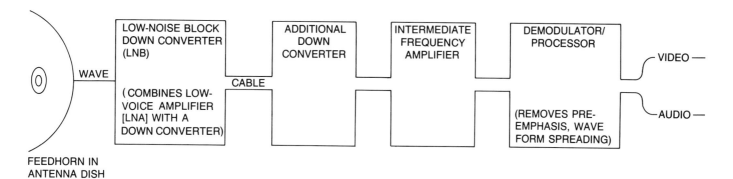

Figure 13–3 The downlink station reverses the uplink process. An antenna gathers the signal being beamed down by the satellite. The feedhorn (or dual feed used by cable TV operators for access to both the horizontally and vertically polarized channels) collects the signals and feeds them via wave guide to the low noise amplifier (LNA), often combined with a downconverter to form a low noise block (LNB). The amplifier strengthens the signal, which has been weakened by its 22,300 mile trip. The downconverter shifts the signal's frequency to an intermediate frequency (i-f). Shifting the wave to cable, the signal moves to a second downconverter, which converts the signal to a still lower i-f, then to an i-f amplifier. The amplified signal then goes through a demodulator/processor, which removes the preemphasis and adjusts the output frequency so that it equals that of the original input. A filter removes any extraneous signals, and the signals are ready for viewing or retransmission.

Because satellites are positioned differently—and their positions can change—antennas must be capable of repositioning. A mechanical arm called an actuator, increasingly with computerized controls, can shift the antennas aim.

DBS follows the same pattern, except that it bypasses the cable company and goes directly to the subscriber. As Figure 13–1 (right) shows, the receiving dish is in the subscriber's backyard, roof, windowsill, or similarly unobtrusive spot.

The birth of DBS to home dishes was not easy. First, the size, configuration, and cost of satellite receiving dishes (antennas) had to change. The first dishes were designed not for home use but for use by television stations. Turner Broadcasting pioneered the use of satellite distribution by superstations. NBC pioneered the use for distribution of network programming to their affiliates. But the dishes local stations used were big. In 1981 the sizes ran about five to ten meters (sixteen to thirty-three feet) and ranged in cost from USA$20,000 to $120,000—not exactly convenient for home use. And the dishes had to be adjusted constantly to aim at the proper satellites for optimal reception. Eventually computer programs were developed to adjust the dishes automatically without having a human on duty—part of the convergence of the media.

By 1990 a ten-foot satellite home receiving dish could be purchased for about USA$3,000, still out of reach for most households, even those with enough space. Some communities passed laws forbidding antennas. But technology never sleeps, and in this area the rest of the world led the USA by a country mile.

Actually, the first DBS system was aimed at the USA, from Canada, but it didn't last. In 1982 United Satellite Communications Inc. (USCI) leased five channels on Canada's Anik C2 satellite and began the first direct-to-home satellite service, serving the northeastern part of the USA. USCI's satellite receiving equipment cost only USA$295 or $395, for a three- or four-foot dish. There was also a monthly fee of USA$24.95. Unfortunately, financial problems closed USCI in 1985.

Even earlier, in 1980, Comsat has asked the FCC for permission to operate DBS in the USA and had some DBS satellites built, but never actually began service. A number of other companies who applied for DBS licenses also let them lapse. By the time the USA got its first all-USA DBS system (K-Prime, later Primestar Partners) in October 1990, other countries had already signed up millions of DBS subscribers. A major factor in the difference between the services is the percentages of cable penetration. A still larger factor—of which cable is one part—is audience demand and how it is dictating transmission techniques.

Chapter *14*

How Viewer Demand Dictates Transmission Techniques

Niche Programming

The industry calls it niche programming—very specific programs aimed at non-mass audiences: sports fans, weather watchers, ethnic groups, opera fanatics, ballet buffs, shoppers, worshippers, credit-seeking students, and so on. But *niche* is more than a buzzword for the 1990s: it is a symptom of how audience demands have forced television—globally—to adapt its methods of transmission to meet audience needs.

One of the reasons *Live Aid* was such a phenomenon was that it spoke the international language—music. Sports and news have also gone global in a major way. Once you get past these three, however, the potential size of each global audience decreases. Languages and audience preferences diversify.

The same is true of domestic audiences. There are mass audiences for first-run movies, the Super Bowl, the World Series and the Academy Awards, and hit television series. The audiences are vastly smaller for dramas, ballets, documentaries, children's programs, college classes, and so on.

Prior to the 1990s, these audiences had to scramble for the available television channels. Public Broadcasting System (PBS) and children's programming and language classes had to fight for space to exist. Narrowcasting in the 1990s will end that scramble, because transmission technology—in the form of cable TV, DBS, PPV, and digital compression—has caught up with audience demand.

For decades, the commercial television networks and local stations argued that they had to serve the mass audience. There was no time to serve all the splinter groups. But as cable expanded, the networks and local stations no longer had a captive audience. The television monopoly ended and the transition to the niches of infinite channel capacity had begun. The networks' share of audience declined by almost 30 percent, temporarily bottoming out in the 1991–1992 season. But 1990s viewers kept going elsewhere, because there were so many elsewheres to go—"boutiques" as well as department stores.

How Cable Television Has Influenced Television Transmission

The shift in audience didn't begin until cable became national in scope and began offering alternative products. When uninterrupted commercial TV started in the late 1940s, there were local stations and only three surviving networks: ABC, CBS, and NBC. Cable television began not as competition to network or local television, but as a method to extend existing television signals to people who could not receive them, such as in the hilly country of Pennsylvania. As opposed to radio waves, which tend to cling to the earth's surface, television signals travel in straight lines and consequently can go only a short distance without being relayed. Any object like a large building or mountain will block a TV signal. Cable—then called CATV (Community Antenna Television)—overcame the mountains. No one expected it to overcome the networks, but it did.

From 1970 to 1980 the number of cable TV subscribers rose from just over 5 million to more than 17 million. In 1975 Home Box Office (HBO) began transmitting its program schedule by satellite. WTBS in Atlanta and other "superstations" like Chicago's WGN and New York's WOR began offering cable audiences many choices they hadn't previously enjoyed.

Niche broadcasting began keeping sports fans happy with sports-only channels. Multiple religious channels became available. Weather buffs could watch weather around-the-clock. Financial news on cable followed the stockmarkets, which followed the sun around the world. Home shoppers never had to sleep, unless they wanted to. Music fans were even further niched into MTV, VH-1, and Nashville. Local news programs provided viewers with home-grown news and cable companies with local sponsors. Where the audiences proved too narrow, competitors merged. HBO's Comedy Channel and Viacom's HA! fused to form Comedy Central. Live courtroom drama appeared, to challenge Perry Mason.

The niches, combined with round-the-clock movies, have been keeping a lot of viewers happy. And cable suppliers like HBO, TNT, and Lifetime have been producing their own movies and specials at an expanding rate, offering exclusive product, often more provocative than what commercial networks provide.

The decade of the 1980s showed the results. Weekly *Variety* (September 17, 1990, p. 34) reported that USA cable penetration in 1980 was 23 percent. By the start of 1990 the figure had jumped to 57.8 percent. And the shift in audience preference was a major event. During the July 1990 sweeps, for the first time in television history, basic cable—as a whole—had more viewers than ABC, CBS, or NBC. Within a decade the previously unchallenged networks had been challenged and knocked onto the rating ropes.

And it wasn't just domestic audiences that had begun to make history. By 1987 the Super Bowl was seen in fifty-five countries overseas. Even though it was on a four-month tape delay, the 1987 Super Bowl was seen by more people in China than in the USA. Suddenly the vast number of potential viewers in Europe and Asia became a factor in worldwide TV production and distribution. In one decade, TV audiences had gone global. Also by the start of 1990 seven of ten USA households owned a videocassette recorder (VCR), another source of alternative programming. Home video became a major challenge not just to television, but to the film industry as well.

Home Video Jolts the Television and Film Industries

It took a long time for television and film executives to recognize and react to the spoiled viewer, who dwells in a marketplace fat with choices, zapping from channel to channel from the remote safety of the sofa. Get up to change a channel? This isn't the dark ages. This is the television age, the age of the image. No still pictures here. A thirty-second commercial may have hundreds of video cuts in it. Viewers want what they want when they want it—and they want it fast! Pizza in thirty minutes, microwaved dinners, cellular phones, answering machines, fax machines. Film scripts have more car chases and bedroom scenes than dialogue. Instant gratification. As "The Haunted Host" said Off-Broadway, "If you stop, you think." Don't stop. Do it now. Get it now. Zap!

Hollywood learned this lesson very slowly. When videocassettes first appeared, film producers fought to have a tax added to blank tapes, claiming that moviegoers would copy films and rob producers of income, destroying the industry. They stopped fighting when they began to realize more revenue from videocassettes than from tickets. Currently movie makers' income from video rentals and sales are double their income from ticket sales. In 1990, *Ghost* not only grossed over $200 million at the box office, it also broke previous home video records. By February 1992, Disney's *Fantasia* (14 + million) has surpassed that mark. By December 1992, *Beauty and the Beast* (20 + million) was the new champion. In 1993 each *Aladdin* cassette advertised the interactive Sega video game. *Aladdin* also went on CD-ROM. Jurassic Park went the 3DO route. The consumer message was coming through loud and clear—TV viewers want to see movies, but they want to see them when *they* want to see them, not when someone else schedules them. Convenience became a theme to add to convergence.

Digital Compression: Channels to Go/PPV

The way to provide viewers instant availability to movies round-the-clock is to program movies to start on viewer demand—or at least every half-hour or every fifteen minutes—but doing that would require a lot more channels than are available today. Fortunately for both viewers and broadcasters, technology has begun to keep pace with consumer demands. An increase in the number of satellites is one reason, but it wasn't enough. Because there is limited space in space, and specifically around the geostationary Clarke orbit, satellite space is allocated. In 1970, the UN allocated the USA slots for sixteen DBS satellites. Each of those satellites can accommodate a minimum of sixteen channels, for a total of 256 available channels—not a lot of channels to cover a country the size of the USA.

Because we can't expand the amount of space available in the Clarke orbit, scientists looked in the other direction: how to expand the use of the channel space already existing on transponders. Their answer was digital compression, a technique of shrinking the amount of transmitted information necessary to reproduce a video image.

Two groups have been charged by global organizations to establish standards for digital compression. Acting for the International Committee on Telephony and Telegraphy (Comité Consultatif International de Téléphone et Télégraphe; CCITT) and the International Standards Organization (ISO), the Joint Photographic Experts Group (JPEG) was charged with setting standards for still video. The Motion Picture Experts Group (MPEG) was charged with setting standards for full-motion video.

Under the JPEG standard, the color image is converted into rows of pixels, each assigned a numerical value representing brightness and color. The image is then broken into blocks, each sixteen by sixteen pixels. By subtracting every other pixel, this block is reduced to an eight by eight pixel block. An algorithm is used to compute varying values for the blocks, leading to further reduction. After transmission, the process is reversed to reproduce the original image. Compression ratios of up to 100:1 are possible. JPEG coded chips have been available for some time and were used as an interim solution for motion video, awaiting acceptance of MPEG standards.

The MPEG-1 standard was approved in 1992. Designed to provide a three-to-one improvement over JPEG compression ratios, MPEG compression requires (and utilizes) prediction of upcoming motion. The June 1991 *Videography* (pp. 38–41) noted that the chair of the MPEG working group cited nine important features in the development of the system: random access, fast-forward/reverse searches, reverse playback, audio-video synchronization, robustness to errors, minimal coding/decoding delay, editability, flexibility of image size and aspect ratio, and affordable real-time decoding.

In terms of television video, these compression techniques have been translated into a number of different processes by a number of companies. General Instrument (GI), the company that created the VideoCipher scrambling technique, developed a DigiCipher digital compression process. It was GI's DigiCipher technique that caused the shift in proposed HDTV standards from analog to digital. Compression Labs Inc. developed a similar system called SpectrumSaver, and Scientific-Atlanta developed a slightly different process. Multiplexing and source-coding based on MPEG-2 standards *could* become interoperable for terrestrial and satellite broadcasts, may be derailed by noncompatible encryption, modulation, etc.

These digital video compression techniques make it possible for multiple full-motion television signals to be transmitted by a single satellite transponder. The compression ratios vary by technique, but they average from 4:1 to 10:1. (The ratios also vary depending on the signal being transmitted. Film can be compressed at a greater ratio than unpredictable live action. Ratios of 8:1 and 10:1 can be realized for transmitting film via DBS. Multiply the 256 DBS channels we spoke of earlier, multiply by ten and you have 2,560 channels. *Now* transmitting film starting on viewer demand seems practical.) See Figures 1 through 3 in the Glossary for examples of compressed video.

Digital compression techniques solved the sending part of the transmission. How about the other half of transmission—receiving? As noted earlier, the price and size of satellite receiving dishes for Ku-Band DBS has shrunk from thousands to hundreds of dollars. If all of this technology is in place, why isn't every television home receiving DBS telecasts? That's another story.

DBS: Around the World

As noted, the world's first DBS system was a Canadian system aimed at the USA, in 1982. Financial problems killed that system as they had halted Comsat's earlier plans to enter the DBS field. Part of the reason for the slow development of DBS has been the pace of the technology. Digital compression systems did not become practical until 1991.

But DBS did not wait on compression technology. While plans for DBS in the USA faltered, other countries moved ahead, however hesitantly, and today there are substantial satellite-delivered programs in the UK (BSkyB), Germany (RTL-Plus), Italy (Telpiu), Scandinavia (TV-3), and Japan (NHK and JSB). But the growth of DBS hasn't been easy anywhere.

DBS in the UK

Probably the most dramatic contrast to the USA lag in DBS systems is in the UK. Rupert Murdoch's News Corporation launched Sky TV in February 1989. By July 1990 Sky TV was serving 1.5 million customers, a combination of cable and dish owners. (The UK had an estimated 21 million TV households at that time.)

Thirteen months after Sky TV was launched, a rival system began in the UK. British Satellite Broadcasting (BSB) was a government-sponsored DBS system with a series of backers led by Granada TV. Other investors included Pearson, Reed International, the French communications group Chargeurs, and the Bond Corporation of Australia.

As a competitor, BSB worked under a double handicap. First, because of problems with its home receiving equipment, it got started a year after Sky TV. Perhaps more important, under orders from the UK regulatory body, the Independent Television Commission, and the EC, BSB was required to broadcast using a new standard called D-MAC, more advanced than PAL's 625 lines and less advanced than MUSE–HDTV's 1,125 lines. D-MAC was not compatible with any of the then-current equipment in the UK. At $575 each, BSB managed to sell only 120 thousand units.

Meantime, Sky TV found that even with a slightly less expensive home unit ($450), subscriber numbers were lagging, so Sky TV began to rent the units. Business boomed, but not enough. The two companies battled until November 1990 when to cut their losses, they merged and became British Sky Broadcasting (BSkyB).

The new company dropped D-MAC, putting it into direct conflict with the EC and its plan to develop a strong competitor to Japan's HDTV. BSkyB used PAL on the ASTRA satellite, which Sky TV had been using. Sale of the D-MAC units was stopped. Despite having placed 1.2 million satellite dishes, BSkyB lost £48 million (USA$81 million) in fiscal 1991. But by the end of 1992 BSkyB was feeding more than three million households by satellite, cable, and SMATV—and was posting steady profits. Estimates place growth in UK DBS at close to 100,000 dishes per month. One issue hovering over BSkyB's future is the EC's plan to mandate transmission via D-MAC and HD-MAC.

This brief history shows that by the time K-Prime/Primestar Partners was launched in October 1990, the UK had already gone through the launching of two companies, testing of the market, consolidation and regrouping, and formation of a merged company. K-Prime/Primestar Partners was formed by a coalition of companies including GE and nine major cable multiple system owners (MSOs). Their test market was in thirty-eight US cities with ten channels aimed at both cable and noncable subscribers. Their home equipment was priced at $550.

Digital Video Compression and DBS: Convenience

Perhaps you noted in the list of backers that there were nine cable companies. Obviously, they saw the real possibility of head-to-head competition between cable and DBS—or the possibility of DBS expanding the reach of cable. Either way, they weren't about to be shut out of direct-to-home broadcasting.

Neither were some other people. Early in 1991, a USA company based in Seattle, proposed SkyPix, a DBS system offering eighty channels. This system, owned by Northwest Starscam, planned to offer so many channels because of the perfection of the technique of digital video compression. By compressing the video, SkyPix—and all that followed—could compress the signal and make one transponder do the work of many. SkyPix pegged the cost of its home-receiving equipment at between $699 and $799 for a dish ranging in size from thirty to forty-eight inches. SkyPix planned what research showed audiences wanted—the convenience of movies they want when they want them. These plans were set back in 1992 when Sky-Pix entered Chapter 11 reorganization, delaying entry into the market to a time that would put SkyPix in direct competition with Hughes Communication's planned Direct TV DBS.

Hughes Communication had been part of an earlier group scheduling DBS. In 1991 a coalition of News Corporation, NBC, Cablevision Systems, and Hughes Communication announced 108 Sky Cable channels to be activated in 1993. Those plans collapsed, but others, like United States Satellite Broadcasting Corporation (USSB) and Hubbard Broadcasting, were waiting in the wings.

In a February 1992 interview with *Satellite Communication* (pp. 14–17), Stanley S. Hubbard, chair of USSB repeated his prediction that USSB's commitment to spend $100 million-plus to buy transponder space on satellites being built by Hughes Space and Communications and another $100 million for a startup will result, finally, in truly successful USA DBS.

Hubbard's prediction is that the combination of a small dish (under twenty-four inches), digital compression, the security of digital scrambling of the signal, and the ability of DBS to transmit HDTV without the added costs that will face terrestrial broadcasters will make the Hubbard-Hughes combination a winner.

Hubbard's prediction echoes others who have prophesied that DBS will outdistance cable systems, which have already begun expanding their use of digital compression techniques. Hubbard thinks DBS will triumph because cable can't go everywhere and, if it could, the cost of fiber-to-the-curb would remain prohibitive. The price of the dishes for the Hubbard/Hughes DBS is scheduled to be under $700.

Perhaps the biggest potential boost for DBS operators came from developments and announcements by equipment manufacturers. In 1990 Matsushita introduced a VCR equipped with a *built-in* satellite tuner. The 1990 version cost USA$1,000. Toshiba's TRX-700 IRD (integrated receiver/descrambler) for home satellite dishes contains built-in PPV ordering capability. In February 1991 RCA announced that starting in 1994 its television sets will be available with built-in DBS antennas. Sony and other manufacturers have predicted that before 1995 large-screen television sets (more than twenty-two inches wide) will come equipped with built-in satellite tuners.

DBS in the Pacific Theater

In other countries, development patterns have varied. Japan has been involved in a sort of double competition: cable versus DBS and DBS versus DBS. At the start of 1991 more than 7.5 million of Japan's estimated 40 million TV households were wired for cable. DBS had fewer viewers but was coming on strong in the face of two competing DBS systems. Japan's semipublic NHK (Japan Broadcasting Corporation) began operating two DBS channels in 1989, in addition to its previously well-established over-the-air channels. NHK's DBS system had signed up more than 2.5 million subscribers before it faced competition in April 1991 from a new DBS system, WOWOW, owned by Japanese Satellite Broadcasting, whose principal investor is the Mitsubishi group. Japan Satellite Broadcasting ran into a major obstacle almost immediately when the satellite scheduled to transmit the system's three channels had to be destroyed before it got into orbit. The satellite was launched April 18, 1991, from Florida but developed engine trouble and was destroyed by technicians just a few minutes into flight. Despite this setback, by August 1992 WOWOW Home Theater had signed up more than one million subscribers.

In Australia, geography played a large part in the start-up of DBS. The population in the Outback is so thin and widely scattered it would be virtually impossible for them to be served by cable. Since the mid-1980s, however, three Aussat satellites have overcome the vast distances with DBS delivery of the regular over-the-air programs of the Australian Broadcasting Company seen in the country's urban areas. In the cities, Bond Media's Sky Channel served commercial subscribers such as pubs, as opposed to private homes.

DBS in Europe

In Europe the situation has varied from country to country, with the continued, if somewhat shaky, advance of the European Community promising an eventual, but very long-term, evening out. The satellites that served the UK, ASTRA 1A and 1B, also served most of the rest of the EC. Individual countries, however, have supplementary satellites. By the start of 1991, availability was as follows.

The united Germany had an estimated 29 million-plus television households (26.6 million in former West Germany and 2.4 million in former East Germany). Cable outran DBS substantially, with almost eight million cabled homes and less than a million home dishes. By 1993 the margin had narrowed with cable's 11 million doubling DBS's 5.5 million. German-operated satellites included Kopernicus and TV-SAT-2. Neighboring Austria, smaller in size, had an estimated three million-plus TV households.

In France, of an estimated 24 million television households, only 600 thousand were cable subscribers. Canal Plus, a pay-TV service, urged the country's three major cable operators (Com-Dev, Générale des Eaux, Lyonnaise-Communications) to join in pushing sales of antenna dishes for the scheduled 1992 launch of the Telecom 2A satellite. Despite the collapse of British Satellite Broadcasting's attempt to transmit in D-MAC, Canal Plus insisted it was committed to the D2-MAC standard. Because of the EC's rules, all channels using the high-powered French TDF satellite were under orders to use D2-MAC, at least until further notice. (A satellite, of course, will send back whatever signal is uplinked to it, whether it's NTSC, PAL, SECAM, DMAC, or HDTV.) France's DBS future was further clouded by Canal Plus's financial problems in late 1991. By 1993, while cable homes expanded to more than 900,000, there were reportedly fewer than 150,000 dish owners in France.

After the UK and Germany, the Netherlands is the third largest market for ASTRA DBS feeds, with an estimated 130 thousand antenna dishes in use.

Italy had virtually no cable penetration but increasing satellite transmission by RAI using the Olympus DBS.

Scandinavia, on the other hand, was highly cabled, with satellite services distributed via cable companies. An estimated 100 thousand dishes had been sold in Scandinavia, largely in Sweden and Norway.

Spain reportedly had no cable systems operating. The coverage (footprint) of ASTRA 1B favored Spain and Portugal more than had ASTRA 1A, so there was a possibility of an increase in sales of dishes on the Iberian peninsula. Those hopes were spurred when Spain's first DBS satellite was launched in September 1992. No national legislation on either DBS or cable existed in Spain as this book went to press.

Television in Czechoslovakia was state-controlled. There was virtually no cable penetration, but there were reportedly some 400 thousand VCRs in use. Late in 1991 privatization of television had begun in Czechoslovakia.

Faced with political unrest and deficit problems surpassing even those of the USA, before its breakup the USSR announced long-range plans to launch a series of up to nine satellites to create a nationwide DBS system. CIS has an estimated 75 million TV households. The breakup of the USSR freed all of the Republics except Russia from the burden of contending with eleven time zones. Russia's satellite leasing went co-op: with Belgium, Combelga; with UK: Constar and World Trade Telcom; with Germany: Santa International. SovCanStar, with Canada, planned a Clarke-orbit sat in 1996.

DBS Systems in Other Countries

In India, as in Pakistan, all broadcasting was government-controlled. Satellite transmission to antenna dishes in villages remained in the experimental stage. By 1992, however, there was movement to loosen state control over TV, especially DBS.

African nations had very active multilingual TV broadcasts, but little cable or DBS. Their broadcast systems were a mix of PAL, SECAM, and even some NTSC. Almost all nations were Intelsat signatories.

Mexico had an estimated 13 million TV households, almost 600 thousand of which had cable. Relatively low income levels slowed the growth of cable and the sale of DBS dishes.

Canada's estimated 12.5 million TV households outpaced even the USA with cable penetration. By 1991 the USA had close to 60 percent while Canada led the way with more than 80 percent penetration. As noted above, with the USCI pioneering in the USA, Canada led the way to DBS with its ANIK satellites.

DBS Versus Cable: Racing Against Time and Technology

However the DBS industry ultimately shakes down, there is no question that DBS will influence cable distribution and vice versa. In the Pacific Rim, the mere threat of DBS caused a five-nation consortium to cancel plans to launch potentially the world's largest cable system. Hong Kong Cable Communications had requested a multiyear ban on competing technologies (DBS), but the government refused to impose the restriction. With a possible 1.5 million households to be signed, the stakes were high, but the cable consortium backed off, faced with the threat of the instant start-up that a DBS system poses. After the Hong Kong Cable project folded, the largest shareholder submitted a new bid to the government's Cable and Satellite Coordination Group, this time for a pay TV license for a service that would combine cable and wireless transmission.

We noted that in the USA the "Baby Bells" have invested heavily in overseas cable companies, largely because in countries outside the USA they have been allowed to integrate data transmission into their systems. The combined concepts of convergence and convenience have convinced the telecom companies that the shift to digital transmission will lead to ultimate transmission of voice/video and data simultaneously. AT&T's Videophone, announced in 1992, although it offered only ten frames per second because fiber didn't go to the curb yet, was another reminder of that. Both politics and technology will influence the rate and the kind of progress that will eventually make two-way interactive global TV standard practice.

The Baby Bells were not the only ones who foresaw the convergence of audio and video via phone lines. Cable companies, broadcasting companies, and newspapers all considered the telephone companies' wired access to almost all homes to be an unfair advantage. A flurry of legal and regulatory decisions left the status of US regional phone companies confused.

On July 25, 1991, the same judge who wrote the 1982 consent decree barring phone companies from either manufacturing equipment or transmitting informational services via their phone lines reversed it. That reversal, however, did not undo the Cable Act of 1984, which, by Congressional rather than judicial decree, prohibited regional companies from using their wired systems to provide cable TV service. In October 1991, however, came an FCC proposal to allow them to do just that. A continued restriction was placed on Baby Bells' *producing* programming; they could only carry others' programs. But this proposal, called *video dial tone service,* would allow the phone companies to carry news, which could eventually make them more of a threat to newspaper publishers than to cable companies. The US Telephone Association (USTA), representing independent phone companies, began lobbying actively to remove the 1984 restriction on ownership of cable program content.

In mid-1993 Bell Atlantic did two things: it sued to overturn the 1992 cable act, which upheld many of the 1984 restrictions preventing telecos from becoming programmers, and used its employees in Virginia to give a public demonstration of its video-on-demand (VOD), Interactive Multimedia Television (IMTV), proving that existing copper phone lines could deliver phone calls without interrupting receipt of film video. However, IMTV let only one TV set at a time receive video—still shy of global TV's need for full two-way fiber. Both phone and cable companies raced to install fiber-to-curb, a costly job.

Fiber-to-Curb Aims Toward Two-Way Television

The real impetus that unleased the electronic rush into our fibered future came early in 1991 when Time Warner, operating five cable systems in New York City, announced plans to double the existing seventy-five channels in its franchises in the boroughs of Brooklyn and Queens by rewiring with fiber optic cable. This rewiring not only increased the system's channel capacity and opened up the possibilities of HDTV and interactive TV, it also shocked all competing businesses. Time Warner's move began to put the convergence/convenience theories discussed in this book into practical action. Competitors stampeded to follow and get a piece of the action. Mergers begat mergers until corporate lines between phone/cable/TV/computer/film/music companies disappeared.

This is just a sampling: In 1993 Time Warner followed its own lead, began installing fiber to its 4,000+ cable subscribers in Orlando, Florida, targeting 1994 to offer 500 two-way interactive channels. Subscribers' converter boxes, in essence computers, put a Silicon Graphics Inc. processor in boxes made by Scientific-Atlanta and Toshiba Corp. Designed to MPEG-1 digital compression standards, the boxes are upgradable to MPEG-2 or other future standards. But note that Scientific-Atlanta also teamed with Motorola Corp. to produce a set-top box with software from Kalieda Corp., itself a joint venture of Apple Computer Inc. and IBM. Also in 1993, Time Warner, the nation's number two cable operator, teamed with the nation's number three regional phone company, US West. US West invested USA$2.5 billion for 25% of Time Warner Entertainment.

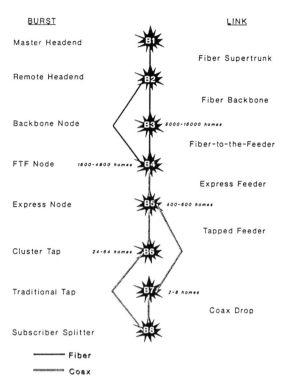

STARBURST ARCHITECTURE
Cable TV System Bursts

BURST   LINK

Master Headend   B1

                                    Fiber Supertrunk

Remote Headend   B2

                                    Fiber Backbone

Backbone Node   B3   6000-15000 homes

                                    Fiber-to-the-Feeder

FTF Node   1800-4800 homes   B4

                                    Express Feeder

Express Node   400-600 homes   B5

                                    Tapped Feeder

Cluster Tap   24-64 homes   B6

Traditional Tap   B7   2-8 homes

                                    Coax Drop

Subscriber Splitter   B8

——— Fiber
∙∙∙∙∙∙∙∙∙ Coax

Figure 14–1 Jerrold Communications' Starburst is a fiber–coaxial cable system. As their diagram shows, from the master headend (B(Burst) 1) through to the B-4 node, the signal is carried on fiber optic cable. (Fiber-to-Feeder; FTF). From that point the current system carries the signal via coaxial cable. A *backbone* is a transmission facility that connects dispersal users or devices. Starburst is upgradable, including for digital transmission. The system's growth plan assumes that HDTV technology will be used first for premium services, that simulcast (as planned by the FCC in the USA) will prevail in the near term, and that all-digital HDTV transmission will become the norm. The provisions are similar to those for compressed digital video and PCN/telephony services. In the UK duplex splitters already allow B8 to feed telephones.

Courtesy Jerrold Communications.

(Time Warner owns 63.5%; US West owns 25%; C. Itoh and Toshiba own 11.5%.) Further, US West was also teamed with Tele-Communications Inc. to test Video-on-Demand in the Denver area and to offer cable/phone service in the U.K.

Meanwhile, Time Warner bought 50% of Metro-Comm, builder/operator of a fiber phone network in Columbus, Ohio, forming a new company, MetroComm AxS. Further, Time Warner met with Tele-Communications Inc., met with Microsoft about forming Cablesoft to manufacture interactive software. Earlier Microsoft teamed with Intel Corp. (chip) and General Instrument (box) to make interactive set-top boxes. General Instrument and AT&T also developed the digital boxes Tele-Communications Inc. used to install interactive service to some of its ten million cable subscribers. With News Corp., Tele-Communications owns fifty percent of Request TV (PPV). Finally, 3DO (backed by both Time Warner and MCA) viewed its interactive multiplayer as the key to interactive TV. Virtually all major firms have so cross-linked themselves they have to be on the winning side. If they're lucky—and interoperatively compatible—maybe all sides will win.

Another sign of the power of digital compression, and of media convergence, was Barry Diller's investment, via Arrow Investments Inc., in QVC Network Inc. Based in West Chester, Pennsylvania, the QVC (Quality, Value, Convenience) cable channel is currently in more than forty million homes. Reportedly Diller got agreements with QVC's largest shareholders (Liberty Media Corp [21 percent] and Comcast Corp [14 percent]) to vote with his stock, estimated below 5 percent. Liberty Media's chairman is also president of TCI. The TCI move to digitally compressed expansion was unquestionably a factor in Diller's decision to invest in QVC and seek the post of CEO. His entertainment background with ABC, Paramount Pictures Corp., and Fox Inc. brings formidable programming skills to what is reportedly the fastest growing home shopping cable network. (TV Macy's got Don Hewitt as consultant.) Diller's investment, Liberty Media's move to buy a controlling share of the Home Shopping Network, and TCI's interest in Request TV presage combined entertainment, merchandising, and interactivity with viewers.

These companies are not alone. According to David E. Robinson, director of cableoptics for Jerrold Communications, "virtually every [cable TV] operator budgeting new, rebuild, or upgrade construction for the early 1990s now incorporates fiber optics." The reason is the decrease in the cost of fiber optics plus the money-saving option of combing fiber optic cable with coaxial copper cable. The traditional tree-and-branch design of coax cabling requires a large number of electronic amplifiers to sustain broad-band video services. The natural urge was to go to all fiber cabling, but the cost is prohibitive: the cost of the main branch isn't great, but getting from the trunk or larger branches to the curb, "the last mile," is very expensive. The solution is to combine fiber with coax, using fiber for almost everything but the last mile to each individual subscriber. Jerrold Communication's Starburst architecture (see Figure 14–1) shows the combination of fiber-to-feeder and coax from the feeder out. Plant improvements like these are accelerating because "amplitude modulation (AM) fiber optic equipment has improved more than 800 percent on a price/performance basis during the last two years [1989–1991]." The best design depends on each individual cable company's existing plant, its geography, subscriber density, marketing strategy, and, of course, budget. "From a global perspective," adds Robinson, "each country's communications regulatory policies, programming availability and local construction costs may lead to somewhat different architectures, even though the 'parts' costs are roughly the same." He agrees with most other experts that fiber-to-the-home will not occur during the 1990s. But systems like Starburst, combining fiber and coax cables, carry provisions for easy upgrade to HDTV, compressed digital video, two-way interactive communications, and narrowcasting *and* personal communication network (PCN) telephony service.

AT&T's similar fiber-to-curb design is called an active double star. The first of the two stars carries signals through fiber from the central office to a remote terminal. The terminal, the second star, routes the traffic through more fiber to curbside pedestals. AT&T is also heavily researching—and applying—analog signal compression to make copper-to-curb support greater signal/interactive capacity.

Equipment supplier Raynet has a system reportedly less expensive than AT&T's, offering a choice of techniques. It can provide a bus system (primary pathway) going by fiber from the central office directly to one or more remote terminals. The bus supports a series of curbside pedestals without any intervening branches. The alternative is to introduce a splitter, either at the central office or anywhere along the bus. As its name implies, the splitter splits the information and relays it—exactly as received—to numerous points, but in a weakened condition that requires more sensitive receivers.

Other companies in the USA and overseas provide similar designs, all providing a transition between the past and the ultimate all-fiber-to-home systems. The technical equipment exists and is increasingly in place for convergence. Planning globalcasts in the future will be less limited by physical forces and more limited only by imagination and budget.

Interactive globalcasting may well get help from even more advanced technology than that available with existing fiber optics. No technology stands still, and the transitional nature of the 1990s will make changes come faster and be less predictable than in the past. We will increasingly integrate electronics and photonics, mixing laser light with semiconductors to increase the speed and capacity of information transmission. Spin transistors (see Glossary) could shrink satellite size from feet to inches.

In Part 5 we discuss briefly the photonic computer being developed by Bell Laboratories and others. Adding the speed of light to computer calculations may produce unimaginable developments. Similarly, fiber optic technology may be advanced dramatically by photonic technology.

Bell Labs is working on optical amplifiers that may replace the current electronic/optical amplifiers used to repeat the signal on fiber. Once they detect a weakened laser, the current amplifiers have to convert the light pulses to electronic impulses to amplify them. The optical amplifiers being tested use "rare earth-doped" fiber and do not have to convert the signal. According to David Robinson of Jerrold Communications, a "co-propagating pump laser" excites the fiber dopant's ions to a higher-energy state. "Through the natural relaxation process, photons are generated (and) combine with incoming photons from the system's source laser, resulting in optical gain." That is only one of many ongoing experiments.

Another futuristic view, already performed in the laboratory, sees a *single* optic fiber able to handle 150 thousand two-way phone calls. Under these conditions, a fiber cable with 144 fibers could handle 21.6 million two-way telephone calls. "The above demonstration," say Vyvx statisticians, "is equivalent to 444 broadcast quality video channels on *one fiber.*"

In still another fiber optic advance, involving color, a Tucson, Arizona, company (Isotec) announced development of "smart glass" devices to

STARBURST EXPRESS FEEDING
2,400 Home FTF Node

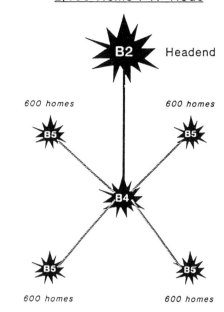

———— Fiber-to-the-Feeder Trunk

············· Coax Express Feeder

Figure 14–1 *(continued)*

expand fiber's signal-carrying capacity. *Popular Science* (March 1991, pp. 64–67, 95) reported that one Isotec device can combine light beams of different color or wavelength and beam them as a single signal through a single fiber to a receiver that breaks the beam back into its original colors. Because it's the frequency of light that determines color in the visible light band, the multiple-frequency signals sent through the fiber could be tuned to, just as you tune to a radio station on a dial. Using color frequencies, a single fiber could simultaneously carry HDTV signals, telephone signals, computer data, and so on. Also being considered is the use of frequencies in the infrared range.

Researchers are constantly testing new methods of transmitting multiple wavelengths via laser light signals. Plastic fiber, easier to work with than glass, is another focus of experimentation. Currently its carrying capacity only matches that of copper wire (ten to sixteen megabits per second), but tests continue. Other researchers are investigating fiber constructed from metal fibers that offer the potential of transmitting signals up to 400 kilometers without repeaters.

In January 1991, the FCC gave preliminary approval to a plan that may make all of the preceding outdated. That form of interactive TV does away with the need for cables of any kind. Tested in Fairfax County, Virginia, in 1990, the system uses two broadcast channels of the radio spectrum. The TV viewer is connected with the broadcaster through a network of small radio towers—no cables. Each customer has a small radio transmitter/ receiver, about the size of a VCR or cable-TV converter, operated with a hand-held remote not unlike a Star Trek ray gun. The system sends frequency microbursts of digitally encoded data back to the signal source for processing by computers. While this system doesn't currently allow viewers to select camera angles during telecasts, like ACTV and some other systems we'll encounter in chapter 20, it does provide interactivity for responding to on-screen questions, ordering PPV movies/specials, and shopping at home, banking, and so on. The FCC notice of proposed rulemaking for the 1/2 megahertz of broadcast spectrum could create a great opportunity for TV Answer Inc., the small Reston, Virginia, company that petitioned the FCC for the spectrum space and the only firm to date to have developed such a radio-based system.

And a development that could catapult the convenience factor to the foreground for any system delivering hundreds of channels came in 1993. Voice Powered Technology had been available earlier for programming VCRs, but the June 1993 announcement of the newest feature of Apple Computer's EZTV (from its Personal Interactive Electronics division) really foreshadowed the convenience future: Casper, a voice-recognition system that allows consumers/viewers to call up any item from an on-screen menu—by voice. Want a program? Want a product? Want a show recorded? Goodbye "computer." Goodbye "remote." Just shout it out. In place for 1994's limited explosion to 500-channel capacity, Casper may have to be redubbed "Genie," replacing keyboards and remote buttons, granting almost unlimited voice-wishes to digitally linked viewers worldwide.

Whatever combinations the future holds—fantastic new fiber or cableless radio spectrum or long-life satellites with vastly increased capacity through digital compression—it's certain that the future will be fast in arriving and exciting, with built-in interactivity. The pace will be spurred by President Clinton and Vice President Gore's following through on their campaign pledge to seek funding (USA$5 billion to be spent over four years) to advance an electronic "information highway," basically an extension of Internet, the nationwide computer information network that links educational, business, and governmental agencies (see Glossary). As these fiber-connected interactive communications links expand, those of us interested in globalcasting will, as always, have to monitor the new developments to evaluate how they can enhance global communication.

Let me reiterate the theme behind all of our discussion of techniques and equipment: the tools of transmission take on meaning only if they carry something meaningful. Transmitting gibberish globally over optical fibers linked with satellite transponders is no more meaningful than gibberish transmitted locally. GIGO is a favorite computer term that we should make part of the convergence of the media. Garbage In, Garbage Out. It's your mind and imagination that have to be sparked by the opportunities the converging media are making possible. Not even the fastest, smartest computer has yet matched Shakespeare or Stravinsky or Michelangelo or Picasso or Costa-Gravas. No image-processing machine has yet brought together the creative process from shooting to editing manifested in films like those of Federico Fellini or Ingmar Bergman or Akira Kurosawa or Martin Scorcese or Francis Ford Coppola.

Chapter 15

Broadcast Television, Media Convergence, and Display Standards

Existing TV Standards

History, the passage of time, economics, and national pride created an international mish-mash of standards for the transmission of television signals—NTSC, PAL, SECAM, and HDTV. To understand the standards, you have to understand the scanning processes they represent.

We've all seen and handled photographs, still pictures. We know that motion picture film strings together a lot of still pictures and projects them at twenty-four frames per second, fast enough to trick the human eye into believing it is seeing uninterrupted motion. The phenomenon called visual retention makes this possible.

You can hold motion picture film up to light and see the still-frame images. Because television signals involve light waves—photons—that have only energy, but no mass or charge, you can't hold or see them the way you can hold and see a frame of film. But the basis of the image-processing is the same. Television cameras transmit enough images per second to trick the eye into believing it sees motion. Originally this was done using a scanning process suggested by G. R. Carey and a mechanical scanning device patented by Paul Nipkow. In the 1920s electronic scanning created the basic system still used to this day.

Electronic Scanning

In black-and-white television, a light-sensitive tube, a form of cathode ray tube (CRT), in a TV camera reacts to the different shades of light and dark light waves it receives from the scene at which it is pointed, and it turns those light waves into electronic signals that can be transmitted. The television camera does not record the scene all at once, the way a still-camera does but rather scans the picture. The basic process used in most of the current standards is as follows: The camera lens focuses on the image. The light waves of that image—the photons—pass through a conductive layer of the tube, which passes those photons to a grid-like photoconductive layer. This grid contains photo-sensitive dots called pixels (for picture elements). The pixels are arranged in a series of lines, with hundreds of pixels per line. Another part of the camera, the electron gun, aims its electronic beam through a ring of deflection magnets, which deflect the electron beam into a scanning pattern that goes both back and forth and up and down. It is the nature of that scanning pattern that determines the transmission standard. There are differences in the way the images are recorded using charge-coupled devices (CCDs), but the end result is the same as using an orthicon or vidicon tube.

Using the NTSC standard as an example, the electron beam in the camera scans 525 lines of pixels, thirty times per second. Every one-thirtieth of a second there is a new 525-line image. In the NTSC standard the beam does not scan the lines in sequence—line one, line two, line three and so on—it scans alternate lines.

The electron beam scans line one, then is turned off by a *blanking pulse,* to create a *blanking interval,* discussed later. The beam is deflected back to the beginning of the lines and simultaneously deflected down to line three. It scans all of the odd-numbered lines, then goes to the beginning and scans all of the even-numbered lines. Each of these scanning patterns is called a *field.* When the two fields—odd and even—are combined, they are said to be *interlaced.* With the NTSC standard, therefore, there are 525 lines scanned at a rate of thirty frames per second. Or if you are talking *fields,* there are sixty fields scanned per second, two fields for each frame. This interlaced scanning becomes important when you begin to compare NTSC or PAL or SECAM with *non*-interlaced systems, called *progressive scanning,* which scans all of the lines in sequence. (See *fields* in the Glossary for a discussion of standards conversion and *VIRS* and *VITC* for use of blanking intervals.)

The electronic information gathered every one-thirtieth of a second is transmitted—either via radio waves, a wired cable system, or satellite—to the television receiving set or monitor, where the image is recreated, line by line, on the screen of the CRT. For color television the process is similar, but there are three tubes that are red-, green-, and blue-sensitive (RGB). (When CCDs are used, there are *triades,* each containing a red, green, and blue pixel.) Because color is involved, the signal carries both *luminance* signals, reflecting the level of brightness, and *chrominance* signals that carry the chroma or color. In a black-and-white (monochrome) receiver, it is the luminance signal that produces the picture.

All of the current transmission standards use some variation of this scanning technique.

NTSC

NTSC, adopted in the USA in 1941, was the first of the three commonly used standards. Because of conflicting systems, the FCC suspended commercial television broadcasting in the USA and appointed a committee—the National Television Standards Committee—to select a standard. The committee selected the system proposed by RCA, but increased the number of scan lines from 441 to 525. It was redesigned later, to accommodate broadcasting color TV. In North America, most of South America, and Japan, the NTSC standard continues to be the accepted standard.

PAL and SECAM

The two systems used in Europe and many African nations, PAL (Phase Alteration by Line) and SECAM (Système Électronique Couleur Avec Mémoire) were adopted after NTSC and were refined to provide a sharper TV image. Both scan 625 lines, at a rate of twenty-five frames per second. In general, PAL has been the approved standard in Australia, England, Germany, Spain, and more than thirty other countries (see the Appendix).

SECAM has been the standard in France, the USSR, and many Eastern European countries. SECAM cannot accommodate some digital special effects that the PAL and NTSC standards can use.

The world has managed to live with these three standards. Converters can be purchased to change signals from one standard to another, and some VCRs have built-in converters. Edit suites and postproduction houses have sophisticated equipment to handle any mixture of the three standards that you may encounter.

And then came HDTV.

HDTV Standards

While no universal global standard has emerged or appears about to emerge, there are four characteristics common to most of the proposed HDTV standards:

* A picture ratio of 16:9 (width to height) as opposed to 4:3 for NTSC, PAL, and SECAM;
* More than 1,000 scan lines (except Zenith's Spectrum Compatible and MIT's ATVA, with 787.5 lines in progressive scan);
* Digital sound;
* Either expanded bandwidth requirements (Japan Satellite Broadcasting uses 8.1 megahertz) or a digitally compressed signal to fit the current terrestrial NTSC bandwidth of 6 megahertz.

The reason for the universal acceptance of the concept of HDTV is readily evident the first time you see it: reception on a home set is as clear as that on a monitor in the control room. The sharpness of Japan's 1125-line picture has been compared to that of 35-millimeter film. And some of HDTV's most important advocates are in film. As early as 1982 Francis Ford Coppola and Glen Larson were involved in experimental productions and demonstrations of NHK's 1125-line system. The Society of Motion Picture and Television Engineers (SMPTE) has also indicated support for the 1125-line standard, although some withhold judgment. Even though the NHK standard could be easily converted to any broadcast format, many feel the production and broadcast standard should be the same.

The interest of film producers and technicians in HDTV is logical on two levels. First, HDTV is seen by many as either a cost-saving replacement for film or a side-by-side alternative for many projects. Second, because they have a picture ratio of 16:9, television films and series recorded on film (rather than tape) are immediate candidates for programming for HDTV transmission. Film distributors are interested, too. While film prints have to be shipped, HDTV motion pictures could be transmitted worldwide by sending scrambled signals to exhibitors, cutting distribution costs dramatically.

But these general characteristics—more lines, improved picture ratio, and digital sound—are the only points of agreement. The rest is competitive, argumentative battle—in the research rooms, in the board rooms, in the marketplace.

Needed: A United Nations for Television Standards

If you've ever been in New York City, you've seen the guy on the corner with a pushcart full of "Gucci" handbags and "Cartier" watches, saying "Hey, have I got a deal for you!" Well, that comes frighteningly close to summing up the situation with new television standards. The

Japanese set up the curb first, in the 1980s, with analog HDTV. "It has 1125 lines! Picture as sharp as 35-millimeter film! Try it!" "Wait a minute!" yell France and the EC. "We've got a better buy—more lines—1250! Buy our D2-MAC!" "Hold on!" screams the USA's FCC. "Don't buy from those guys! They've got shoddy merchandise! They can only send out their HDTV signals by satellite. And some of theirs aren't compatible with existing sets. Buy theirs and what happens to your local eleven o'clock news? What happens to "Wayne's World"? Buy ours and you get it all. We're compatible! Of course, we won't have ours ready until after 1993." Just as you're ready to throw up your hands, you hear "Allow me to introduce myself. I've invented a number of advanced techniques in video. And I'd like to suggest your consideration of Super NTSC. Available now! Sets only cost $300 more than your present set. And if you don't have an adapter in your set, you still get the picture on your old TV. Think Super NTSC."

What can we do to get out of this maze? Let's analyze the situation.

MAC and Analog HDTV While there are a number of proposed advanced television standards, until 1993 the two standards with which globalcasters had to deal were the EC's D2-MAC and Japan's MUSE-HDTV, both analog transmission systems.

D2-MAC was established as an interim standard leading to high definition. MUSE-HDTV was established as a final product, but the expectation of a USA digital standard has put MUSE-HDTV into questionable perspective as a broadcast standard. As you will see, it is already firmly entrenched in a number of areas outside of broadcasting.

The MAC Standards: D-MAC, D2-MAC, HD-MAC The MAC standards have 1250 scan lines, scan time of fifty frames per second, progressive scanning, analog signal, and distribution by satellite. The EC's D-MAC standard was envisioned as a series of compatible transitions, from existing systems to a totally high-definition standard, HD-MAC, to be in place by 1995–1996. Virtually all of the nations in the EC supported the concept of a European HDTV standard, but they disagreed on how to reach their goal. The conflict between the UK and the EC concerning the mandated use of D2-MAC for satellite transmission raged well into 1993. In June 1993, reacting to the USA's decision to regear its three contending HDTV standards into a single digital standard, the EC shifted into high gear on its HDTV plans. Philips Electronics, Thomson Consumer Electronics, Germany's ARD and the British Ministry for Trade/Industry joined other EC interests to form a group to coordinate a shift from HD-MAC's analog signal to digital.

A third voice—the equipment manufacturers—entered the fray. With millions invested in research and development, companies arrived in 1991 at the point of offering home receivers for sale. Needless to say, the manufacturers were in favor of everyone's using D2-MAC on the way to HD-MAC as a spur to the sale of new TV sets. And, of course, advocates of the D2-MAC standard saw the sale of home sets as a spur to production and acceptance of the standard. But like the adapters for existing sets, the new model sets were expensive.

The first Thomson sets put on sale, dubbed "Space System," were aimed at the top of the market and cost about 35,000 French francs (USA$7,000, £3,500). Less expensive sets were released by the end of 1992.

The Thomson interim HDTV sets had two major advantages over the Japanese sets. One early advantage was the price ($7,000 versus $30,000), partially overcome by Sharp's 1992 price cut. The other major advantage was that—like the proposed USA standard sets—Thomson sets were capable of receiving the interim D2-MAC and the proposed HD-MAC but were also compatible with existing standards *and* offered better reception of those existing standards.

Furthermore, these top-of-the-line thirty-four-inch diagonal sets had a number of future-looking features. They featured the 16:9 aspect ratio screen, of course, but could also use the old 4:3 ratio. When the 4:3 ratio picture was being shown, it could either fill the screen or shrink to allow a picture-in-picture (PIP)-like service, displaying three additional smaller pictures at the same time. The top-of-the-line sets also featured built-in satellite equipment for receiving DBS. And, of course, all of the sets offered one of HDTV's advantages: digital sound.

Non-EC Support for MAC Standards

The EC is not alone in recommending and using the MAC standard. For a number of technical reasons, including its ability to reach a high level of resolution by separating the video components (multiplexing), a number of countries consider MAC a better standard than the 1125/60 standard.

MAC stands for Multiplexed Analog Component. Time multiplex techniques were developed in the UK (by the Independent Broadcast Authority), originally to separate monochrome and color components during transmission and later adapted to separate RGB signals in color transmission, as well as to transmit audio in multiple languages. Currently four varieties of MAC standards exist, one of which, A-MAC, has bandwidth limitations that restrict its use.

B-MAC B-MAC is the standard used at the Captain New York production/transmission center in New York City. It is also used for satellite transmission in Australia and is compatible with Scientific-Atlanta's HDB-MAC, which it says "is designed to deliver a secure high-definition signal, along with six encrypted audio channels, through transmission channels not limited by regulation to 6 megahertz of bandwidth" (i.e., not by terrestrial broadcast). It is, however, "ideally suited to satellite, fiber, coax, or microwave" transmission. Like D2-MAC, HDB-MAC is a compatible standard. Those who have HD receivers get a 16:9 picture, with 1,000 lines of horizontal resolution, while sites that have only B-MAC equipment will receive the same signal, but process instead a 4:3 picture with NTSC resolution. Scientific-Atlanta's B-MAC system uses digital processing to enhance video quality, but converts the signal back to analog before transmission, thus assuring compatibility with existing TV systems. Audio, data, and control information are delivered during the horizontal and vertical blanking intervals.

C-MAC Wide bandwidth limits the use of C-MAC for cable transmission. It is, however, used in Norway for satellite transmission.

D2-MAC D2-MAC, of course, is the standard, developed by the Eureka 95 consortium of manufacturers championed by France and the EC. Delivered a blow by the UK rebellion against mandated use and by pending digital standards, D2-MAC's bandwidth compatibility, high-resolution, and cost-effectiveness still make it a contender in the battle for a single world standard, or one of several world standards.

EC industry is doing its best to promote the D2-MAC standard. Numerous organizations have been formed to stimulate D2-MAC production. Vision 1250 represents twenty-four members and six associate members, including broadcasters, hardware manufacturers, producers, and telco organizations. For an admission fee plus an annual fee, members of Vision 1250 have access to the use of HDTV equipment, including six-camera mobile units, postproduction facilities, transmission/reception equipment. By mid-1991 EC producers using Vision 1250 facilities had reportedly turned out more than seventy major HDTV productions, including Wimbledon tennis, the Football World Cup Final, Tina Turner's London Concert, the London Symphony. They also used the 1250/50 standard at the 1992 Barcelona Olympic, the Winter Games in the French Alps, and World Expo92 at Seville. A special group, Savoie 1250, was set up to coordinate HDTV coverage of the Olympic games. With more that one thousand 16:9 monitors placed for public viewing, Savoie offered ten to twelve hours of HDTV coverage every day for the big screens. There were, however, some complaints from Germany about the technical quality of the satellite feed, including blurred video with quick movements on screen, a complaint sometimes lodged against MUSE-HDTV.

There was dissent in Europe, as in Japan, claiming that a successful digital HDTV system from the USA would render both the 1250/50 and the 1125/60 analog standards obsolete. Even before the EC's digital study group formed, a report released by the French foreign ministry called for EC manufacturers to form a consortium with USA companies to coordinate development of an improved standard.

NHK's MUSE-HDTV The standards for MUSE-HDTV are 1125 scan lines, scan time of sixty frames per second, progressive (noninterlaced) scanning, analog signal, and distribution by satellite. First demonstrated publicly in 1985 at the World's Fair at Tsukuba, Japan, HDTV shifted rapidly from being "an interesting technique" to being "the wave of the future." In 1989 in New York City, when NHK televised *Our Common Future* in MUSE-HDTV, I directed using separate HDTV cameras alongside NTSC cameras. That meant clearing a few more seats to make room for the HDTV equipment and installing more lights for the more sensitive HDTV cameras. Because special receivers are required to receive MUSE-HDTV, the HDTV portions of the program were sent to limited locations—Japan and two public demonstration areas in the USA.

With typical Japanese thoroughness, NHK researched human vision before designing its 1125-line/5:3 (later 16:9) aspect ratio standard. With the 4:3 aspect ratio of pre-HDTV, you have to sit at a distance approximating seven times the height of the television screen to avoid seeing grain in the picture. With HDTV's 16:9 ratio, you can watch from a distance just three times the screen's height. At that distance watching television transmitted by NTSC, PAL, or SECAM, you would see the motion of the lines of pixels as they are being scanned. Double the number to approximately 1050 or 1100, and at the same distance you would not see the scan lines. And so in the 1970s—with the first experimental broadcast in 1978—NHK developed MUSE-HDTV. (MUSE, originally called MUSE-E, stands for Multiple Sub-Nyquist Sampling Encoding. See *sampling* in the Glossary.) Because MUSE uses a bandwidth just over 8 megahertz, greater than the NTSC broadcast bandwidth of 6 megahertz, it is transmitted only by satellite or cable. It also requires unique HDTV camera equipment and can be received only on special TV receivers designed to accept MUSE-HDTV. It is not compatible with existing sets or existing standards. NHK's submission for FCC testing in the USA was called Narrow-MUSE, having been compressed further to fit into the 6 megahertz bandwidth.

The International Battle Over HDTV Standards

MUSE-HDTV came close to being accepted as an international standard but was blocked at a 1986 meeting of CCIR (Comité Consultatif de Radio Communication), an organization that sets international radio/television standards. It was representatives from Europe who blocked the acceptance of 1125/60 as a global standard, even though the EC didn't present its own version of HDTV until two years later, in 1988. One reason was the compatibility issue. Another was money—$144 billion per year, according to the 1988 report from the USA's National Telecommunication and Information Administration (NTIA), which suggested this figure as potential sales of HDTV equipment, including home TV receivers and VCRs, by the end of the 1990s. It is widely believed that it was this CCIR battle that prompted the television industry in the USA to begin thinking about a USA standard other than Japan's 1125/60. The compatibility issue may have been the more powerful for the USA, because only one manufacturer (Zenith) remains active in the global television market.

MUSE-HDTV for Broadcast

The future of Japan's MUSE-HDTV as a broadcast standard is uncertain. If the USA's FCC tests result in a successful USA digital standard, MUSE-HDTV's analog signal could suddenly become antiquated for broadcast use. There have been numerous reports that NHK is readying a digital version of MUSE, but its submission of Narrow MUSE for FCC testing used an analog signal.

The loss of planned satellites slowed the schedule of HDTV broadcasting in Japan, and the cost of HDTV receivers, while diminishing, remained comparatively high in Japan. In 1991 the cost was in the range of three million yen (circa USA$27,000). (USA$17,000 was inexpensive, USA$30,000 was top line.) Only a few hundred sets were sold. In February 1992, Sharp Corporation lowered the price of its thirty-six-inch screen to one million yen, about USA$8,000. The price reduction reportedly resulted from using highly integrated semiconductor chips to replace a large number of unintegrated chips and from simplifying the circuitry. Some manufacturers complained that it was not true HDTV. (In 1991 a Japan-USA team was formed to develop semiconductors specifically for use in HDTV systems. Japan's Hitachi, Sony, and Fujitsu joined with USA's Texas Instruments to develop TV sets, MUSE decoders, and so on.)

In Europe, Thomson's interim HDTV sets were priced close to USA$7,000, but Thomson listed its late 1992 price as circa $6,000, and its possible year-end price for a 16:9 wide-screen HDTV-compatible set for as little as $2,300. The NHK's 1991 schedule of one or two hours daily of HDTV telecasting expanded to eight hours a day in early 1992. The daily telecasts were seen mainly on several hundred large screens located in public areas such as shopping centers.

While it is probable that Japanese set costs will decrease enough to stimulate increased production and to satisfy the increased viewing audience, it's less certain what effect the 1991 shake-up in NHK's heirarchy will have on the 1125/60 in the long term. NHK chair Keiji Shima was the driving force behind the creation of the Japanese company MICO (Media International Corporation), funded by forty-seven leading Japanese corporations. When MICO was formed, there were charges that it was a ploy to monopolize television rights to films, among others. Shima weathered that storm but was forced to resign on July 15, 1991, after giving conflicting testimony to a subcommittee of Japan's parliament. His replacement Mikio Kawaguchi, chairman of the NHK Symphony Orchestra, indicated that there would be few changes in the overall direction of either NHK or MICO, but other sources indicated that there might be a slowdown in the startup of the international news network, Global News Network, to compete with CNN. Shima, now CEO/Chair of Shima Media, may revive the concept.

The second major shift envisioned by Kawaguchi was redirecting the development of MUSE-HDTV as a broadcast standard. As a nonbroadcast standard, it has been well entrenched for some time in business and industry.

HDTV for Nonbroadcast

Motion Pictures   Even if MUSE-HDTV fades as a broadcast standard, it may have a thriving second life in nonbroadcast fields. Some estimate that broadcast use will account for only one-third of the standard's use, with the other two-thirds divided between business and computer applications. We mentioned support from the Society of Motion Picture and Television Engineers for the 1125/60 standard, and there is a growing amount of motion picture product being done with HDTV. The PBS feature *The Ginger Tree* was an HDTV production. Viacom teamed with MICO to cofinance and coproduce a *Perry Mason* telefilm shot in Japan using MUSE-HDTV. And when Sony bought Columbia Pictures, the company allocated space for the Sony Advance Systems' demonstration studio to show how HDTV can save production time, speed up editing, and still produce film-quality images as well as the incredible special effects already incorporated into film production. Other HDTV systems may compete in production, but at least until a USA digital system is established, MUSE has the field to itself, except in very limited areas of the EC.

NHK Enterprises USA established an HDTV production center using the facilities of the Kaufman Astoria Studios in New York City. A major focus of the center was reported to be computer graphics, working in cooperation with Shimax, a USA subsidiary of the Shima Seiki Corporation. High-definition computer graphics have become a major focus of NHK and other Japanese companies. (A number of them are working with Carnegie Mellon University in Pittsburgh, Pennsylvania, to develop HDTV animation.) One goal is to enhance the three-dimensional (3-D) effect that people claim to see in HDTV with actual 3-D visualization.

There are three other HDTV production facilities in the USA, all in New York City: Captain New York Inc. (a commercial production and transmission facility, the first such combined facility in North America, which uses the HDB-MAC system developed by Scientific-Atlanta), REBO High Definition Studio (REBO also has a Tokyo office and is reported to be planning an office in Hollywood), and Zbig Visions, Ltd. There are reportedly more than ninety HDTV production companies in Japan.

Individual cinematographers are beginning to react to videotape and, specifically, HDTV in motion picture production. In late 1990 a coproduction team involving Japan, Italy, and the USA shot an experimental eleven-minute short subject called *Pacific Coast Highway,* about a marathon bicycle race. Using all Panavision/Sony equipment, the project was recorded on both half-inch and one-inch digitalized tape. *Daily Variety* (January 9, 1991, pp. 28, 35) quoted cinematographer Steve Poster as saying that rather than trying to make video look like film, it may be desirable to consider that "these are two separate mediums and should be thought of as such . . . There is a need for both of these styles," which "can work side by side without one trying to dominate the other."

Even some non-HDTV video has begun to intrigue film cinematographers. A 1991 article in *Videography* (April 1991, pp. 34–36) reported that producer/director/cinematographer Al Giddings, worked with Sony to adapt a BVW-400 Betacam SP camera and a Fujinon 14X zoom lens to his needs to tape *Water: Gift of Life* for the Nature Company and *A Celebration of Trees,* a Discovery Channel special. Gidding is also using the video equipment, with specially designed housing, for underwater shooting. The advantages include lighter equipment and being able to shoot for thirty minutes (more than three times longer than with film), cutting out processing costs. Manz estimated that a film production that would cost $75,000 to film, process, and print could be turned out on video for about $3,000. When asked about the future of video and film, Giddings said the transition to video may be slow but that many film traditionalists will begin to see advantages in the three-chip CCD video cameras.

Showscan Located in Culver City, near Sony's HDTV studio, Showscan developed and patented a motion picture format using 70-millimeter film projected at sixty, rather than twenty-four frames, per second. The resulting image is reportedly four times brighter and larger than standard projection with an almost 3-D effect similar to that some ascribe to HDTV. Showscan executives have been quoted as saying it was not coincidence that the Showscan process matches that of MUSE-HDTV—it was designed to be compatible.

Television Commercials The first area to succumb to the lure of HDTV's enhanced qualities was—and will probably continue to be—commercial shoots. Non-HDTV videotape invaded the commercial field long ago, and the superior images produced by HDTV provide an irresistible attraction.

Computer-Aided Design (CAD) CAD allows physicians, scientists, and industrial designers to manipulate visual elements it is impossible (or too costly) to manipulate with actual objects or patients. Industrial designs that once were expensive and time-consuming to construct are now produced by rapid manipulation of computer images. This process existed before the advent of HDTV, but HDTV has added a touch of virtual reality—the ability to integrate computer-generated design with real images. For example, automobile design images are plunked down into real-life traffic situations. And CAD allows designers to disassemble and reassemble cars however they wish, without damaging an actual vehicle or an actual model. Three-dimensional computer images eliminate the need for models until the final stages. Some reports estimate that HDTV-CAD may be able to cut as much as a year off the design time for an automobile and to reduce costs while doing it. Ford and Toyota are two companies reported to be using HDTV-CAD in their car design.

Medicine   Because of its high resolution and clarity, especially on a large screen, HDTV has found a unique place in medicine. Videoconferences that allow surgeons, in particular, to view distant operations have found the sharp detail of HDTV to be a godsend. As virtual reality techniques increase in use, allowing physicians at remote locations to interact, the need for HDTV will increase.

The Military   Everyone who watched television coverage of the Persian Gulf War in 1991 gained a full appreciation of the value of TV and computer images in fighting modern battles. It comes as no surprise, then, that a study by the USA's General Accounting Office (GAO) on behalf of the House of Representatives Subcommittee on Telecommunications found that HDTV has more uses for military defense than for any other purpose. Industry officials reportedly told the GAO that having a single universally accepted production standard would not influence the development of nonbroadcast uses for the standard.

The USA's Defense Advanced Research Projects Agency awarded a million dollar contract to New York-based Projectavision to develop and produce liquid crystal display (LCD) projectors for high-definition displays. Apparently, Projectavision's series of patents solves many of the problems with light levels and color saturation and phasing that have kept LCD projectors from equalling the output of cathode ray tube (CRT) projectors. JVC (Matsushita Electric Industrial Co.) and Hughes Aircraft Co. (General Motors) joined to form Hughes-JVC Technology Corp. to produce liquid crystal light valve displays, reportedly bringing almost-film quality to high-resolution projectors.

Equipment   No new system comes cheap, including high-definition systems. Estimates are that current HDTV equipment runs a little over three times the cost of the best available NTSC equipment. As with any new product, prices will decrease as demand and production increase. In 1992 the price for a high-definition camera or video tape recorder ranged from USA$325,000 to $350,000. But new equipment is being developed and produced constantly, including portable cameras. And if the cost of replacing NTSC with HDTV equipment seems astronomical, remember that commercial TV equipment does not last forever. Cameras have an average life span of about ten years, so the cost of replacement will be three times the cost of NTSC equipment minus what the NTSC replacement would have cost. Fox Television Stations, Inc. was the first to move: in early 1992 Fox signed with Harris Corporation to build HDTV transmitters, based on whatever system the FCC approved, for seven affiliate stations.

Pending Television Standards

The world waited, not always patiently, for the FCC test results for an HDTV standard. As noted, the fear for the EC's HD-MAC and for Japan's MUSE-HDTV, both analog, is that a digital USA standard will antiquate their systems in terms of quality, capacity, and equipment.

To make a decision, the FCC followed the pattern it set in 1941 and 1952. It called into being an industry group to perform the tests and make recommendations. FCC chair Albert Sikes and Richard Wiley, chair of the Advisory Committee on Advanced TV Systems, oversaw the operation.

The FCC set down requirements that any contending standard had to meet: it would have to fit the 6 megahertz bandwidth and be compatible with terrestrial broadcast of NTSC signal so that existing TV sets would

Figure 15–1 This Ikegami HL-1125 high definition portable television camera uses a 2/3-inch Harpicon tube and a 2-2/3 inch HDTV lens. It scans 1125 lines and has interlaced (2:1) scanning and an aspect ratio of 9:16. The camera head weighs 14.3 pounds, excluding the viewfinder and lenses. It provides a high signal-to-noise ratio (–45 decibels).

Courtesy Ikegami Electronics, Inc.

receive video of at least equal quality. The committee envisioned a system capable of simulcasting on two 6-megahertz bands, one carrying the HDTV signal and the other carrying the NTSC signal.

Starting in July 1991, the industry group began putting the six accepted systems through a three-phase test program: objective testing at the Advanced Television Testing Center in Alexandria, Virginia, from July 1991 through May 1992; subjective testing, including consumer reactions, at the Advanced Television Evaluation Laboratories in Ontario, Canada, during May and June 1992; and over-the-air tests beginning in June 1992. The recommendation to the FCC, first scheduled for September 1992, was advanced to February 1993, the FCC's decision to mid-1993. If the FCC approved none of the standards accepted for testing, it could theoretically have accepted one of the other two global systems or called for more testing. Of the six standards scheduled to be tested, two were analog systems and four were digital, as follows.

Analog Systems
July 12 to September 3, 1991:
1. ACTV (Advanced Compatible Television), submitted by the Advanced Television Research Consortium, representing NBC, the David Sarnoff Research Center (Princeton, NJ), Philips Consumer Electronics Co. (Philips, N.V. of the Netherlands), Thomson Consumer Electronics Co. (Thomson S.A. of France); 525 lines, 59.94 cycles per second, progressive scan (1:1) (Withdrawn in 1992)
September 10 to October 24, 1991:
2. Narrow MUSE, submitted by Japan Broadcasting Company (NHK), compressed from MUSE 1125/60 to fit the 6-megahertz bandwidth, interlaced scan (2:1) (removed February 1993)

Digital Systems
November 14, 1991 to January 7, 1992:
3. DigiCipher, Submitted by General Instrument Corporation and ATVA (American Television Alliance, representing General Instrument and Massachusetts Institute of Technology); 1050 lines, 59.94 cycles per second, interlaced scan (2:1)
January 14 to March 2, 1992:
4. DSC-HDTV (Digital Spectrum Compatible HDTV), submitted by Zenith Electronics Corporation and AT&T; 787.5 lines, 59.94 cycles per second, progressive scan (1:1)
March 9 to April 22, 1992:
5. ADTV (Advanced Digital TV), submitted by N.A. Philips Consumer Electronics and Advanced Television Research Consortium; 1050 lines, 59.94 cycles per second, interlaced scanning (2:1)
April 29 to June 15, 1992:
6. ATVA Progressive System submitted by Massachusetts Institute of Technology and ATVA (General Instrument, MIT); 787.5 lines, progressive scan (1:1).

In early 1992 the Advanced Television Research Consortium withdrew its ACTV submission and in February 1993 the FCC advisory panel rejected Narrow MUSE because it created too much interference with NTSC transmission, leaving only four systems, all digital, in contention.

As the tests progressed, digital compression began to spread in other areas. The "big picture" began to indicate a rapid digital evolution coming close to a revolution. One possible indication of this trend was the mid-1992 announcement by two of the HDTV rivals of a plan to share royalties. General Instruments and MIT agreed tentatively with Zenith and AT&T to share royalties if any of their systems were selected. The announcement also hinted at technical cooperation in perfecting any chosen standard. The door was left open for the Philips/ATRC consortium to join in such an investment/profit-sharing arrangement. In the end it was the FCC group's recommendation to combine technology from the four surviving systems into a single digital standard. Computer-compatible (progressive scan) but able to be transmitted in either progressive or interlaced scan, the standard will use square pixels as opposed to inter-laced systems' rectangular pixels.

With FCC approval of one system, theoretically there will be three standards vying for global acceptance: Japan's 1125/60, the EC's 1250/50, and the USA system.

Two Standards Detailed

To offer a sense of the challenge faced in selecting an advanced standard, I will detail two of the contending standards, both digital, one offering interlaced scanning and one offering progressive scanning. Note that both specifically addressed the problems inherent in the FCC's requirement for simulcasting NTSC and HDTV signals.

To accomplish the simulcast, today's NTSC television stations will have to broadcast as they have done, with the HDTV signal broadcast simulta-neously on currently unused channels of the spectrum. These "taboo" channels have historically been left vacant to avoid interference between channels. Note the words *spectrum* and *spectrally* in the two presentations.

ADTV: Advanced Digital Television ADTV was the surviving system submitted by N.A. Philips Consumer Electronics and the Advanced Television Research Consortium. One reason I include it here is to demonstrate how current hardware is being designed with built-in flexibility, reflecting both the transitional nature and the convergence of global television and the rest of the multimedia. We all know everything is coming together, but no one knows exactly how, so industry is hedging its bets by designing equipment with flexibility.

The following is an excerpt of the Advanced Television Research Consortium's description of ADTV.

Advanced Digital Television (ADTV) is a fully digital system that delivers high definition television (HDTV) in a 6-MHz channel. To achieve its goals, ADTV has made significant improvements to proven digital video compression and digital transmission techniques, and molded them into a single cohesive system. There are three key elements in the ADTV system.

First, ADTV's video compression, called MPEG++, is based on a specific implementation of the MPEG (Moving Pictures Expert Group) compression approach. MPEG++ upgrades the standard MPEG performance level and incorporates a video data prioritization layer that allows the most important video data to be transmitted with the greatest reliability.

Second, ADTV incorporates a Prioritized Data Transport (PDT) layer. PDT is a cell relay-based data transport layer that supports the prioritized delivery of video data, thus providing the feature of graceful service degradation under impaired channel conditions. PDT also offers service flexibility for a wide mixture of video, audio, and auxiliary data services, and compatibility to broadband ISDN (integrated services digital network).

Third, ADTV applies spectral-shaping techniques to Quadrature Amplitude Modulation (QAM) to carefully minimize interference from and

Figure 15–2 Researchers evaluate the high-definition picture performance of the Zenith-AT&T Digital Spectrum Compatible HDTV system. In the foreground is a live digital HDTV image from a special HDTV camera displayed on a fifteen-inch flat tension mask high-definition monitor. In the background are computer-simulated television images: today's (left) and those of HDTV (right). See Figure 15–3.

Courtesy Zenith Electronics Corp./Charlie Westerman.

to any co-channel NTSC signals. The result is an extremely robust data transmission system, known as the Spectrally-Shaped QAM (SS-QAM).

Although submitted as a 1050-line system, ADTV is designed to provide flexible support for a wide range of services and future media formats. The initial hardware implementation for the FCC tests will use the interlaced scan format for video source and display (1050/59.94/2:1), with a 16 × 9 aspect ratio and more than twice the NTSC resolution. Selection of this initial format was based on current camera and display technologies, and does not preclude a future adoption of other video formats, consistent with the evolution of studio equipment and production standards.

In summary, ADTV is an integrated system that uniquely incorporates elements of digital video data compression (MPEG + +), data transport (PDT), and digital transmission (SS-QAM) into a high-quality HDTV system that meets the needs of the broadcasters, receiver manufacturers, and the consumers. By adapting proven, widely-accepted techniques, ADTV provides an open door for industry acceptance and future ease of integration with other video and multimedia equipment and services. Together, these important attributes of ADTV form the basis for a successful development of an HDTV industry in the United States.

Zenith/AT&T: Digital Spectrum Compatible One of the points Zenith and AT&T stressed in their HDTV entry was their progressive scanning. Both this system and MIT's ATVA called for 787.5 lines progressively scanned. Zenith noted two strengths of progressive scan: progressive scan is also used to create computer display images (making it easier to interface with computer workstations); and comparative studies have shown that the quality of 787.5 progressively scanned images is greater than 1,575 interlace scanned images. Aside from the possible blurring and flicker that can occur on the horizontal edges of interlaced systems, the numbers alone show one reason for the strength of progressive scanning:

Scanning Format	Lines per 1/60 Second	Lines per 1/30 Second
787.5 progressive	787.5	1,575
1,125 interlace	562.5	1,125
525 progressive	525	1,050
1,050 interlace	525	1,050
525 interlace	262.5	525

Figure 15–3 This is an enlargement of the background shown in Figure 15–2. These computer-simulated television images contrast the high-definition images of the Zenith-AT&T Digital Spectrum Compatible HDTV (left) quadrupling the video information transmitted via current television broadcast signals (right).

Courtesy Zenith Electronics Corp./Charlie Westerman.

Zenith/AT&T also noted their approach to solving the interference problem inherent in simulcasting NTSC and HDTV signals, a solution reflected in the name of their proposed system—Digital Spectrum Compatible. The proposed solution, the use of a low-power transmitter and signal processing for the HDTV signal, avoids interference into NTSC channels. To keep the stronger NTSC signal from interfering with the weak HDTV signal, they proposed digital filtering of both signals, a new technology centering on a unique digital filter at the HDTV transmitter and a complementary filter in the HDTV receiver. The technique allows error-free reception of HDTV signals by eliminating interference from the standard television signal. An added advantage is that ghost images, common in some broadcast areas, are also eliminated.

One final note on this proposed system relates to an "up-conversion" process developed by Zenith/AT&T, which would allow stations to switch to HDTV without having to replace all of their equipment. Using a Digital Spectrum Compatible simulcast transmitter and antenna, stations could "pass through" network HDTV broadcasts. For locally produced programs, they would use the up-conversion process to broadcast high-quality NTSC images on the digital transmission system.

The up-conversion process starts with the regular interlaced 525-line NTSC signal. In the first stage of conversion, each interlaced line is doubled, resulting in a 525-line progressively scanned signal. In the second stage, the line rate is increased from 525 to 787.5, also progressively scanned—a digital match for the HDTV signal. This technique would allow stations to phase in the more expensive HDTV camera and other studio equipment gradually over a period of years, rather than having to invest in new equipment all at once.

In developing their video compression algorithm they gave special attention to minimizing the decoder circuits that will have to be part of every HDTV home-set receiver. Some elements of the encoding algorithms that affect picture quality can be altered without requiring modifications to the decoder. This feature provides an opportunity for future improvement without affecting the decoder in the home sets.

Still Needed: A United Nations for Television Standards

From a globalcasting and a technician's point of view, a single world standard that does not require conversion anywhere would be ideal. Unfortunately, globalcasting and technical considerations are not the only factors involved, and it appears that time alone will dictate who the HDTV winner(s) will be. It might seem that, having gone through the NTSC/PAL/SECAM standards conflict, the world would welcome an opportunity to form an international consortium to work together on a single, unified HDTV standard. Apparently it is as difficult for television industry executives and government officials to coordinate communication techniques as it is for politicians to solve tariff and border disputes.

In any event, whether or not the countries of the world unite on HDTV standards, interactive global productions will have to be sure that both the standards and the methods of transmission and distribution for each country are known and compensated for.

Technology Providing Interim Help With Standards/Conversions

Global HDTV is a few years off, whatever form it takes. In the meantime, globalcasts dealing with multiple world standards might reduce their costs in postproduction by having a universal VCR. Panasonic offers the AG-W1

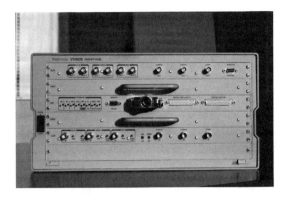

Figure 15–4 This format converter manufactured by Tektronix, Inc. for the Advanced Television Test Center (ATTC) was created to tape specially designed test pictures so comparable picture evaluations could be made for the competing HDTV systems. Absent this equipment, use of live-action scenes or computer-generated images would have delayed testing by months. Cablelabs, a research and development consortium of cable TV operators, developed additional test equipment to test the high-definition standards for cable use.

Courtesy of Tektronix, Inc.

Universal VCR, which automatically determines the standard (NTSC, PAL, SECAM) used on any inserted tape and adjusts for playback. You can also record a program in any system.

Similar equipment is available from the Advanced Television Test Center (ATTC). Created to perform the tests for the FCC Advisory Committee, the Tektronix/ATTC Format Converter (Figure 15–4) was designed to record the four different HDTV scanning formats on the only commercially available tape machine, itself built for only one of the formats.

The Tek/ATTC Format Converter enables digital recording and playback of all four digital HDTV scanning standards proposes to the FCC, using a Sony HDD-1000 DVTR. RGB source material in any of the proposed standards can be edited and copied in the same manner as 1125-line material. On playback, the encoded standard is automatically recognized by the Format Converter and converted back to its original RGB form. A digital output signal indicating the playback standard used is also provided. The likelihood of artifact generation or masking is reduced because the Tek/ATTC Format Converter does not employ techniques such as digital resampling of pixels or lines. The converter was offered for sale through the Center.

In October 1992 NHK announced development of an in-receiver converter that allows a TV set to receive any of the existing HDTV systems—MUSE's 1125 lines at 60 Hz, the EC's HD-MAC (1250 lines at 50 Hz), whichever system is adopted by the USA. Weighing in at 440 pounds, the converter will require serious miniaturizing. NHK estimates a manageable size by 1994–1995.

Other transitional, mutually compatible digital/analog equipment:

- W-VHS (Victor Co. of Japan): JVC's still-analog VCR records any digital/ HDTV signal *after* the TV set has decoded the signal. Others (Toshiba, Hitachi, Sony, Goldstar) propose digital recording of compressed signal.
- Digital Betacam BVW: Digital *but* still tape-linear, not computer-random. Tape players (circa $30,000) playback analog or digital. A VTR or studio camera is circa $50,000+. Cf. Panasonic's D-3 format.

Computer, Television, and Film Standards

It has been technically possible for some time to convert interlaced TV signals for progressive scan display on computers, resulting in a sharper picture, but that technology has remained in the world of computer screens. Until the proposed digital TV standards emerged, not much attention was focused on creating a single compatible standard.

While the advance of computer technology is not a model of orderly progression, the competing systems have at least been fairly agreed on goals. Microsoft's Multimedia version of its Windows software is recognized as a very basic tool that integrates video, voice, graphics animation, and music. The Apple/IBM pact announced July 3, 1991, included plans for collaboration on standards for new multimedia software. They cannot afford to do less than Microsoft has done. Unfortunately, the advances in computer technology have not yet hit TV receivers.

Separating Television and Computers

The remaining differences between television and computer displays are in scanning methods, scanning rates, and bandwidth. Computer monitors— frequently called video display terminals—win out on all counts.

Most personal computer (PC) graphics standards use progressive scanning. NHK's MUSE-HDTV and several of the systems submitted for FCC testing use progressive scanning, but the NTSC, PAL, and SECAM standards still in use all have interlaced scanning, despite the fact that progressive scanning tends to produce sharper, more stable images.

Another variable between TV and computer standards is the scanning rate. The scan rate of computers, called the *vertical refresh rate,* is double the scan rate of the NTSC standard. The third major difference is in bandwidth. Computer terminals have higher bandwidth, as much as four or more times greater than current television standards.

Given the proper added equipment, a computer video display terminal can easily be transformed into a receiver for current television signals. In fact, many of the large (twenty-five- to thirty-five-inch) presentation monitors are NTSC-compatible and have direct inputs for composite signal. Some are also PAL- and SECAM-compatible. Even a PC monitor—as opposed to a presentation monitor—can be rigged to accept NTSC signals (*PC Magazine,* May 14, 1991). And some Super VGA monitors, like the Mitsubishi Diamond Scan 14, accept NTSC signals. These monitors are extremely useful for multimedia presentations integrating inputs from different sources. Because of their scan rate and interlaced scanning, however, NTSC signals are not as sharp as computer images in such presentations.

If computer monitors can display TV images—as they shall increasingly, as Windows technology expands—the reverse is not yet true. Normal TV monitors cannot double as computer terminals. But there is interim technology that shows the way.

Sony's VBox controller is a computer/video interface that lets PC users control a series of Sony video peripherals while working within their multimedia software. VBox uses Sony's Video System Control Architecture (ViSCA) to synchronize control of peripherals such as camcorders and VCRs. With RGB signals converted to NTSC signals, the system lets the computer output to video.

Redlake Corporation's SPECTRUM-NTSC+ and SPECTRUM-XVA offer further examples of what you can accomplish. SPECTRUM-NTSC took video from any video source, added it to text and graphics developed by a video graphic adapter (VGA), and displayed it on a VGA-compatible monitor. SPECTRUM-NTSC+ with the tape caster option (see Figure 15–5) does all that SPECTRUM-NTSC does and adds the capability of PAL video input and the ability to output mixed VGA and NTSC (or PAL) video to NTSC or PAL videotape.

SPECTRUM-NTSC+ offers color frame digitization to 16.8 million colors. It uses an eight-bit, color GRGB digitization method to achieve full-color digitization and display of images. The NTSC or PAL image can be

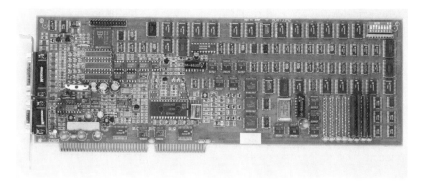

Figure 15–5 This Spectrum NTSC+ board features special synchronization circuitry that enables it to synchronize to different video input sources, such as video cassette recorders and hand-held RS-170 output still picture cameras. Two Spectrum NTSC+ cards can be used to provide simultaneous overlay and display of two live videowriters on the same VGA monitor. Additional cards can be added to the system to increase the number of video windows.

Courtesy Redlake Corporation.

sized to fit small video windows on the display and can be positioned anywhere on the screen under software control.

SPECTRUM-XVA is an advanced video overlay controller and frame digitizer. It can also display video from any NTSC or PAL source on a screen with graphics and text generated from a VGA. Digitized resolution is 640 × 480 for NTSC output, and 640 × 512 for PAL. (These are greater than the SPECTRUM-NTSC+, which has digitized video at 512 × 480 resolution and 512 × 512 in PAL.) SPECTRUM-XVA can accept input from four Gen-Locked video sources simultaneously. Add an optional multiple video input module (MVI-XVA) and it can accept up to four *a*synchronous video sources at the same time. With two megabytes of memory, the board can input video into any or all of up to 256 video windows, which can be panned on a two-pixel boundary and zoomed at noninteger zoom factors. With the optional Tape Caster output module, the combined VGA graphics and video windows can be recorded on regular videotape. SPECTRUM-XVA's Digital Video Bus can be used with digital video coming from Intel DVI equipment.

And one step further: In early 1992 AT&T launched Acumaster Management Services, leasing/linking phone lines worldwide to tie together computers and telephones operated by multinational organizations. The system both creates worldwide connections for telephone/computer systems and interfaces different makes of computers.

As the signals grow increasingly digital and some form of HDTV becomes the norm, television and computer monitors will be one and the same. And it may also be "Lights! Computer! Action!" The "camera" documentary producer/director Scott Billups used to create "A Day in the Life of Melrose Avenue" went directly to the future: all computer. Using Macintosh with adapted Betacam equipment, Billups went straight from image to digits, ready for computer editing, manipulation, compression. We have met our globalcasting future. It is digital and it is here.

Equipment That Bridges the Gap in Standards

In addition to equipment designed to integrate existing and proposed TV standards, there are transitional technologies that integrate film, TV, and computers. We noted film's readiness for transmission via HDTV, but there are more artistic reasons that film will remain a significant medium and will not be replaced by HDTV or any other television recording standard, for the foreseeable future.

The main artistic reason is what might be called a patina—a finish that seems to soften scenes on film, as opposed to the stark reality of videotape. On screen, unless there are glitches or dropout, it's impossible to distinguish tape from live. Film has a sheen, an added dimension that creates undefinable but real reactions in viewers. This difference is what will probably dictate that film coexist with tape in our video future. But even as they coexist, these media are also converging as film images travel to tape for computer manipulation en route back to film.

Two of the most highly visible examples of these converging media are Eastman Kodak projects. To combine the qualitative differences of film with the manipulative potential of electronic imaging, Kodak developed its High Resolution Electronic Intermediate System (Cineon). With Broadcast Television Systems (BTS) of Darmstadt, Germany, Kodak also developed a telecine to translate the quality of film into *any* proposed HDTV standard, from any country.

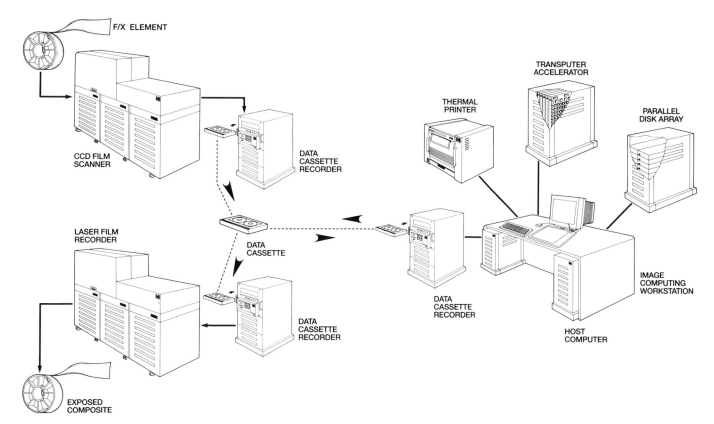

F/X ELEMENT

CCD FILM SCANNER

DATA CASSETTE RECORDER

LASER FILM RECORDER

DATA CASSETTE

DATA CASSETTE RECORDER

EXPOSED COMPOSITE

THERMAL PRINTER

TRANSPUTER ACCELERATOR

PARALLEL DISK ARRAY

DATA CASSETTE RECORDER

HOST COMPUTER

IMAGE COMPUTING WORKSTATION

Figure 15–6 shows how the High Resolution Electronic Intermediate System translates optical images to/from high-resolution digital images. It scans the film images at twice the resolution of HDTV and converts them to digital bits, thus making them capable of computer manipulation. After the digital effects/imaging is completed, the digital images are processed back into film form. A new job title will probably emerge for whoever works these systems. More than an editor, different from a director, the operator will be a digital manipulator, able to combine ten or fifteen layers of imagery as easily as compositing a simple foreground or background. Prototypes of these systems were used to develop the effects in *Willow, Die Hard 2,* and *Terminator 2.*

The high-resolution *film scanner* uses a Kodak-made CCD sensor with three linear photosite arrays (RGB) of 4096 pixels each. A xenon light source (see *atom switcher* in the Glossary) provides high-power, diffuse illumination for the film. The sensor, signal-processing electronics, datapath, and framestores are all digital and capture a frame every three seconds. This digital information is interfaced to external storage in a data cassette recorder (DD-2 standard). The scanner is designed to support the scanning and storage of half- and quarter-resolution frames, which correspond roughly to HDTV and conventional TV resolution, thus requiring only one-fourth or one-sixteenth the storage space. The transport system is also based on a proprietary Kodak design.

The *film recorder* uses three visible gas lasers, RGB. The red laser is a common 633-nanometer helium neon laser; the green is a 543-nanometer helium laser; and the blue is a 458-nanometer argon laser.

The image *composing workstation* is based on a SUN platform with VME bus and UNIX operating system. A transputer-based image-processing accelerator provides high-speed image manipulation for such tasks as

Figure 15–6 Kodak's High Resolution Electronic Intermediate System represents the increased marriage of film images with computer-controlled/created digital images. Input images of the system are stored on random-access disks. The video disk recorder can make a video print of the digital image for en-route evaluation before the computer images are recorded to film (or downloaded from disk to tape for later recording). See text for a fuller description of this system.

Courtesy Eastman Kodak Company.

compositing film resolution images or scaling down film resolution images to best-quality HDTV or TV resolution images.

The software lets the operator interactively manipulate operational settings on key frames to define a sequence of operations that will be applied automatically to other frames. In addition to color balance and grading, matte generation, painting or touchup, and image-processing functions such as filtering, blurring sharpening, resizing, repositioning, the operator can create custom functions as needed. Settings can be fixed, choreographed, or time-varied.

In general, the throughput of this system is expected to be similar to that of conventional digital video. Kodak worked with the SMPTE Ad Hoc Group on Digital Pictures to develop a file format to be used for the exchange of images between workstations or facilities. Increased use of systems such as this one will sharpen viewer expectation, especially in music and action video, and will put a greater burden on globalcasts to match film techniques. If global audiences are to be lured and held, integrating live images with digitally manipulated images may have to become standard practice.

Helping to ease concern about transmitting any global HDTV standard, Kodak, with partner BTS, developed a telecine that will output to any proposed HDTV standard—in Japan, the EC, or the USA. Figure 15–7 shows the telecine in its experimental stage. It employs two new CCD linear array sensors (different from those in the Intermediate System) developed by Kodak to scan and digitalize the film output. Kodak also added an advanced illumination system (a patented integrating cylinder) to improve the quality of light.

All-in-all, these two pieces of equipment stand as models of adaptability to the new world of hybrid digital imaging.

Figure 15–7 Kodak's experimental charge-coupled device (CCD) HDTV telecine operates at any proposed HDTV standard—1050, 1125, or 1250. It was the first HDTV telecine to take advantage of advanced CCD sensor technology. It assures that film programming—including even the new high-resolution 16-millimeter film, or super 16-millimeter—will be compatible with and can provide adequate image quality for any currently proposed HDTV standard.

Courtesy Eastman Kodak Company.

Waiting for Full-Wall Television: Thin Panels for HDTV

One reason for the fast approach of the day when one Vidtalk-Stereofax will assimilate our daily communications is the video screen itself. Many believe that HDTV and other advanced systems will not be accepted fully by the consumer until the long-promised thin-TVs and wall-monitors become realities. To a limited extent, they already have. Figure 15–8 demonstrates that though the size may be small, the direction is clear. Wall-TV is the future. Your Vidtalk-Stereofax won't be simply a piece of equipment, or even furniture—it will be part of your house. And as the Sharp models displayed in the illustration show, chances are it will be designer-inspired. The size of the screen may currently be limited by technology to 8.6 inches, but those 8.6 inches can be placed in sophisticated environments.

Current liquid crystal display (LCD) technology will follow you out of the house, too. In 1991 Sharp Corporation introduced a new four-inch color LCD TV for outdoor life. This can be combined with a kit for installation in your car, with an optional antenna system that automatically selects the best signal while your car is moving. Once installed, the set is detachable for recreational use or security purposes. (See the Glossary for other LCD-inspired products.) Look for new uses for LCD flat screens in computer products. In spring 1992 Apple Computers signed with the Sharp Corporation to develop new uses for their LCD screens.

How LCDs Work

Other promising, easier-to-produce products are on the horizon, such as Alpine Polyvision's flat panel displays that use a solid silk-screen film in lieu of liquid crystal. But until the mid-1990s, LCD will be the available flat television screen. And its technology is expanding.

Liquid crystal is almost a halfway between a liquid and a solid. It is a liquid whose atoms arrange themselves such that they approach the order of solid crystal and thus have many of the characteristics and properties of solid crystals. Because the molecular order of the liquid crystal is not so firmly fixed as that of solid crystals, however, liquid crystal can be modified with relative ease by the application of a number of outside forces, such as temperature, mechanical stress, or—in the case of electronic/photonic communication—magnetic radiation. It is this ability that led to the use of liquid crystal in battery-operated calculator readouts, clocks, watches, and computer and television video screens.

The concept was developed in 1965 by RCA researcher George Heilmeier. RCA didn't think much of Heilmeier's invention and abandoned efforts to use it in their products. Japanese engineers, however, were intrigued with the idea and foresaw potential that RCA did not see. (One of the characteristics that differentiates USA industry from Japanese industry is the Japanese willingness to invest in long-range thinking, research, and development. Even if MUSE-HDTV is never accepted as a universal HDTV standard, all of the standards under consideration were motivated by the 20-plus year investment made by Japanese industry in developing the concept.) Similarly, as a result of those short-term, long-term decisions made in the 1960s, if you want liquid crystal technology in the 1990s, you go to Japan. Companies like Sharp invested in years of research that led to the first commercial uses of LCD.

Japan also developed the technology that seems to be on the brink of expanding the use of LCD screens dramatically—the active-matrix. The active-matrix display is accomplished by combining liquid crystal with a

Figure 15–8 Technical and manufacturing limitations have kept these thin LCD TV screens from matching the size of large-screen TV, but LCD screens offer the lure of complete color control not possible via CRT technology. Working independently of Sharp, whose LCD screens seen here are already available, NHK's Science and Technical Research Laboratories are aiming for thin screen gas-plasma flat TV screens, in which varying voltage levels excite gas (brought to gas-plasma stage by initiating voltage), causing RGB coatings on attached glass to glow. Color and brightness are controlled by varying the voltage to individual cells, each containing RGB phosphor coatings.

France's Alpine Polyvision Inc. is working on a thin screen that uses a solid, an electronically sensitive film that is silk-screened onto the glass. Currently used in calculators, Polyvision electric flat panel displays for television are not expected until the mid-1990s.

Photo reproduced with the permission of Sharp Electronics Corporation

large semiconductor containing more than 1.5 million transistors, each of which controls a single pixel on the display screen. This point-by-point control is leading gradually to larger screens and to much improved color images, a lack of which has tended to keep LCD TV screens from overtaking the sharp color images of the CRT.

LCD Screens with Projection Television

Regardless of the application, LCD projection has three advantages that make its ultimate triumph predictable: the light weight and thinness of LCD screens, often called flat panels; its sensitivity to low-power energy, resulting in lower power consumption; and the ease with which the signal is transmitted in LCDs.

The unchallenged consensus has been that LCD technology has not reproduced color as accurately as CRT technology and that the pixel quality in LCD projection doesn't equal that of CRT projection. The complaint is that LCDs show visible pixel edges. Projectavision's system reportedly overcomes both the visible pixel effect and the color problems. If those reports are accurate and the potential technology is used industrywide, LCDs will conquer the CRTs because of their convenience.

We described earlier how CRT technology uses an electron gun to shoot the signal at the CRT. In most projectors based on CRT technology, there are individual CRTs for red, green, and blue. The light for each of the RGB

Figure 15—9 Ikegami's TPP-1500 Super Projector has three nine-inch high-luminance, high-resolution liquid-cooled CRTs and fl.1 large glass lenses (optically coupled). It achieves a light output of more than 1,100 lumens and a horizontal resolution of more than 1,000 lines. LCD projectors utilize a single lens that simplifies operation but cannot yet match the quality of CRT.

Courtesy Ikegami Electronics, Inc.

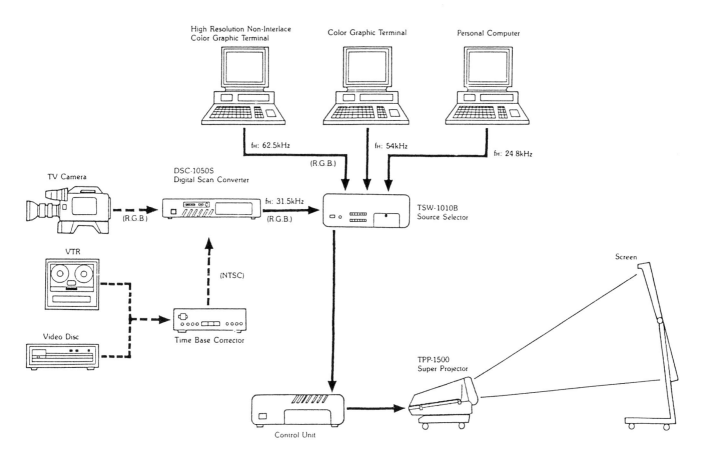

High Resolution Non-Interlace
Color Graphic Terminal

Color Graphic Terminal

Personal Computer

fн: 62.5kHz

fн: 54kHz

fн: 24.8kHz

(R.G.B.)

TV Camera

DSC-1050S
Digital Scan Converter

fн: 31.5kHz

(R.G.B.)

(R.G.B.)

(R.G.B.)

TSW-1010B
Source Selector

VTR

(NTSC)

Screen

Video Disc

Time Base Corrector

TPP-1500
Super Projector

Control Unit

Figure 15–9   *(continued)*

tubes is projected through a separate lens. There are therefore three tubes and three lenses, all of which have to be focused together to produce an accurate image. The LCD projectors, on the other hand, develop their color matrix by sending light through a series of color-sensitive mirrors that focus the image *within* the projector, so that it transmits its image through a single lens. No setup time is required—you plug in the LCD projector and it's automatically adjusted.

Current Uses of CRT/LCD Technology

CRT projection is basic to many of the increasingly popular videowalls. A number of companies now offer this technology, with stacks of sets that can be arranged to cover a thousand or more square feet. As noted in discussion of *Earth 90* and *The 1990 Goodwill Games,* these can be extremely effective production tools for globalcasts.

In-home projection TV sets, fifty- and fifty-three-inch screens are now available, some including PIP technology when linked to a VCR. And at the other extreme, you can get a three-inch active-matrix LCD screen on a Sony Color Watchman portable TV that has built-in AM/FM stereo tuner, weighing 2.5 pounds, including batteries. There is no question that flat, thin screens are in our future, in the television studio and at home. They may be LCD or solid panels, but they will be.

Part *Five*

Global Trends

Chapter *16*

Trends in News

In these pages we have covered extensively the globe-circling power of satellite/wired television. Inherent in power, however, is the need to exercise that power responsibly. In no other area of global television is the relationship of power and responsibility so clear or so perplexing as it is in global news. Because of conflicting rights, often both justified, the role and responsibility of global TV news are still blurred. The role is clearer than the responsibility.

Global television's role was thrust upon it by its mere existence: satellite TV signals circle the globe, leap borders, cancel geography, influence people. No one could have foreseen the impact of adding global pictures to global sound.

Until satellites soared to the Clarke belt, the dissemination of news had developed relatively slowly. The Chinese invented printing, and the Koreans were using movable type as early as 1400 A.D. Johann Gutenberg, who was apparently unaware of the discoveries in Asia, is credited with creating movable metal type, and the Mazarin Bible was the first book printed in England, in 1455. Gutenberg's invention caused revolutionary changes: it made books—and therefore education—previously available only to the select few in the clergy, royalty, and government, available to millions. It wasn't until the early 1600s that the first daily newspapers were printed in Europe, and England didn't have a daily paper until 1702. It was another 200 years before newspapers became part of the daily life of large masses of people.

The telegraph sped up news reporting, and then radio sped it up even more, until finally video news replaced print as the main source of news for most people. Some media critics (Marshall McLuhan among them) have suggested that, paradoxically, the visual medium of television would drive us out of the age of the visual into the age of the aural. To the extent that television, with news images that are both brief and fleeting, is replacing the hard copy of newspapers, McLuhan's prediction was accurate. He was able to foresee that the video/audio medium would create a new environment, altering the way people experience the world around them. What he did not foresee was that the *content* of the medium would have as dramatic an effect as the medium itself. He did not foresee what I call *teleplomacy,* the use of television by heads of state to carry on the diplomacy that was once restricted to tuxedo-filled and smoke-filled rooms. That phenomenon was dramatized most vividly by another phenomenon called Cable News Network (CNN) and its coverage of the 1991 Persian Gulf War.

Teleplomacy: Global Television in War and Peace

The Persian Gulf War was the flashpoint focusing political and military attention on the power of global television news, but it was far from the first world event to reflect this power. As far back as the televising of President John F. Kennedy's funeral in 1963, individuals around the world began to sense that personal communication that had once been limited to their neighborhood had spread. There had been global radio, but the words had been remote, detached, reportorial. With television the inanimate words became living pictures and entered people's living rooms. Mourning became black clothes and flag-draped coffins. Tears became visible. With the death of a young president, the world became a neighborhood.

More powerful images followed. The on-camera murder of President Kennedy's accused assassin, Lee Harvey Oswald, stunned the nation and the world. Films of the Vietnam War televised worldwide altered the views of millions of people in many countries. The 1968 on-camera assassination of Senator Robert Kennedy seemed like a grotesque electronic punctuation to the off-camera assassination of civil rights leader Martin Luther King, Jr. In that same year, television cameras carried to the world live images of tanks from the USSR invading Prague to put down First Secretary Dubcek's attempt to democratize Czechoslovakia. In 1969 the world watched as a single audience as Neil Armstrong and Edwin Aldrin walked on the moon—on live television. Later, when the Challenger spacecraft exploded only seconds after being launched, once again the world watched as one, on live television.

Perhaps the most dramatic demonstration of the power of global television news came in 1989. In that one year a non-Communist regime was elected in Poland; the Berlin Wall came down; Communist leaders were peacefully overthrown in Bulgaria, Czechoslovakia, East Germany, and Hungary; the Romanian regime was overthrown; Chinese students, apparently on the brink of effecting peaceful change in their government, were slaughtered in Tiananmen Square; and Manuel Noriega was overthrown in Panama.

While images are burned into viewers' minds, like that of the lone young Chinese man defiantly halting a line of tanks in the Tiananmen Square confrontation, the true testament to the political and even military power of television news is that it was part of the *process* of change in countries like Romania and Lithuania. In December 1986, the transitional government in Romania housed itself inside the television station, using it to maintain contact with the outside world and with the people of Romania. After the overthrow of Premier Ceausescu, his trial and execution were both televised. Having learned from this, when troops of the USSR later entered Lithuania, some of their first targets were communications centers—the phone exchange and the main printing plant. Throughout the weekend of January 12, 1991, hundreds of pro-Independence Lithunanians stood vigil outside the nation's main television station. When the troops moved in and captured it, the Lithuanian Congress was able to continue televising their message of dissent because they had kept control of a mobile TV unit. The wars of independence in two nations were fought in and around the television stations.

None of these stories, however, dramatizes the power and responsibility of global TV news to the extent of the Persian Gulf War coverage. And no single-source coverage highlighted it more than that of CNN.

Global Television News: A State of Readiness

Cable News Network (CNN) was not the first television news organization to distribute its reports around the world. Visnews—a joint effort involving the BBC, Reuters, and NBC—sent out daily multidestination satellite news feed as early as 1975. CNN wasn't launched until 1980. What then prompted its ascendancy?

I have alluded many times to the need for preplanning, for always asking "What If?" I believe preplanning explains CNN's global position, and perhaps I can explain why by recalling my experience televising the late Premier Tito.

When I televised Premier Tito in Yugoslavia, I was told that the route he would take was definite and would not be changed. Instinct told me not to count on that. My background in sports broadcasting helped me plan for contingencies. When I began directing the Kentucky Derby, the first thing I did was to add a second camera on the finish line. I did that because if you direct the Derby and—for whatever reason—you do not show the horses crossing the finish line, you would have no reason to report back to the network for another assignment. So in Yugoslavia I did what I had always done at the Derby: I gave myself a back-up. I imagined myself as an aide to Premier Tito and asked myself what route I would take. I chose what I considered the best alternate route, and put cameras there, too.

And they used the alternate route. I never found out whether the change was for security or other reasons. On meeting Premier Tito, I never asked. I didn't care. I got the coverage, and that's what counts. Preplanning is what got it for me. If there is a secret to preplanning news coverage, it's the same secret I've stressed for every live telecast, for every live globalcast—and for every budget. *Never stop asking, "What if?"* What if the weather goes bad? What if a camera goes out? What if we lose audio? What if you foresaw the possibilities and set up an alternate communication system that would keep you on the air?

Many newspapers and magazine accounts of the phenomenal growth in audience and stature for CNN claimed that growth was due largely to CNN's ahead-of-the-pack coverage of the Persian Gulf War. That's only surface truth. CNN led the other networks and news organizations in covering the Persian Gulf War because CNN was in position and ready to cover the war.

Long before Saddam Hussein invaded Kuwait, Ted Turner was one of the founders of the Better World Society. He was known globally as a broadcaster interested in using communication to unify the world. Turner banned the use of the word *foreign* on his airwaves. This was one world, and he meant to prove it to the best of his ability. He proved it by reaching out to the world. CNN invited the world into the news mix by starting the *CNN World Report*, airing uncensored news clips from around the world to countries around the world, covering whatever subject those countries wanted to cover.

And CNN reached out to seek audiences—in the world's capitals, in the world's homes, in the world's hotel rooms—long before Saddam Hussein became a screaming headline.

1980 to 1989: CNN Sets the Stage

CNN started slowly. It was the butt of industry and columnists' jokes: Turner was "The Mouth of the South," with a "Chicken Noodle Network." For years the network lost money, but Turner Broadcasting persisted. By the start of 1985, CNN began to show a profit, and it also began to be a major factor well beyond the boundaries of the USA.

In 1986, CNN's role in teleplomacy began in earnest. When president of the Philippines, Ferdinand Marcos, felt he was not getting a fair hearing in the US Congress and the White House, he asked CNN to carry a speech he planned to make on the subject. In that same year, Libyan leader Muammar Gaddafi insisted that CNN be present at one of his news conferences and that a CNN reporter ask him a question during the televised conference. These world leaders knew where their target audience was. They knew that in the USA the Pentagon and the Congress watched, because TBS had furnished them with satellite receivers so they could watch. They knew that two years earlier a microwave service had made CNN available to the US State Department and to the world embassies in Washington, D.C. They knew the world's diplomats would get their message.

By 1987, two years before Iraq invaded Kuwait, in addition to its diplomatic audience, CNN was being watched in fifty-four countries and in sixty thousand hotel rooms around the globe. In 1987 China signed on to have CNN news made available to that country's millions of television sets.

In short, by the time the crisis in the Persian Gulf erupted, CNN had long-established worldwide contacts in place, immediately responsive to whatever news broke, wherever news broke. Like most "overnight sensations," CNN became a star after paying a decade of dues.

Globalcasting the Persian Gulf War: Four-Wire to the World

CNN was not the first to break the story that the UN coalition forces had begun their strike on Baghdad. The first report came from ABC's Gary Shepard, who was in Baghdad. What made the initial difference—and what directed the world's attention to CNN—was that when reporters for all the other networks lost contact with the world outside Iraq, CNN's team was able to stay in touch using their four-wire telephone setup.

As we have noted, live television programs—especially globalcasts and sportscasts—have historically used four-wire communications. When you need instantaneous communication and cannot run the risk of not having it when you need it, you need such a dedicated phone line that does not go through a switchboard and that has four wires, two outgoing and two incoming. This guarantees you two things: instantaneous, uninterrupted communication with no delay (no dialing, no operator, no competition) and the opportunity to have two communications simultaneously. If you're working with talent—as CNN was in their Persian Gulf globalcast—the four-wire lets you have a conversation with the talent and to have the talent speak on the air. CNN's four-wire ran from the Al-Rashid Hotel in Baghdad across Iraq to a microwave transmission dish in Jordan. The signals were sent via satellite from Amman, Jordan, to TBS's dish farm outside Atlanta, Georgia. From there the signals were sent via fiber optic cable to the control room.

CNN spokesman Steve Haworth said their four-wire was not only a communication godsend, it was also a bargain. From August 2, when Iraq invaded Kuwait, through the January 16 start of the UN action, CNN's coverage of the war cost approximately USA$15 million dollars, a third to a half of which was for satellite rental. The four-wire, which became active in November 1990, cost $15,000 a month. The other networks reportedly had applied for four-wire service from Iraq, but never got final approval. It took CNN two months, from September to November, to get permission to have the lines installed. (See *PTT* in the Glossary.)

Viewer Reaction to the Gulf War Coverage

There were literally millions of viewers clinging to their television sets, awaiting word of the latest developments. Oddly, it was a virtual 1991 interpretation of Marshall McLuhan's prediction, for the reporting was audio-only out of Baghdad. On television it was the power of words, not pictures.

On a normal night, CNN's USA audience is just shy of 600 thousand viewers. The night of January 16, CNN's audience soared to almost 11 million cable viewers. And that figure does not come close to representing reality, because hundreds of noncable television stations, including a number of network affiliates, incorporated CNN coverage into their programming.

CNN was able to keep their four-wire communication intact until Iraqi officials forced them to shut down live reporting at approximately 11:00 a.m., January 17. Gradually, then, the focus shifted from live reporting from within Iraq to live reporting via portable satellite uplinks, called *flyways*, by all of the networks. For a number of reasons, members of the Arab State Broadcasting Union did not send reporting teams to the battlefields. The main source of information for the Middle East remained CNN throughout the war. Even Iraq, picking up signal illegally with satellite dishes, was tuned in. Foreign Minister Tariq Aziz said they watched CNN, and unofficial reports indicated that Saddam Hussein himself watched.

Note that CNN's overnight success could not have happened if its signal had not previously penetrated to all those nations, all those war rooms, all those diplomatic departments, all those hotel rooms. The intelligent planning that went into having the four-wire and into having access to the only twenty-four-hour satellite transponder out of Israel were just the last steps in a decade of dedication. Showbiz "luck" has been described as being in the right place at the right time with the right stuff. For the six weeks from January 16 to February 26, 1991, the duration of the Persian Gulf War, CNN had it all.

Questions Raised by Gulf War Coverage: Power Means Responsibility

> No newspapers. You are parasites and gossips, no better than spies. You disclose all our plans to the enemy. . . This is one military operation that is going to be carried out without meddle of the press.

That is not a quote from President George Bush, when his generals decreed that reporting of the military action in the Persian Gulf would be restricted to authorized pools of reporters, under supervision/censorship of a representative of the UN force. General William Tecumseh Sherman uttered those words 129 years before the bombs fell on Baghdad, but the echo was distinct and clear. Sherman was approaching Vicksburg on December 19, 1862. President Bush and the UN coalition were approaching Baghdad in 1991, and the President sent out the edict in response to complaints by the news media and others over the total news blackout that had been imposed during the Grenada invasion and the questionnable arrangements during the invasion of Panama. Many US officers still smarted over what they considered biased reporting of the war in Vietnam in the 1960s. In Vietnam, the military learned along with everyone else that television reporting could influence and change a nation's will to wage war.

Figure 16–1 This CNN slide supported many of the audio-only reports from Baghdad during the 1991 Persian Gulf War.

A study of school students ranging from grade 4 to grade 12 was undertaken by the University of Pittsburgh. Half the students viewed broadcast segments on drug busting. The other half listened, with the video turned off. The results confirmed those of earlier British research: listeners remembered more of the news than did viewers who had the "advantage" of both audio and video. Does this mean that with young viewers that most prized of television news products—video—may actually block communication?

In tests of 400 college-age students, the University of Pittsburgh's School of Education, Department of Psychology in Education, found that more men than women found TV news to be entertaining and would choose to watch it "for the purposes of seeing violent and tragic events." Women, however, were more likely to say that TV news made them worry. The students in the survey split evenly on the question of whether there is too much violence shown on television news. On the other hand, almost all agreed that watching television news "makes them worry."

Finally, the survey respondents "agreed that television news does represent the real world, but they disagreed that it represents their [the audience members'] real world."

Figure 16–2 Taken four days later that the video in Figure 16–1, this off-screen image of CNN's Charles Jaco reporting live from Saudi Arabia symbolizes the shift from audio-only coverage from within Iraq to audio/video coverage from outside Iraq. Reporting on live Scud missile attacks raised questions of censorship and responsibility still not fully answered by either government or the media.

Figure 16–3 While other USA network news departments have been shrinking, CNN has continued to expand. The simple pressure of filling twenty-four hours each day with news redirected the world's attention to global rather than parochial coverage, thrusting television news into higher levels of both influence and responsibility.

The Vietnam War has often been called the first television war, as action scenes were air-freighted out to be transmitted to the living room. We progressed far from the days when the telegraph clicker announced the beginning of the USA's Civil War and made possible daily news from the battlefield, and even from the days when Edward R. Murrow's radio reports from London brought World War II vividly into homes. But however much these media sped up the transmission of news or brought the sounds of war closer to home, none of them had the sight-plus-sound impact of television reporting—and certainly not of *live* television reporting. During the Persian Gulf War, when viewers saw reporters called off the air because a Scud missile had been detected en route, millions of hearts skipped a beat, and millions held their breath until the missile landed and everyone knew where it had landed and what damage it had caused. As viewers, we were drawn into the war, through television.

But the immediacy afforded by portable satellite uplinks and the personal involvement felt by reporters and viewers alike created a potentially dangerous situation and a major unanswered question: How do we control military news and still maintain a free press?

Censorship Versus Responsibility in Live Battlefield Reporting

When one television reporter reported on live television that an Iraqi Scud missile had landed just a mile from Saudi Arabian military headquarters, he was in effect functioning as a forward observer for the Scud missile launchers. There *must* be military censorship—self-imposed or forcibly imposed—to prohibit even accidental transmission of military information that can endanger the lives of the men and women involved in the conflict. And this need becomes more critical as global television equipment becomes more portable. CBS got the first news crew into Kuwait City using a truck that carried a portable satellite transmitter, providing instant, portable communication. And the networks generally agreed that lightweight cameras, such as High-8 ENG equipment, made it easier for news crews to be fast and mobile, even under battlefield conditions. This equipment, however, places an increasing burden on our judgment. Each reporter, each network, each station has the solemn responsibility to honor essential military censorship.

But flip the coin. In a free society the citizens have a right to know what their political and military leaders are doing. And they have a right to know in time to judge whether their leaders' actions are proper and justified. During the Persian Gulf War, reporters were actually arrested in Saudi Arabia for trying to initiate their own reporting outside the authorized pools. The four USA television networks, joined by the Reporters Committee for Freedom of the Press and the American Civil Liberties Union, filed a protest with US Secretary of Defense Richard Cheney. Reporters also protested restrictions on unscheduled interviews and on communicating the sights and sounds of casualties. A federal lawsuit was filed—by representatives of the print media, not television—challenging the military restrictions on reporting on constitutional grounds.

Then there was the bizarre predawn landing of USA Marines in Somalia in December 1992. As the lead troops of the UN's Operation Restore Hope, they were almost outnumbered by TV crews using portable lights to get better video than that supplied by the night-vision cameras used to cover the prelanding operations. The open invitation to cover the landing came from the Pentagon; the excess seemed to come from overzealous TV crews. The issues involved in defining battlefield coverage have obviously not been resolved.

News Responsibility Off the Battlefield

Let me cite one more incident for you. Not once, but numerous times before Saddam Hussein's forces blew up and set fire to Kuwait's oil fields, TV news programs enumerated—with appropriate real and graphic visuals—the disastrous effect such devastation would have on the water supplies and air quality in Saudi Arabia. The effect, it was stressed, could be both militarily and ecologically disastrous. Did Saddam Hussein plan the explosions and fires in the Kuwait oil fields independent of what was shown on television? Did reiteration of this point carry a message to Saddam Hussein's staff that the oil fields were a uniquely vulnerable point? In other words, did the reiterative use of a potential military tactic on television news programs affect the outcome of the war and the post-war world? On the January 27, 1991, episode of *60 Minutes,* correspondent Steve Kroft asked a scientist how he would use poison gas and biological weapons. The scientist replied, "I don't want to give Saddam Hussein any ideas. I prefer not to answer that question."

Solomon-like Wisdom, Where Are You?

We need a balance in the shared world of television and war. It is as great a sin for government and military leaders to mislead free citizens as it is for the television spokespeople for those free citizens to endanger the lives of men and women on the battlefield. Global television coverage of the world's wars will continue. How do we attain a balance?

If I had the answer to that question, it would have been on the first page of this book. I don't know what steps we need to take to attain that balance. What I *do* know is that we need a lot of discussion—away from the heat of battle—so that thinking men and women can focus on these issues and guide us to a workable solution. If we citizens of our global village are to make proper decisions about the future of our increasingly endangered planet, we need to have as many facts as possible—including unpleasant facts like battle losses and body counts. We need to know that modern wars mean modern warfare and that just as satellite TV signals cross national borders at will, so do many of the effects of battle. The oil fields of Kuwait that smouldered for months after the war ended provided just one symbol of that significance. Damage to the environment is not limited to the participants. The world pays.

The War of Words and Pictures

In sports, live television's greatest strength is exercised by immediate decisions on the air and in the control room. In global newscasting we are not playing a game. We have to guard against a producer's or reporter's lack of restraint in airing material, against giving way to live coverage that may result in feeding frenzy to the global village. As any number of columnists pointed out, there was no question that television was used as a weapon during the Persian Gulf War. It was also used for teleplomacy, to communicate and make points off the battlefield. The TV camera can win the propaganda war.

Even when the time comes for "newspaper" writers and editors to communicate their columns electronically to your Vidtalk-Stereofax, they will have a span of time, however brief, in which to digest, analyze, weigh the words. Live television does not have that leisure.

The job, then, is to find a way in which critical news is not withheld from viewers at the same time that information that could tilt the outcome of the conflict is not conveyed. If we could successfully outlaw war, our problem would be solved. We cannot even solve the much simpler determination of a global standard for HDTV. How much harder, then, must everyone involved in communicating news work to simultaneously protect truth and protect lives. The dialogue *must* go on, inside and outside the television industry—before the next "live" war.

What Direction Is Newscasting Taking?

As a result of ten years of dogged determination and the overnight success of its Baghdad reporting, CNN International is now seen in 125 nations. In the USA, its cable subscribers total nearly 60 million households, close to 100 percent of all cable households. (And the official statistics don't tell the whole story. Fidel Castro has announced that he is a fan—watching a pirated signal—and it's estimated that as many as 20 thousand European hotel rooms receive a pirated version of CNN.)

The impact of CNN's Persian Gulf War coverage sparked a reported 18 percent increase in its cable base, adding close to half a million new subscribers. And future generations of subscribers are in training. A 1990 national survey found that half of the USA school districts contacted used CNN for instructional purposes.

Other Global News Services Challenge CNN

As we noted, other television news services fed their product worldwide before CNN was started. And since the Persian Gulf War, many twenty-four-hour global news services have been either started or proposed, and a number of existing services have joined to widen their mutual coverage. Here is a summary of some of the major organizations.

Regional/International by interlocking

- Asiavision: Asia-Pacific Broadcasting Union
- Eurovision Network: EBU (European Broadcasting Union), serving Western Europe and North Africa
- Servicio Iberoamericano de Noticias: Latin America
- Servicio Intervision: Afghanistan, Cuba, Eastern Europe, Nicagargua, CIS, Vietnam

These regional services exchange coverage with each other and—by providing their service to subscribers like the USA and other networks—with the world. There are also a number of private pacts that tie the globe together. Tokyo Broadcasting System, one of Japan's five commercial networks, and CBS signed a long-term agreement that will eventually encompass sports and entertainment programs as well as news. NHK has ties with ABC, separate from possible incorporation into MICO's Global News Network. (ABC also has a nonexclusive agreement with Fuji TV.) NHK has ties with the BBC, and both NHK and TV Asahi have rights to CNN news. Nippon Television has an agreement with NBC.

International by Design

Existing

- CNN: Turner Broadcasting International
- USA Armed Forces Radio and Television Service: via satellite or videotape to more than fifty countries
- Visnews (BBC, Reuters—NBC had part ownership, withdrew, still uses Visnews): includes VisEurope, a satellite news channel.
- World Television News (WTN): owned largely by ABC
- Worldnet (US Information Agency): to more than fifty countries

Proposed or under consideration (becoming or expanding to twenty-four-hour service)

- BBC World Service: The famous global radio service transformed into a twenty-four-hour television service
- BSkyB: The realigned satellite service in the UK: linked with two other continents via Hong Kong-based Star TV, Fox Network in the USA
- Euronews: start–up date 1992, satellite-fed, using PAL standard, broadcast in five languages simultaneously (English, French, German, Italian, Spanish); endangered by the EC's ruling that the EBU/SkyTV's Eurosport channel was a breach of the Treaty of Rome and by the possible EC regulation requiring transmission by D2-MAC
- Global News Network (MICO/NHK): proposed rotating eight hours ABC, eight hours Europe, eight hours NHK
- Newsworld: Canadian Broadcasting Corporation
- RTL News: Luxembourg

The pattern seems to be clear, although how much success will result from cluttered twenty-four-hour competition remains to be seen. Whatever the outcome, the beginning can almost certainly be traced back to the success of CNN's Persian Gulf War coverage.

Let's Make a Deal: Some Networks Pull Back from the Power

Just as global news appears to be expanding, USA domestic news coverage appears to be shrinking. The overall audience share of the "Big Three" USA television networks has diminished in recent years, and so have their news audiences gotten smaller. No one talks any more about expanding the half-hour evening newscasts to an hour. Instead, the Big Three are shifting their focus from that time period to other time periods—and cutting staff in the process. CBS News has gone through a series of staff cutbacks and has been in discussion with Turner Broadcasting System about establishing a working arrangement between CBS and CNN. While the other networks were cutting news staffs, CNN was expanding. News bureaus opened in the fall of 1991 in Amman, Jordan; New Delhi, India; and Rio de Janeiro, Brazil. CNN's executive vice president of news, Ed Turner, no relation to Ted Turner, told weekly *Variety* (July 22, 1991) that the new overseas bureaus brought the network's total to twenty-seven.

In July 1991, both ABC and NBC announced plans to cut approximately 100 people from their news staffs, both domestic and overseas, at the same time that they announced plans for overnight news programming.

In the same month that it closed news bureaus in St. Louis and Rome, made cuts in its bureaus in Washington, D.C., Miami, and Paris, and announced a cutback of fifteen to twenty people in its London bureau, ABC announced it would launch an overnight news service, from 1:00 to 6:00

a.m., starting in January 1992. Entitled *World News Overnight,* the show is presented in half-hour segments, featuring both domestic and global news. Prior to the July cuts, ABC had shut down operations in Budapest, Frankfurt, and Prague, consolidating them into their Berlin bureau.

NBC's announced cuts came in the same month. It closed news bureaus in Miami and New York. At the same time, the network's news division announced it would continue production of *Exposé* and *Real Life with Jane Pauley* and add two daytime news-oriented shows, *Cover to Cover* and *Close-Up.* Later, Jane Pauley shifted to the prime-time news magazine *Dateline NBC,* and Tom Brokaw began *The Brokaw Report.* NBC also began offering its affiliates an all-night news service in 1992.

The cutbacks have not been limited to the USA. In the UK, also in July 1991, ITN announced it would cut staff. Also in that month, obviously a bad month for television staff workers, the BBC announced plans to cut one thousand workers in addition to the previously announced cut of three thousand. Not all of those cuts were to be in news, of course. At the end of 1990, the Canadian Broadcasting Corporation (CBC) cut more than 10 percent of its staff and closed eleven local stations (again, not all of the cuts were in news).

Global News from Local Sources

While domestic news services cut staff and farm-out coverage to independent agencies, a new—perhaps ultimate—source of news has been introduced: The Video Journalist.

The charge-coupled devices (CCDs) that made possible the Sony High-8 and JVC Super VHS equipment also made possible the concept of the video journalist—a one-person operator who carries his or her own camera and field edits the material. Some National Association of Broadcast Employees and Technicians (NABET) contracts with the major networks already include limited video journalist clauses. And video journalists are not limited to the professional ranks.

Perhaps the most celebrated amateur video journalist is George Holliday, who in March 1991 taped Los Angeles, California, police officers beating apparently defenseless Rodney Glen King, who the police report maintained had been apprehended after a high-speed car chase and resisted arrest. Holliday offered the tape to Channel 5 (KTLA) in Los Angeles. The station paid Holliday $500 for the use of the tape. Channel 5 aired the tape March 4, 1991, and gave a copy to CNN, with whom it has an ongoing agreement. CNN, and later countless network and local programs, aired and reaired the graphic tape. As a result of the incident, the taped evidence of it, and the subsequent inquiries, members of the L.A. police department were suspended, a lawsuit was filed, and L.A. Police Chief Daryl F. Gates was pressured to announce his retirement. The April 29 jury verdicts in the case triggered riots and demonstrations in Los Angeles and in a dozen or more cities around the USA. No riots followed the federal retrials on civil rights charges.

There have been numerous other cases in which hand-held minicams have "made news." A few, such as ex-police officer Don Jackson's being shoved into a plate glass window by Long Beach, California, police, have been pre-planned and taped by professionals (an NBC camera crew had followed Jackson to record whatever happened when he entered Long Beach). Most, however, have been unplanned tapings, and they have become a factor in news coverage.

Perhaps the most famous homemade video preceded the minicam, in 1963, when Abraham Zapruder filmed President John F. Kennedy's motorcade in Dallas, Texas. But film was complicated, required processing. The new, inexpensive camcorders are lightweight and virtually defy error. A number of police departments are mounting cameras in patrol cars so that officers can have visual evidence of a car search or arrest, and the FBI has been using video cameras for "sting" operations for years.

As camcorder prices continue to drop, more and more video journalists are expected to become a factor in how news is discovered and recorded. Privacy rights will undoubtedly become an issue. (A couple in Florida made love with their curtains not fully closed. Neighbors filed a complaint—and a videotape—with the police.) Unquestionably the phenomenon of the ubiquitous camcorder will spread. With it will spread one more avenue of capturing and distributing news worldwide.

From the Globe to the Village

A more likely outlet for home video news may well be the increasing number of local programs. We mentioned the probability that some of the hundreds of cable channels made possible by digital video compression technology would be dedicated to local news. But the trend preceded digital technology. In 1986, Cablevision began *News 12,* a twenty-four-hour cable news channel seen strictly on Long Island, New York. In September 1990, Freedom Newspapers launched a twenty-four-hour news service in Orange County, California (Orange County News channel). In the summer of 1991 Fox TV stations and Tele-Communications Inc., one of the leading cable operators in the USA, started a joint twenty-four-hour news channel. Other similar automated camera, high-graphic operations are planned at stations around the country. The Sony "video jukebox," long used to automate the running order of station breaks/commercials, now automates news video. Producers' personal computers can access/alter the running order of stories taped by video journalists in the field.

How many of these efforts will result in producing global news is hard to predict, but the first news crew to deliver video of the 1990 crash of an Avianca jet near Kennedy Airport, New York, was from Cablevision's *News 12.* Neighborhood news is something to keep an eye on.

Summary: News/Global News

By its nature, news will probably provide the single greatest number of globalcasts for the foreseeable future. Multiple live satellite pickups and feeds are increasingly characteristic of global news services, as well as of domestic evening news programs in individual countries. Live news discussion programs and programs incorporating live segments, such as *20/20, Prime Time,* and *This Week with David Brinkley,* provide a second tier of globalcasting possibilities. Non-news programs that incorporate news segments, such as *The Today Show* and *Good Morning America,* also frequently include live satellite pickups.

As Eastern Europe and Asian countries expand their news capabilities, chances are that globalcasting will gain increased meaning and depth. The Pacific Rim is one of the fastest growing television markets. There are millions of viewers in Hong Kong, Singapore, Korea, Taiwan, Indonesia, Thailand. News and sports are particularly popular in Asia because on-air talent can provide local voice to the video.

Despite many ups and downs, politically and economically, the former USSR's combination of restructuring (perestroika) and openness (glasnost) resulted in opening up potentially significant news channels. While some of the smaller Soviet republics have had local television channels, it was only with great effort from reform-minded Boris Yeltsin, president of the Republic of Russia, that Russian Television came into being. And in Leningrad, even before the breakup of the USSR, the city council voted to create the Independent Leningrad Television and Radio Company, a move to free the station from domination by the Kremlin. It's estimated that 70 million Soviet citizens view Leningrad TV.

Even in countries like India, where government control over communication channels is only beginning to loosen, there are signs of an emerging global future. Several years ago private videotape news magazines began to appear in India, some produced by newspapers. Barbara Crosette noted in the *New York Times* (January 2, 1991) that tapes like *Newstrack,* a monthly video magazine similar to *60 Minutes,* have helped fill the news void in the nation. She also noted that in this country that barred private broadcasting and international satellite networks for decades, the number of illegal satellite dishes increased steadily as more and more people discovered that they could receive signals from CNN. In 1993 India announced that far from continuing to oppose DBS, the government will add its own programming. And as Hong Kong's Asiat satellite feeds increase, the ability of any Asian country to remain isolated from world news will diminish significantly.

One final thought on the future of news programming, relating to the theme of convergence. Convergence of computer technology with television technology may finally result in newspapers being delivered by way of the Vidtalk-Stereofax. Most of the videotex experiments—like those of CBS and Knight-Ridder, which offered news and other video services—have failed. Despite all the features it offers computer readers, Prodigy struggled to become profitable. But as digital technology takes over, it will become possible to program the Vidtalk-Stereofax to download and print out whatever news you are interested in.

Newsrooms in newspapers and magazines have long been invaded by computer technology. In 1974 *U.S. News & World Report* got rid of typewriters and installed video display terminals (VDTs). Out went the Linotype machines, in came the photographic processors. Computers took over. The incredible speed and accessibility of computerized printing lets magazines publish late-breaking news that previously would have had to wait for the next edition. Leap ahead to 1992 and the requirement that new TV sets contain a decoder for text on the vertical blanking intervals (VBIs). What's to stop *U.S. News & World Report* or any other magazine or print medium using VDTs from hooking up with a transmitter to VBI?

In fact, experimental electronic publishing has already begun blurring the multimedia lines. In November 1992 NBC joined with IBM and NuMedia Corp. to transmit news feeds via personal computers. Called Desktop News, the video news wire was announced as a six-month trial, aiming to add information from sources in addition to NBC news. In the same month, *Daily Variety* (November 16, 1992, p. 84) announced Monday-to-Friday electronic publication, including full text and advertising plus availability of archive material, via Baseline entertainment information service to PCs. And still in November 1992, *TV Guide* announced its 1993 Denver-area test of a video version of the magazine (*Daily Variety,* December 3, 1992, pp. 1, 10). Many see electronic listings as the only way to cope with the hundreds of channels available via cable and DBS/DTH.

Almost no one reads all of a newspaper or magazine. Millions of trees lose their life for sections that go unread. Digital technology will allow viewers to order just the section they want. This may reduce the peripheral learning that occurs when readers skim through other sections en route to their particular interest. But it may also increase newspaper readership, which has been declining for years. In any event, it's important for people working in television to know that newsprint will soon be in video competition with telecasts. "The tube" will not be a synonym for television much longer. The Tube will encompass all the incoming signals: very possibly within the 1990s, those signals will begin to include your daily newspaper.

Chapter **17**

Trends in Sports and Pay-Per-View

Ahead in sports: (1) Sports will continue to go global, in both participation and globalcasting. (2) As sports coverage by USA networks is increasingly will sports continue to go to pay per view (PPV)?

Sports Will Continue to Go Global

After news, sports events probably offer the richest field for globalcasting. Major events like the Olympics, Wimbledon, The World Cup, The Goodwill Games, and The Pan American Games bring together athletes from different countries. *The 1990 Goodwill Games* demonstrates how international "athletes' villages" like these present the opportunity to weave together different cultures and different countries to produce a secondary level of audience involvement—a powerfully personal and emotional involvement—that adds a new level of depth and significance to the sports events themselves.

But it's not just events like these, with automatic worldwide audiences, that are thinking global. The same is true of professional sports. Baseball has become as big a favorite in Japan as it is in the USA. More Chinese than American viewers see the Super Bowl. World League American (Style) Football survived into its second season before being suspended by the NFL. Many teams in baseball and football regularly play preseason games overseas. Soccer is already gigantic virtually everywhere outside the USA.

I've already stated my prediction that USA networks, faced with unrecoupable costs, will increasingly forego sports events or coproduce them with sports franchises. The staff cuts in news will be echoed in sports. Despite the long symbiotic relationship between sports franchises and commercial TV revenues, sports events will continue to go global, no matter what the networks do.

One example of how both sports leagues and sports broadcasters have begun to look beyond borders is basketball. Kim Bohuny, International Events Manager for the National Basketball Association (NBA), did not shy away from the nickname some sports writers have assigned to the NBA: "Basketball Ambassadors." The Association's goal, she said, is to "work closely with FIBA [the International Basketball Association] to make basketball the number one sport in the world." Some of the ways the NBA is working to attain that goal include holding NBA clinics abroad, licensing NBA products worldwide, and aiding in popularizing annual global basketball events.

The NBA runs about ten clinics abroad each year, in which four NBA coaches visit a country and meet for three days with coaches from the host nation. Annual international events include the 1992 Olympics (NBA and collegiate players); the 1993 McDonald's Open Club Championship (featuring an individual team, such as the L.A. Lakers), seen in 109 nations; the 1994 World Championship, in which the competing USA team is composed of both NBA and collegiate players; another McDonald's Open Club Championship in 1995; another Olympics in 1996, and so on.

The teams around the world seem to be doing a pretty good job of making the sport international. The major Spanish league, for example, permits each league team to have two nonnational players. 1991 teams featured players from Lithuania, the USA, the USSR, Venezuela, Yugoslavia. Europe's highest paid basketball player plays in Italy for Messaggero-Roma. His name is Dino Radja; he's from Yugoslovia. And 1990 saw the first Russian player signed to an NBA team. As NBA Commissioner David Stern told The Sports Summit, "The global decade of the '90s is upon us."

Sports Will Continue to Go to Pay-Per-View

Any global enterprise has to include global numbers. And sports statistics not only fill record books, they also fill the eyes of franchise owners with dollar signs in the form of PPV.

I said that outside the USA soccer draws gigantic audiences. *The Guiness Book of Records* shows just how gigantic: the largest cumulative audience (thirty-one days) for television is listed as the 1990 World Cup in Italy. From June 8 to July 8, 1990, an estimated 26.5 billion people worldwide watched soccer. Add to that the PPV income statistics for boxing and wrestling. *Hollywood Reporter* (December 13, 1991, pp. 1, 53) stated that in 1990 PPV grossed USA$135 million for thirty-two boxing and wrestling events. That figure nearly doubled in 1991. With numbers like that staring them in the face, it's no wonder sports franchise owners are looking increasingly to worldwide TV and global PPV as a lucrative future.

Boxing and Wrestling on PPV

What is the future for other PPV events is past and present for boxing and wrestling. The World Wrestling Federation (WWF) first offered *Wrestlemania* on PPV in 1985, grossing USA$300 thousand. The March 1991 take from *Wrestlemania* was more than $20 million. Twelve of the top twenty top-grossing PPV events have been WWF events.

Boxing has done well too, but it appears to be more dependent than wrestling on the quality of the match. For example, the Mike Tyson–Michael Spinks match in 1988 grossed more than USA$20 million; Sugar Ray Leonard–Thomas Hearns in 1989, more than $20 million; Evander Holyfield–James "Buster" Douglas in 1990, well over $30 million; Mike Tyson–Razor Ruddick in March 1991, more than $30 million; and Evander Holyfield–George Forman in April 1991, more than $30 million. Mike Tyson's preconviction draw resulted in a PPV series on Showtime Event Television. Time Warner developed their TVKO PPV series. An analysis of one fight card reveals a number of the variables PPV promoters have to deal with.

Steve Wynn: Knockout at the Mirage

Steve Wynn paid $32 million to get the James (Buster) Douglas and Evander Holyfield heavyweight championship. The fight was held at his Mirage Hotel in Las Vegas, Nevada, long-time US gambling mecca. He also got the rights to televise the fight via PPV—and a lot of laughs from a lot of skeptics who said his bid was outrageously high and that he'd lose a lot of money.

Prior to this, most major PPV boxing matches had been franchised out to local/regional promoters who marketed the fight right to their area's cable systems. Ever the gambler, Steve Wynn kept the 10 percent he would have had to pay to regional promoters and produced and marketed his own PPV telecast. In 1990 there were an estimated 14.8 to 14.9 million cabled TV households in the USA capable of ordering the fight. The price was pegged at $34.95.

The question was how many of the potential 14.8 to 14.9 million potential viewers would ante up. PPV promoters talk in terms of "buy rates." For the sake of convenience, say you have 15 million potential homes and a price of thirty-five dollars. If *every* home bought in, you would have a 100 percent buy rate:

$$15,000,000 \times \$35 = \$325 \text{ million}$$

The only event that might draw that size audience would be Michael Jackson's marriage to Madonna. It will never happen. So you lower your sights—a lot.

Music shows—rock concerts—on PPV have low buy rates. The highest up to 1991 was reportedly the Rolling Stones concert, which came in with a buy rate just over 2 percent. Boxing and wrestling matches do better, usually somewhere in the range of 4 to 6 percent. If Steve Wynn got a 5 percent buy rate (.05 × 15 million = 750,000) at thirty-five dollars per household, that would total

$$750,000 \times \$35 = \$25,250,000$$

That total would, of course, have to be split with the cable companies. There were reports that some of the Mirage deals with cable outlets had a 70/30 split, with Mirage copping the 70 percent. But let's average it at 60/40, with 60 percent to Mirage and 40 percent to the cable companies:

$$60\% \times \$25,250,000 = \$15,150,000$$

Fifteen million is still shy of $32.1 million, but Steve Wynn had other potential income from the fight: $9 million from Mirage's 16,200-seat stadium; $5 million from delayed television rights; $1.5 million from closed-circuit TV. That's $15.5 million to add to the potential PPV income of $15 million, a total of $30.5 million, still shy of $32 million—but we haven't counted in overhead to present and promote the fight, an added cost of at least several million dollars.

Betters Counted the Mirage a Loser/Winner

The consensus was that the Mirage would lose money on the Douglas and Holyfield fight but would recoup most of the loss by increased revenue from the gambling tables at the casino. With deliberate foresight, the fight had been scheduled for a Thursday night, creating a long weekend, Saturday historically being the biggest income-earning day of the week at the casino. That gave the Mirage the best chance of making up potential losses on the fight itself.

Wynn's Win Wins Converts

In the fight in the ring, challenger Holyfield knocked out title-holder Douglas (who had upset Mike Tyson to win the championship). In the fight for PPV dollars, the Mirage knocked out the skeptics. The PPV buy rate was a reported 7 to 8 percent. Not only did the Mirage not lose money on the fight telecast, it made money. And whatever added income came to the casino tables was pure profit.

With that gamble having paid off so well, speculation and predictions about future PPV flooded the air, with claims and counterclaims feeding off each other in news report after news report. Everyone was convinced it was a turning point.

The Olympics and NBC and PPV

One group that was certainly convinced was NBC, who conceived triple-cast plans for the 1992 Summer Olympics from Barcelona, Spain. NBC sports president Dick Ebersol said their research showed that 69 percent of the viewers of the 1988 Summer Olympics had expressed a desire for greater in-depth coverage of the events (*Daily Variety*, March 28, 1991, p. 8). That's what NBC planned for 1992. Their coverage spanned fifteen days, with twelve hours of programming per day. The PPV coverage in the USA supplemented—did not replace—"free" TV coverage. Europe had extensive HDTV coverage. And even though Eastern Europe's OIRT was not incorporated into EBU until the 1993 Summer Olympics, portable ground stations downlinked the signal for Eastern Europe.

The advertising slogan for the PPV portion of the telecast read: "Catch the Fire Behind the Flame . . . A Whole New Way To See the Games . . . Three full-time cable channels, live 12 hours each day, then replayed in entirety . . . uninterrupted showing of Gymnastics, Track & Field, Swimming, Diving, Basketball, Volleyball and Boxing . . . Commercial free pay-per-view . . . To complement the extraordinary NBC Network Broadcast . . . Convenient Bronze, Silver and Gold Packages for your choice of subscription level."

The $95 Bronze Package bought seven weekend days and nights of viewing (opening Sunday and two three-day weekends). The $125 Silver Package offered all fifteen days of PPV coverage plus a 50 percent discount on merchandise. For $170, the Gold Package offered all of the above plus a videocassette and a hard-cover commemorative book. Added later was a daily package, available for $29.95. An abbreviated form of interactive TV offered PPV customers a chance to call a "Viewer Request" phone number on the last weekend only to access some of the less popular events not included in the regular coverage.

NBC used Panasonic's new half-inch digital compact video format for their coverage, including a super slow-motion replay device. They aired some 150 hours of "free" coverage and some 600 hours of PPV coverage. They also paid USA$401 million for the rights to televise the 1992 Summer Olympic Games. The outcome of this PPV experiment may have even longer-lasting significance for PPV than did the Douglas–Holyfield fight.

To recoup that investment, they had to reach beyond normally addressable cable homes. The term *addressable homes* refers to those TV households fitted with the technology that allows the signal transmitter (the

Figure 17–1 This Jerrold IMPULSE 7000 impulse pay-per-view addressable converter symbolizes the in-home link to interactive communication, which has only begun to revolutionize the entertainment and sales components of television. The success of commercial telecasting of global events like the Olympics will depend increasingly on the expansion of both DBS and fiber optic cabling to addressable homes.

© Todd A. Trice, 1988.

cable company) to distribute PPV events on a selective basis and to bill the receiving household for them. Logically, the larger the number of addressable households, the greater the potential income from PPV events. The two largest suppliers of PPV in the USA have been Viewer's Choice and Request TV; their combined outlets totaled an estimated 15 million households in 1990, 18 million in 1991, 20 million in 1992. But with hundreds of local cable companies accessing those PPV services, there are addressable households and addressable households—some inconvenient and some convenient. Guess which system resulted in more PPV activity? The inconvenient systems required viewers to spend half the evening punching numbers on a touch-tone phone: first the phone number of the PPV service (seven numbers), then the numbers of the viewer's account (as many as nine numbers), then the number for the event code (four numbers). It takes dedication well beyond impulse to order an event. A more convenient system allows the viewer to push one or two buttons at the most. Figure 17–1 shows a Jerrold IMPULSE 7000 pay-per-view addressable converter, an example of the type of on-off device that cable companies will gradually supply to their subscribers.

NBC's triplecast had to reach beyond the addressable households to recoup its $400 million-plus investment in the Summer 1992 Olympics, simply because 20 million addressable households would not be enough to pay back their investment. So NBC set a trap for the estimated 40 million cabled households that did not have addressable cable converters—an updated version of a device used by PPV boxing promoters during the 1980s as a one-time-use access adapter. Because NBC's cable coverage was designed as a triplecast, offering PPV viewers a choice of any of three channels, round the clock, for fifteen days, Cablevision Systems supplied a three-channel access adapter that was either installed by the local cable service or shipped directly to the viewer's home for do-it-yourself installation. For a small fee (in the six to seven dollar range) plus a deposit that was refunded when the trap was returned and a shipping and handling charge, nonaddressable households were temporarily lifted into the hallowed circle of the addressables. This upgrade charge, however, was peanuts compared to the fee for the triplecast itself, also three-tiered: $95 for weekends only (or for the first seven days of coverage, viewer's choice), $125 for the entire fifteen-day 500-plus hours of coverage, or $170 for all those hours plus prizes, such as books and videocassettes.

Unfortunately, the traps didn't trap enough extra viewers to pay off NBC's investment—doubtless closer to $500 million than the $400 million it paid just for the rights to televise the games. Was the prestige gained worth the loss? One NBC answer: After paying $456 million for rights to the 1996 Summer Olympics: no PPV. You can be assured, also, that others will reflect on the 1992 outcome long and hard. Also being watched will be the ABC Sports/Showtime Event Television teaming to offer three PPV college games per weekend at $8.95 for the first game plus up to three more for an additional dollar. Seasonal charge: $59.95. Since games are already being televised there's little chance to lose.

Politics and PPV

When you look at the records of sports events on TV, it's easy to see the PPV temptation—for politicians as well as franchise owners. Seven of the ten most-watched programs in the USA through the 1980s were NFL Super Bowls. Viewership of the Super Bowl has been declining, but not a lot. It's still the USA's biggest sports draw, followed by the NFC and

AFC championship games, followed by the World Series, followed by the NCAA Basketball Championship game, followed by *Monday Night Football*. The Super Bowls draw more than 30 million viewers. *Monday Night Football* averages more than 10 to 12 million. You can see why everyone is wondering when the other shoe will drop.

The first shoe will be the NFL tests in the PPV market. In an interview on ESPN's *Outside the Lines/The Commissioners' Report,* NFL Commissioner Paul Tagliabue said that the league's television contracts provided that the NFL could run PPV tests in the 1992 and 1993 seasons. That announcement prompted letters from Massachusetts Congressman Ed Markey, chair of the US House of Representatives Communications Subcommittee not just to Commissioner Tagliabue, but to the commissioners of the NBA, the NHL, and Major League Baseball asking for clarification of any plans for PPV.

In April 1991, Amos B. Hostetter, Jr., of Continental Cablevision, actually urged that Congress rule on which sports events should remain on "free" television and which could go to PPV. He didn't have long to wait. In July 1991, Pennsylvania Congressman Peter Kostmayer introduced the Fairness to Fans Act of 1991, which aimed to take away the sports leagues' federal anti-trust exemption if they allowed teams to shift local games from "free" TV to PPV. Another bill was introduced to require that all major league championships be televised on either "free" TV or basic cable—not PPV or premium channels. Neither bill was given a chance of passing in the near future, but the stirring of the PPV hornet's nest did draw promises from the commissioners of baseball, basketball, and football that league championship games will remain on "free" TV through the end of the decade.

PPV Sports Results Vary

Another PPV caution for franchise owners is that the results of past PPV events have varied. When the Philadelphia Phillies announced that their opening game would be available only on PPV, fans rebelled and that PPV test was cancelled. On the other hand, the Minnesota North Stars scored big on PPV telecasts of their 1991 home games of the NHL Stanley Cup Playofffs. A team no one had expected to survive beyond their first playoff games, the North Stars went on to be in the finals against the Pittsburgh Penguins. The North Stars ultimately lost the series, but they reportedly grossed $600 thousand for each home game, a surprisingly good income for a regional telecast.

With the championship games protected from PPV until at least the year 2000, will the leagues go for the frequently mentioned season ticket idea? Could baseball, with games played every day, top the four-year $1.06 billion deal it got from CBS (along with $400 million from ESPN for cable rights)? The six-year ABC/NBC post-CBS (1993) deal had no upfront money, only a percentage of sales to the teams, mostly regional. Sound like global PPV?

Sports Worldwide

The picture in European sports telecasting remained muddled as this book went to press. In February 1991 the EC ruled that the ownership of Eurosport, a European satellite TV channel, violated EC law. In early May, Eurosport went off the air. In late May it resumed broadcasting under the aegis of the private French network TF-1. That same month, a French consortium that included ESPN agreed to buy the W.H. Smith

Group's satellite and cable TV company, WHSTV. WHSTV's holdings included 75 percent ownership of the European Sports Network, the parent company of Screensport in the UK, TV Sport in France, Sportnet in the Netherlands, and Sportkanal in Germany. In the meantime, Rupert Murdoch, freed from his involvement in Eurosport, launched a new sports channel, Sky Sports, on BSkyB. It was announced, however, that Sky Sports signal would be scrambled so that only UK audiences could receive it. While that situation appears muddled, it certainly indicates much interest in televised sports outside the USA.

On the Pacific Rim, sports programming Prime Network International signed a partnership agreement with Hong Kong's Hutch Vision, the satellite broadcasting division of Hutchison Whampoa. Prime Network reportedly produces/distributes some three thousand sports events worldwide annually. Meanwhile, ESPN continues its global efforts. ESPN is shown on six satellites around the world and is seen in more than sixty countries.

PPV Beyond Sports

Following the Douglas–Holyfield match, various prognosticators foresaw expanded production well beyond sports events: entertainment specials, rock concerts, and so on. But the unwritten law for PPV is: "PPV is event-driven." That law prompts the question: "What is an event?" And the answer prompts the financial and promotional battles being waged.

If *event* is defined as the most financially successful PPV telecasts, the answer is big-time boxing and big-time wrestling, accounting for more than 90 percent of PPV revenues to date. According to PPV sports promoters, another unwritten law of PPV is that big-time sports are EVENTS and everything else is a nonevent. They have a point.

Sports provide the biggest PPV draw because the events are one-time-only, they contain the drama of conflict, and their outcome is unpredictable. Further, the audience for sports events tends, by its nature, to be worldwide and to span age brackets. As a result, boxing and wrestling have been the mainstay.

But there is PPV beyond sports. In fact, one of the nonevents—theatrical films—is in a knock-down-drag-out battle with Home Video rentals and sales.

PPV: Events Versus Nonevents

The difference between an EVENT and a nonevent is the difference between a 2 percent and a 6 percent buy rate. Many promoters say that difference is too great for them to spend time on musical events or films, or even minor sports. Big leagues, big money—the rest is not happening, as far as they are concerned.

There are, however, other promoters who think they can live with a 2 percent buy rate. If the event charge is only ten dollars, a 2 percent buy rate of 15 million homes brings in $3 million. Split 50/50 between signal carrier and promoter, that's $1.5 million for the promoter. So the same phenomenon is gradually developing in PPV that is developing in—or because of—cable channel proliferation: niche telecasts. The Stones did it. Tina Turner did it. Rod Stewart niched a Valentine's Day special in 1992. The audience for PPV tends to be young. There may even be a niche audience for rap. There could be a niche for country and western. But the Metropolitan Opera's Twenty-fifth Anniversary PPV special lost money.

To turn a musical nonevent into an EVENT requires some quality of uniqueness. If some rockers are full of conflict and unpredictability, generally rock concerts aren't, but there are fans willing to pay. If a promoter can come up with a one-time-only never-to-be-repeated concert, that might become an EVENT. With global PPV, a one-time-only reunion of the Beatles—every promoter's dream—could become an income-producing bonanza and a long ancillary life.

Films released in the theaters are considered nonevents: they are not premieres, not unique, not one-time-only. That doesn't mean that films are not a major factor in the future of PPV—far from it.

Home Video Rentals Versus PPV: Windows

The second factor in the future of PPV is the battle with home video rentals and sales. While there is symbiosis as well as a battle between PPV on DBS and PPV on cable, there has been nothing but pure antagonism between PPV and the members of the Video Software Dealers Association. Despite cable operators' claims that PPV and home video outlets can and will complement each other, not kill each other, some members of the Dealers Association went so far as to threaten boycotts against selected films if producers let PPV precede or move simultaneously with video store releases.

The producers apparently agree, and indications are that they will continue to grant the video outlets a window—a period of time, customarily six weeks—during which the video outlets will have exclusive rights. Only after that window will films be released for use on PPV or any other TV source. (Producers' other windows have wavered, however. For some time there was another unwritten law stipulating that theatrical films would not go to home video until six months after theatrical release and would not go to television until a year after theatrical release. Both parts of this law have frequently been broken.) Maintaining the exclusive window for home video makes sense from the producers' point of view, however, because they get two sources of income. Big hits in the theaters tend to produce big hits in the rental/sales market. But hundreds of non-hits in the theater return big profits for both video dealers and producers. Just the continuing return from home video rentals of martial arts and horror films would be enough to convince producers to keep the video rental stores alive.

It's easy to see why some members of the Video Software Dealers Association feel threatened. At the fourth annual Western Regional PPV Conference of the Cable TV Administration and Marketing Society in December 1992, Pete Warzel, president and CEO of United Artists Theater Circuit, predicted a decline in event PPV coinciding with strong growth in film distribution via the pay-TV route. At the same meeting 20th Century Fox president and CEO Strauss Zelnick predicted that in ten years PPV will be the chief medium to distribute films on TV. Robert Friedman, New Line TV president, echoed similar thoughts in July 1993 at a National Academy of Television Arts and Sciences luncheon. There was a near panic when Time Warner's plan to double capacity on two of its cable franchises in New York City took these predictions closer to reality.

What concerned the video retailers, of course, wasn't niche channels dedicated to foreign languages, sports, music videos, or home shopping. What upset the retailers was knowing that half of the new channels would be following the same pattern SkyPix planned on DBS: major PPV feature films with start times every half hour or on-demand, around the clock.

And the Time Warner move in Brooklyn/Queens was just the tip of the iceberg. In May 1991 Tele-Communications Inc. joined with AT&T and U.S. West for an eighteen-month market test in the Denver area of Colorado called "Viewer-Controlled Cable Television." The service offered first-run films at selected times plus some one thousand other film titles available at any time to subscribers. It was at the end of this market test that TCI announced its plan to install digital cable boxes in the homes of one million subscribers.

At the same time, Home Box Office began "multiplexing" tests on its product on both HBO and Cinemax, effective August 1991. The multiplexing test—on cable systems with unused channel space—consisted of offering programming on three different channels (at no additional cost). No new programs were introduced; the move was not to expand product but to increase access to it. With each channel programming the same product but at different times, the viewer would have three choices instead of one at any given hour. HBO's competitor, Showtime, also reportedly planned a new low-cost supplement to both Showtime Channel and the Movie Channel. Both the HBO and Showtime efforts were seen as possible reactions to reports that subscription rates had leveled off and that the percentage of cable subscribers paying extra for pay cable services had begun to decline. Where was the blame placed for the decrease in cable subscriptions? On the increases in basic cable rates, which multiplied rapidly after the deregulation engineered in 1986 by President Reagan's administration, on the increased penetration of VCRs (very close to 80 percent; more than 70 million households), and on—you guessed it—PPV.

Everyone seemed to be staring at the same wall, aware of the handwriting but not sure of the message. The two words that seemed to be visible to all, however, were *convergence* and *convenience*.

Copy Protection for PPV

The "Captain Midnight" syndrome plaguing cable also plagues PPV—on cable and on DBS. Copy protection has become a sore subject between film producers and PPV outlets. Reasoned the producers: "You show our film on PPV, everybody and his brother tapes it on the VCR and there go all our home video rentals and sales!" Now reply PPV transmitters like Northwest Starcan's SkyPix: "We have you covered. Viewers accessing the film just to watch pay one fee. The feed to them is transmitted using copy-protection technology. They watch, but they can't copy. If they want to copy, they pay an extra fee. Our system bypasses the copy-protection system electronically. They tape. We split the extra fee with you, the producer."

No marriage has been announced yet between film producers and PPV outlets, but they're living together. As in most relationships, communication and negotiation are the key ingredients to success. So whether you call it Video on Demand or Home Theater or Home Video Store or—as Viewer's Choice calls it—"Continuous Choice," keep your eye on those elements of the Age of Transition: convergence and convenience.

Summary

Even before the Douglas–Holyfield fight, there was speculation that with the growth of PPV in the EC and the Pacific Rim, a major fight would eventually produce a gross of as much as USA$500 million worldwide. The

NFL's announcement, later modified as noted, that it might begin experimenting with PPV in 1993 prompted predictions of a PPV Super Bowl that would bring in hundreds of millions globally. The broadcast networks and elected officials countered with their vows that the major sports events would never go to PPV. But the jurisdiction of the US Congress doesn't extend much beyond the fifty states and the Philippines. And in the end, maybe both sides will be right. Perhaps the Congress will be able somehow to legislate continued "free" TV coverage of the Super Bowl and the World Series in the USA with globalcast rights outside the USA going to PPV.

Once the world gets fully equipped for DBS and wired with fiber optic cable, those who say the economics of the world will force these major events into PPV will probably be as smiling—and wealthy—as Steve Wynn after the Douglas–Holyfield fight.

The need to summarize all of these conflicting, converging factors prompts me to issue two globalcasting proverbs:

Globalcasting Proverb 1: "It isn't what you do, it's the way that you do it."
Globalcasting Proverb 2: "It isn't the way you do it, it's what you do."

By way of amplification: proverb 1 says—truthfully—that it isn't whether you broadcast a good special of a good film on cable or DBS, it's "the way whatcha do it"—*conveniently,* available with no-effort access and, if it's a film, available every half-hour or every fifteen minutes or continuously, around the clock. Do it that way, bingo! Success.

Proverb 2 says—truthfully—that it isn't how conveniently you broadcast an event (be it wrestling, boxing, a film premiere, a rock concert), it's what you broadcast. Is it a unique, exciting program people want to see, or is it a dull match between unknown contenders, a bomb, or just one more stop on a global rock tour?

The rules of globalcasting, I suggest, come down to the same common sense as all other programming. You need a good show at the right time for the target audience, delivered in a manner that requires little or no effort on the part of the viewer. To find significant rating numbers, you need to find your niche, your "names," your time slot, your couch potato-proof one-button-one-voice availability and proper promotion. End of PPV proverbs.

Chapter *18*

Trends in Entertainment

Trying to predict trends in entertainment programming—globally or domestically—is like trying to thread a needle in the dark, blindfolded, with one hand tied behind your back—and no eye in the needle. The graveyard of former programming vice presidents is filled with living corpses. If ratings didn't kill them off, the pressures did.

Programmers are a strange breed. They have tremendous power, and so they crave power, so much that it's difficult for them to accept outside judgment, except for statistics compiled by an outside research group that can be blamed and fired for failure. It's grown almost impossible for a network programmer to follow his or her own instincts or judgment. The days when an innovative master like Sylvester "Pat" Weaver could conceive *The Today Show* and *The Tonight Show,* thus assuring network control of programming that had previously been controlled by ad agencies, seem so remote now that it's ancient history. And since it was forty years ago that Pat Weaver wove his magic into the business, maybe in television time it qualifies as ancient history.

In any event, I'm not going to predict a single thing here. I'm just going to indicate what is going on and some directions in which programming seems to be heading, with one qualification. Because of the proliferation of choices available to viewers and because of the ever-diminishing share of audience tuned to past programming leaders, network programming has little or nothing to do with taste or quality or imagination. As this decade neared its halfway mark, programming was a struggle for survival, and the networks seemed to be loading up with scattershot, hoping to hit *some* target—anything—even if it wasn't the target they were aiming for.

Reality Rears Its Many Heads: Television on Trial

An occasional sitcom, such as *Cheers, The Golden Girls,* or *The Cosby Show,* seems to last forever, to rise, fall, be reborn, gain new heights. Most new shows introduced on television fail. A few individuals, like Norman Lear, with *All in the Family* and its many spinoffs, and Susan Harris, with *Soap* and its follow-ups, change the way people look at television and at themselves. The rest is a crapshoot with one-parent families and troubled teens and strange animals populating the small screen. In desperation, then, the USA networks and their counterparts around the world have been turning increasingly to programs that are scripted by the headlines. "Reality" has become the refuge of the program director.

60 Minutes started a trend nobody recognized as a trend. Real stories, real people, top-notch talent and producers. Keep it simple. Tell a story. *60 Minutes* and its creator/executive producer Don Hewitt deserve all the awards—and audiences—they've won. And the best part, from the networks' point of view, is that programs like *60 Minutes* are cheap compared to dramatic or comedy shows that cost $.5 to $2 million for a half-hour or $1 to $2.5 million for an hour production.

The other networks have tried to duplicate CBS's incredible ratings success with *60 Minutes*. No one has matched it yet, but some solid programs—*20/20, Prime Time Live,* and *48 Hours*—have developed niche audiences. Some, such as NBC's low-rated *Real Life with Jane Pauley,* gave way to new attempts (*Dateline NBC*). A trend that, happily, faltered tried to integrate drama with news, the so-called docudramas. Programs like NBC's *Yesterday, Today and Tomorrow* and Connie Chung's Saturday night show on CBS (before she anchored "Eye to Eye with Connie Chung" and co-authored with Dan Rather) found audiences and critics alike rebelling against half news/half supposition.

The entertainment side of network programming picked up on the idea of using real news as real drama, lured by the same relatively inexpensive production costs, no cast to worry about beyond a narrator, and an endless supply of story lines. Programs like *Top Cops, Unsolved Mysteries, America's Most Wanted,* and *Rescue 911* searched the news for plots and occasionally made news themselves. A crew for *Rescue 911* was taping a police crew in Boston when they were called to what turned out to be a headline-making murder of a pregnant wife. *America's Most Wanted* made news because of viewers' responses to video of wanted criminals.

The interactive nature of the reality shows, as they are called, is a characteristic of many of these shows worldwide. Germany developed *Aktion Zeichnung, X.Y. Ungelost,* asking for viewer calls to solve crimes. Italy came up with *Chi L'Ha Visto?* In the Netherlands a similar program, *Opsporing Verzocht,* sought audience help. A number of countries have also adopted the *America's Funniest Home Video* format (itself based on a segment of a Japanese program), *Candid Camera,* and *Rescue 911.* In England the BBC went one step further, with the series *Video Diaries,* by providing hand-held cameras to individuals to produce their own stories on tape. The BBC helps with the editing, but all of the taping is purely personal.

Perhaps the extreme in this category is the new Courtroom Television Network in the USA. Co-owned by a group of MSOs (TCI/Liberty Media Corp., Time Warner, Cablevision Systems Corp.) and NBC, Court TV is a 24-hour cable operation that televises live courtroom cases from around the USA, interspersed as needed with taped segments. Off to an iffy start in July 1991, the new network's executives no doubt cheered silently when the judge in the William Kennedy Smith, Palm Beach, Florida, trial ruled that the camera would be allowed and the trial could be televised. Postponed until early 1992, the trial's combination of the Kennedy name and the sensationalism that surrounded the original accusation of rape guaranteed a substantial boost in the fledgling network's viewing audience. As a follow-up, Jeffrey Dahmer's sex-slaying trial in Milwaukee, Wisconsin, added to the impetus. After the pretrial publicity by prosecution and defense in the Kennedy Smith trial, some in the news media wondered if the court of public opinion hadn't become more important than the courts of law. Why have a trial when you can go to a 900 number for audience call-in? Fifty cents a call: Guilty or Not Guilty? Cheaper and quicker, just like reality TV.

Whatever else it was, because of its low cost and interactive nature made reality TV a trend. It slowed when sponsors began to balk at the unpredictability of "reality," but low cost is alluring. Will it ever go global? I promised no predictions, just observations.

Global Game Shows: Call 900-Interactive

In France it's called *Le Juste Prix.* In Germany it's called *Der Preis Ist Heiss.* In Spain it's called *El Precio Justo.* Bob Barker calls it *The Price Is Right.* It's just one of the game shows that have gone global.

The number one rated game show in the USA, *Wheel of Fortune,* latched onto the interactive lure by offering a chance to win prizes to home viewers who called a 900 number. Proceeds from the show's first call-in went to Toys for Tots. *Variety* pointed out, however, that the profits from the phone calls don't have to be donated to a charity. In the future they could add to producers' profits. They estimate 45 percent of the revenue goes to the phone company, plus another 10 percent to the company providing the system. The other 45 percent could end up in the syndicator's profit column. Producers claim that profit is less a motive than establishing the interactive contact with viewers. (Programs like *Inside Story* and *Hard Copy* have often used 900 number phone interactivity to poll their audiences on "hot" subjects, such as messages to hospitalized stars (Elizabeth Taylor) and stars being badgered (Pee Wee Herman).

In the CIS, *Wheel of Fortune* is called *Field of Miracles,* and the show reportedly receives 40 thousand letters a week from viewers, many of them hoping to become contestants. The CIS also teamed with Phil Donahue for interactive global talk shows, with q-and-a between nations.

Se Habla Español

One exceedingly positive trend in television programming is expansion into multiple language feeds. For example, in January 1991 PBS's highly respected news program *The MacNeil/Lehrer Report* began a test with interpreters simulcasting the program in Spanish in more than thirty major cities in the USA. In the same month, using the B-MAC satellite signal, Turner Broadcasting's TNT Channel began offering its service in three languages—English, Portuguese, and Spanish—in Latin America and the Caribbean. In June 1991 Venevision International, a Venezuelan company, announced plans to offer a new Spanish-language cable network in the USA, based in Miami, Florida. There are an estimated 3 million-plus households in the Spanish cable market in the USA, already served by Univision, Telemundo, and Galavision. In 1993 the Fox Latin American Channel opted for Spanish/Portuguese/English. There was HBOle. *Oprah,* often dubbed, was in more than 60 countries. Barry Diller's QVC with BSky B brought Home Shopping to UK/EC and, with Groupa Televisa, S.A. de C.V., to Latin America, Mexico, Spain.

As the multiple channels open up with increased digitalization, watch for more and more multilingual telecasts. "Local" television may edge ever so slowly toward globalcasting.

Let Me Infotain You

There are two trends in USA entertainment programming to watch, both related to commercial interests. As other countries deregulate state-owned television, this phenomenon could ultimately become a global problem. With luck, the USA experience will alert other countries.

Among many deregulations under President Reagan's administration in the USA came deregulation by the Federal Communications Commission of the number of commercials allowed in any given time slot on television. This anything-goes deregulation resulted—inevitably, one might suggest—in programs dubbed *infomercials*. Programs such as *Amazing Discoveries* pretend to present amazing discoveries that are actually commercial products. An inordinate number of the products seem to remove stains or cure baldness. Despite the fact that the entire program consists of pitching commercial products disguised as "discoveries," most of these infotainment vehicles take commercial breaks for undisguised commercials. The total program, therefore, consists of commercials disguised as programming interrupted by commercials posing as commercial interruptions. For this scientists sweated to create radiovision.

A second trend, not quite so appalling perhaps, but disturbing nevertheless, has to do with news program tie-ins to programming. If there is a network special about child abuse, the late local news show follows the special with a report on a local case of child abuse. The early news show can plug the late news report—and, coincidentally, of course—the network special. If there is no local tie-in, frequently the network will provide a taped interview with one of the stars featured on the special. Sometimes both elements show up on a newscast. Shrewd programming? Misuse of news programs? Borderline?

In an almost parallel, all-network effort, NBC ran a miniseries called *Drug Wars: The Camerena Story*. It covered a powerful story about the murder of a USA Drug Enforcement Agency agent that suggested possible implication of Mexican authorities. The incident created long-lasting problems between USA and Mexican officials. The questionable tie-in was between *NBC Nightly News* and the miniseries. For the week preceding the miniseries, *NBC Nightly News* carried a lot of drug-related stories. After each episode of the miniseries, *NBC Nightly News* anchor Tom Brokaw hosted a fifteen-minute special report about the Drug Enforcement Agency. In addition, the miniseries incorporated actual news footage from *NBC Nightly News,* featuring Tom Brokaw. Shrewd programming? Misuse of news programming? Borderline?

Let Me Infotain You.

"I Surrender, Dear"

At a July 1991 news conference ABC-TV's president of entertainment programming, Robert A. Iger, offered a possible solution to all of the concerns just raised. Eventually—although not now—the USA networks may cut back from the current twenty-two-hour prime-time programming they supply to affiliates. They would return some of the unprofitable time to the local stations. The FCC recently defined a network as a service providing fifteen hours of prime-time programming per week. If the networks cut back enough, they could cease to be defined as networks and deregulate themselves. Could they then consider doing infomercials?

TV Inside TV: *Dream On*

One new trend has shown up in two programs, one on cable and one on both cable and an over-the-air network. The program idea often produces interesting results, but the fact it reflects is even more interesting.

Dream On is an award-winning HBO presentation, a half-hour comedy aired twice a week. Its premise has its hero's thoughts revealed in fast clips of old movies and television shows. Conceived by a pair of off-Broadway playwrights, Marta Kaufman and David Crane, the show creates a comic situation for its book-editing hero, then expands the comic moment by accenting it with the flashback clips.

Hi Honey, I'm Home, a half-hour sitcom produced by and aired on the cable channel Nick at Nite (Nickelodeon), also aired on ABC. Actually, it aired first on ABC, then was repeated—twice—on Nick at Nite. However brief its lifespan, the show went into the television history books as the first program to air on cable and noncable at the same time. The plot point in *Hi Honey, I'm Home,* however, ties it to that of *Dream On. Hi Honey, I'm Home* plopped a TV sitcom family from the 1950s into a sitcom of the 1990s, part of a "Sitcom Relocation Program."

The hook in both, of course, is that television has reached such an age that it can feed on its own history. A trend? Two programs probably do not indicate a trend. But two programs are one more than one program.

TV Guide Guides Us

Do viewers want to watch television programs about old television shows? Or news programs that plug nonnews programs? Just what do viewers want to watch in the 1990s? In 1990, *TV Guide* (January 20, pp. 11–15) reported on a telephone survey of more than one thousand people that it commissioned by the Roper Organization. The results?

- Three out of four interviewees said they want more movies, and over half said they would prefer to see first-run movies at home rather than in a theater. Only 38 percent, however, said they'd be willing to pay to see such first-run movies on TV. Almost as many said they'd also be willing to pay to see Broadway shows, special concerts, and so on.
- A whopping majority wanted sports championships kept in the land of "free" television, but about one-third of those interviewed said they'd be willing to pay a little bit if they had to.
- More than half expressed a desire for more news and reality programming. Obviously the program directors read this part of the survey.
- Only 24 percent expressed a desire for more game shows.

Probably the only certain thing one can say about TV programming is that it, like everything else in the industry, is in a state of transition. Cable has wooed viewers away from over-the-air networks and local stations. DBS and cable channel proliferation will alter existing patterns even more. The number of available programs—good, bad, or infomercials—will be staggering.

Trying to predict trends in entertainment programming—globally or domestically—is like trying to thread a needle in the dark, blindfolded, with one hand tied behind your back—and no eye in the needle. So much for reruns

Chapter *19*

Trends in Coproduction

How Quotas from the European Community Sparked Coproduction

Behind all of our discussions of advanced and advancing technologies, one element has hovered: The economic and political rules and restrictions of individual nations and groups of nations that will, in large part, determine the directions in which globalcasting will develop in the future.

Three major forces have driven television production, especially global television production, to coproduction: politics, financing, and the need for global audiences.

Quotas and limitations established during the official formation of the European Community (EC) had a great impact. The trend set by the EC rippled out to create additional quotas and restrictions by other nations. Simultaneously, the financial recession in the USA, the Soviet Republics, and elsewhere pushed more and more productions into cooperative efforts to cut costs. A need for income fueled the outreach for broader, more global audiences. "Will it play overseas?" became a critical part of production designs.

These three forces, combined with the new relationships being forged in Europe as a result of the unification of East and West Germany and the gradual opening of Eastern Europe, are bound to have a profound effect on globalcasting, on program and film distribution, and on coproductions in general. Similarly, dramatic economic and political pressures are being felt along the Pacific Rim and worldwide.

How the EEC Became the EC

The European Community, in its 1992 formation, consisted of twelve countries with an estimated total population between 325 and 350 million and a collective gross national product just under $5 trillion. The twelve members were Belgium, Britain, Denmark, France, Germany, Greece, Ireland, Italy, Luxembourg, The Netherlands, Portugal, and Spain. Just as the coming together of the original thirteen states in the USA was not an easy transition, the formation of the twelve-member EC followed a long and arduous path leading—so far—only to economic, not political, union. Eleven different currencies, ten different languages, age-old trade barriers—these and much more had to be overcome to arrive at a community with a common currency and no internal trade restrictions. As we'll see, the *external* trade restrictions between the EC and rest of the world are many and complicated.

The transformation of twelve nations into the EC began a forty-year germination in 1952, when six nations established the European Coal and Steel Community to unify their labor market and products. In 1958, under a group of economic and trade policies set up by the Treaty of Rome, another Pre-European Community organization was chartered. It was officially called the European Economic Community (EEC). Unofficially it was called the Common Market. It also included only the six member nations: Belgium, France, Italy, Luxembourg, the Netherlands, and West Germany. In that same year, one more part of the developing EC was created: The European Atomic Energy Community.

In 1958, when these groups were formed, Europe's main economic competition was seen to be with the USA. Japan was not then recognized as the trade giant it would become. And for the most of the three decades from 1958 to 1992, most of the news about the EEC seemed to relate to agricultural trade barriers. ("If you'll remove your restrictions on our citrus crops, we'll remove our restrictions on your dairy products.") GATT, the General Agreement on Tariffs and Trades, was the focus of ongoing discussions and negotiations, as it still is. Gradually, through the years of bickering, the structure of the European Community emerged. In 1967 the EEC joined the Coal and Steel Community and the Atomic Energy Community to form the nucleus of today's EC. The coalition created a Council of Ministers, an Executive Commission (sometimes called the European Commission), a European Parliament, and a Court of Justice. In 1973 Britain, Ireland, and Denmark joined. Greece joined in 1981, and Spain and Portugal joined in 1986. During all of these years, television did not appear as a major issue. That situation changed dramatically at the end of the 1980s.

In 1988, the EC set the target date of 1992 for the member nations to complete their conformation to the economic and trade practices outlined in the Treaty of Rome. One of the categories covered under the treaty was the broad category of services. Broadcasting was defined as one of those services and was also defined to include all television—terrestrial, cable, and satellite.

A Quota Is a Quota Is a Quota

In 1989, the EC ruled that its twelve members had to reserve most of their transmission time for European-produced programs. News and sports were excluded from this specific restriction, but films, sitcoms, dramas, animation, and so on were included. Hollywood and the rest of the non-European television community shivered. They shivered more when the individual twelve-member nations announced their specific restrictions within the EC guidelines. The quotas varied from nation to nation, and many went far beyond the EC ruling. Some examples follow.

France

- Sixty percent of network prime-time programs have to be produced by Europeans.
- To be classed as a European production, the production company has to be incorporated in Europe.
- To be classed as a European production, 50 percent of the production cost has to come from European sources.
- Half of the sixty percent that has to be European-produced has to be original French-language programming.

- Two-thirds of the actors, screenwriters, and technicians have to be European citizens.
- Two-thirds of the production's final cost has to be spent in Europe.

There are also union restrictions in France, not related to the EC ruling, such as a stipulation that crews work overtime only if they choose to. I can say, however, from experience with the production crew on *Earth 90*, that the French staff was extremely cooperative under tremendous deadline pressures.

Italy

- Forty percent of television broadcasts have to be produced in Europe.
- One-half of that 40 percent must be produced by Italian producers.

The good news is that the number of national TV channels in Italy more than doubled, from seven to fifteen, by the end of 1991. The number of local stations will *de*crease from 900 to 800, clearing airspace. (A number of nations in Europe face a shortage of available frequencies.)

Spain

- Forty percent of television broadcasts have to be produced in the EC.
- Twenty percent have to be produced in Spain.

A note here about barter and how it can be a factor along with coproduction. Spain has only recently begun authorizing privately owned and operated national television channels. With the private enterprise system came the barter system, used frequently and effectively in the USA and elsewhere. Barter is just what its name implies: swapping commodities considered to be of equal value. In television that means a producer provides a TV station with a program at no charge in exchange for part of the advertising time during that program, also at no charge. The producer makes money by selling the advertising time, and the station makes money by selling the remaining advertising time during the show—and gets the show for free. Everybody wins. How does barter relate to globalcasting and EC quotas? All I can tell you is that if one group of lawyers can write rules, another group of lawyers can interpret them. As part of a barter deal with Lever Brothers, one of Spain's private channels (Antenna 3) acquired "ownership" of *Wheel of Fortune,* within Spain only, of course. Since the Spanish channel owns the show, it becomes a Spanish production within the EC regulations. It is reported that Antenna 3 is negotiating with another international giant, Procter and Gamble, to acquire ownership of another game show. Pick a cliché: Where there's a will, there's a way . . . Le plus ça change, le plus ça reste le même . . . My lawyer can beat your lawyer.

Portugal   As this book goes to print, the future of Portugal's deregulation is not clear. It appears that the long-standing monopoly of the two national state-controlled channels (Radiotelevisao Portuguese) will face competition from one or two privately owned and operated national channels. Reportedly, 10% of the broadcasts of the new channels would have to be produced in Portugal while one-half of all programming would have to be in Portuguese. While this quota is similar to that in France, Portugal makes an exception for programming produced in Brazil, where Portuguese is spoken.

There is a strong possibility that other nations may join the EC. But even if other countries do not immediately join with the twelve, every country will be affected dramatically by the changes within the EC. One potential roadblock to coproduction with EC countries developed with that body's adoption of new copyright legislation that included the concept of "moral copyright," inalienable rights of writers, directors, actors, and musicians in their contributions, leading to an "equitable share" of revenues generated within the EC. This revenue-sharing concept may make it rougher for European countries to pull productions in via coproduction. The new legislation was mandated in all member states by 1994.

In a parallel move, there will be a television jointure formed separately from the EC proper. In Western Europe the European Broadcast Union (EBU) has represented some thirty-nine member broadcasters in thirty-two countries. In a series of meetings held in 1991, a merger plan was set in motion to join the EBU with its Eastern European counterpart, the International Radio and Television Organization (OIRT, for Organisation Internationale de Radiodiffusion et Télévision). The OIRT represents thirteen member broadcasters in six countries (Bulgaria, Czechoslovakia, Hungary, Poland, Romania, CIS). 1993 was set as the target date for the OIRT to be absorbed into the EBU, sharing technical standards, satellite facilities, legal rulings, and so on.

Quotas Not Limited to the EC

Possibly in keeping with the concept of merger, quota limitations in Western Europe quickly extended beyond the EC. Eastern European countries began to follow suit. In Hungary, for example, new broadcast legislation, in the formative stage as this book goes to print, called for 35 percent of broadcasts to be produced in Hungary. Barter has also found its way into Hungarian TV programming, with *Disney Presents* airing on Magyar TV's Channel 1 in return for airtime. Poland and Czechoslovakia are edging their way toward some form of private television channels and can be expected to follow the EC's pattern, but Czechoslovokia's split into two nations created an unpredictable situation. In June 1992, Czechoslovokia Television split into Czech TV (in Prague) and Slovak TV (in Bratislava), but beyond that lay political uncertainty.

The trend toward quotas to limit the influence—cultural and commercial—of the USA by limiting exposure to USA programming went well beyond Europe also, especially in the English-speaking countries. In 1991 Australia required that 40 percent of television programming be local, with the percentage scheduled to rise to 50 percent starting in 1992. Canada has gradually been cutting back on USA-produced series. Despite the fact that the federal agency that provides funding for Canadian production (Telefilm Canada) had its funding level frozen through 1996—at the 1989 level of just over Can$145 million annually—in that same year Canadian broadcasters made some of their deepest cuts yet in USA programs.

Sharing Costs and Audiences

What follows is in no way comprehensive or even representative. It only hints at the many areas outside of entertainment that have been involved in mergers and joint ventures for years. These examples of coproduction offer a glimpse of its scope, nature, and intent. There are literally thousands of "deals" in existence, pending and in the take-a-meeting stage. The deals, participants, and outcomes fluctuate daily. Take this section, then, as an indication of what is "out there." The only guarantee you can take to the bank is the fact that, however many coproduction deals exist now, more will exist in the future. As technologies converge and nations deregulate themselves, everyone will be seeking a safe haven that guarantees an audience and a chance for future production.

The most publicized events in the entertainment world in the USA in recent years have been mergers and takeovers: Time and Warner Brothers became Time Warner; GE bought RCA/NBC; France's Thomson bought RCA; Capital Cities bought ABC. Even more in the spotlight were the takeovers by the Japanese: Sony bought CBS Records; Sony bought Columbia Pictures; Matsushita bought MCA, which included Universal Studios, Universal Pictures, MCA Records. A number of voices screamed, "Beware the invasion!" But the sky didn't fall. Perhaps the calls for panic didn't work because so many USA citizens were driving Toyotas, Nissans, and Hondas.

But wait! Are cars bearing the names of Japanese manufacturers the only Japanese cars being sold in the USA? Where is the Dodge Stealth built? The Eagle Summit? Where is General Motors' Geo Metro made? These are all built by the Japanese, mostly in Japan. Pontiac LeMans is made in Korea. So is Ford Festiva. Mercury Tracer is made in Mexico. So is the VW Golf. Until some Japanese officials began disparaging American workers, nobody except the automobile unions took much notice of who builds what cars where. Coproduction had become so much a way of life that it was invisible, except to the people who lost their jobs and a few bookkeepers who kept pointing at the USA's ever-increasing national debt. But even if the US Congress finally admits that the national debt has never been cut and is always increasing (they've cut the rate of increase, never the actual debt) and that automobile sales are a factor, it might not make any difference. Coproduction in the automobile industry is so ingrained it can probably never be changed. Will the entertainment industry go the same direction? Will globalcasting be the television equivalent of the Dodge Stealth? In a way, yes. And the reason is basic and important.

Let's look at a few of HBO's coproductions and—more important—the philosophy behind their coproductions. It is fundamental to understanding what is driving the economic *and cultural* patterns of global productions. (And keep in mind that HBO is part of Time Warner.)

HBO has partnerships with a number of people on a number of projects, all with the same guiding philosophy. *The Tragedy of Flight 103: The Inside Story* and the series *Women and Men: Stories of Seduction* were coproduced with Granada Television. The film *Fellow Traveller* was coproduced with BBC. *The Josephine Baker Story* was coproduced with Anglia Television. *Kids in the Hall,* the ACE award-winning comedy series, is coproduced with the Canadian Broadcasting Corporation (CBC). The link tying all these together? The partner in each case is what HBO's executive vice president, Lee deBoer, calls "end-users"—companies that have an audience to satisfy and so a vested interest in the quality and success of the production.

(This point also underscores the difference between *cofinancing*—in which production control is not shared—and *coproduction* in which the minds as well as the money have to meet.)

The key, then, to expanding into world markets is to join with people already in those markets: "You scratch my back, I'll scratch your back." Together, carefully, we will choose subject matter that will appeal to both of our audiences, and then together we will create products to satisfy the audiences. In coproduction, marketing is the message for the future.

Examples of Coproduction

And the message has come through loud and clear. Read this list of just a few of the elements joining together in today's global market.

- HBO's parent, Time Warner, has entered into a joint cable TV project with United International Holdings in Hungary.
- Ted Turner's interests go beyond CNN. TNT Pictures coproduced *A Season of Giants* with RAI: shared expenses; a shared international "flavor" in the subject matter (the careers of Michelangelo, da Vinci, and Raphael); outlets in both markets.
- *The Global Family* (a thirteen-show series on how the human care or lack of care of the environment can affect animals) was coproduced by TVONTARIO and NHK Japan.
- The Walt Disney Company raised more than half a billion dollars to finance films from a Japanese limited partnership, Touchwood Pacific Partners I. In television, the Disney Channel coproduced two series with CBC: *Avonlea* and *Danger Bay*.
- CBS's long-planned pre-Letterman late night programming turned out a virtual catalog of coproductions to be aired on CBS in the USA and in the coproducing countries. The five one-hour shows aired across the board on CBS, Monday through Friday, included *Dark Justice,* combining Lormar, Magnum Productions, and Spain's TV3; *Sweating Bullets,* linking Kushner-Locke with Accent Productions in Canada; and *Silk Stalkings* (also airing on the USA cable network), a coproduction of Cannell Entertainment Inc. and Stu Segall Productions. From the Monte Carlo TV Market in February 1992 came reports of discussions between Leo Kirch's Beta-Taurus and Silvio Berlusconi Communications about a series of coproductions with CBS that could involve advertising revenue for the investors. A number of shows, all live action-adventure and aimed at Saturday night audiences, would be expensive, about USA$1.5 million per hour. To recoup the huge investment, the suggestion was made to give the overseas investors a share of the advertising revenue once the USA ratings reached an agreed-upon level. Reportedly, ad spots in all of the series, premiering in 1993 or later, would be sold on a global basis rather than country by country. CBS is also coproducing a half-dozen films with Great Britain's Granada Television.
- USA Network's adventure series, *Counterstrike,* mixes Grosso-Jacobson Productions with Alliance Entertainment in Canada. *The Hitchhiker* comes from the joint efforts of France's Atlantique Productions and Canada's Quintana Productions and Lewis B. Chesler Productions, shot partly in Toronto and partly in Paris.
- Australian producers, hard hit by economic setbacks in Australia's television industry (two networks, Seven and Ten, ended in receiverships) are looking overseas for partners. The networks cut their pay scale for telefilms by one-fifth to one-third. The ceiling for a series drama is reported to be about Aus$150 thousand. The Nine Network reached out to the UK and the BBC to coproduce the series *Flying Doctors.*
- Hearst Entertainment and France's Ellipse Programme coproduced the made-for-TV film *Fatal Image,* a story set in Paris.
- France's Quinta Productions signed a coproduction agreement with Michael Mann (executive producer of *Miami Vice*) for the NBC miniseries *The Medellin Cartel.*
- Coproduction doesn't always involve cash: there can be in-kind payments. Germany's Ravensburger Film and TV reportedly coproduced a series of children's programs with Czech TV. The series was shot in

Czechoslovokia with Czech TV's contribution coming largely from supplying the facilities for the shoot.

- For the Japanese Satellite Broadcasting's WOWOW Channel (partially owned by Mitsubishi), Mitsubishi and Group W News Service agreed to coproduce a series of monthly shows—entertainment, not news—in the USA, in Los Angeles, with Japanese-speaking hosts.

- A fascinating coproduction involved Japanese Satellite Broadcasting (JSB) with Broadway. Weekly *Variety* (April 1, 1991) reported that JSB put up $2 million of the $6.25 million capitalization for the musical *The Will Rogers Follies*. Japanese investors were involved in backing Broadway shows prior to *Will Rogers* (the recent productions of *Gypsy* and *Fiddler on the Roof* were partially financed by Japanese companies). For its investment, JSB got exclusive electronic rights for Japan and Southeast Asia, including a limited number of telecasts of a taped version of the Broadway stage show, plus a potential home video, plus first option on a stage tour of the region. Is this a mini globalcast, aiming a non-TV event in one part of the world at TV in another part of the world? If there is no film deal for *Will Rogers,* will the JSB tape end up as a worldwide globalcast? Keep your eyes on the bouncing ball called coproduction.

- Perhaps the most dramatic example: in 1990 MICO (Media International Corporation) was born. MICO involves forty-seven Japanese investors, including the five biggest banks in the world. MICO defines itself as "The Direct Link Between You and NHK" and as "The comprehensive media trading house for the world market." It lists as its work "international coproductions, production investment, special events promotion, software purchasing and sales." One early coproduction arrangment joined MICO with the UK's Primetime Television Ltd. Primetime's distribution arm, RPTA, had been involved for years as distributor for NHK Enterprises, prior to the 1991 formation of Primetime/MICO, described as a joint venture company. The first joint effort announced by the company was the miniseries *Iran,* an extended coproduction involving Primetime/MICO with Turner Broadcasting, Consolidated Enterprises, Tele-munchen (Germany), and Antenne 2 (France).

  The most ambitious of MICO's announced list of ventures is the 20th Century Project, scheduled as a ten-year operation involving documentaries and miniseries. Each of these projects will be developed separately, with no overall partnership arrangments. Potential partners include the successor to the USSR's Gostelradio, the UK's BBC, France's Antenne 2, the USA's ABC.

Again, this list of coproduction efforts is in no way inclusive, but it demonstrates the scope and nature of the beast: money that travels, product that travels, both based on ideas that travel well. Symbiosis.

Coproduction: Elements on the Fringe

Although they are not directly related to globalcasting, some areas of coproduction come close enough to have an effect on it. Be aware, then, of the following trends.

- An April 1991 FCC ruling gave USA television networks (defined as a network that distributes fifteen or more hours of prime-time programming) 100 percent of "foreign" syndication rights. A December 1992 federal court gave the FCC 120 days to redesign its 1991 ruling—or leave the networks totally free from Finsyn regulation. (CNN obviously had to use a different word in reporting this story, called *Finsyn,* short for Financial Interest and Syndication rules.) Distributing rerun programming overseas may not specifically qualify as globalcasting, but it reestablishes the USA networks in another relationship with countries outside the USA.

- Weekly *Variety* (February 11, 1991, p. 37) reported that the number of non-Americans attending the annual convention of the National Association of Television Program Executives (NATPE) increased from 1,076 in 1990 to 1,269 in 1991. In 1992, overall attendance decreased, but there was greater attendance by European representatives—a trend of the times.
- The twelfth annual New Music Seminar in New York City in July 1991 reported more countries represented than ever and the availability of simultaneous Japanese translations. Nations exhibiting at the convention included Australia, Canada, France, Holland, Ireland, and Spain.
- Advertising agencies have been expanding operations in the Asian market. At the end of 1990 D'Arcy, Masius, Benton & Bowles announced creation of a network of agencies there, including joint ventures with existing agencies in India and Pakistan. J. Walter Thompson added staff in Asia. Other agencies heavily involved are Bache-Spielvogel Bates Worldwide, McCann-Erickson, and Ogilvy & Mather. Motion Picture Association of America president Jack Valenti indicated that the Association has been working to establish relations with Indonesia, a large market still closed to USA product. Protecting copyrights and preventing piracy have historically been major Association goals in the Asia market.
- Time Warner formed a coventure with Japanese supermarket chain Nichii to construct thirty-five multiplexes in Japan during the 1990s and announced plans to expand, by coventure, in Denmark, Germany, Holland, Italy, Portugal, Spain, the UK, and the Soviet Republics.
- Perhaps the most tantalizing coventure is the plan announced by Jameson Entertainment Corporation to build a series of large multiplexes, with eight or more screens, in China and the Soviet Republics. The multiplex concept would allow presentation of feature films, PPV events, non-PPV spectaculars, even videoconferencing. If these China/Soviet projects are successful, the plan would expand, ultimately, to more than 100 thousand screens on five of the world's seven continents. The plan will have local governments or business coalitions providing the sites and bankrolling the buildings, with Jameson providing management and booking.

What makes the plan intriguing, however, is the third part of the coventure. The equipment for the theaters would come from Japan's Ikegami Electronics, and that equipment would be electronic. The plan is to project the films using high-definition laser discs. Aside from fitting into the themes of transition and convergence, this move could profoundly influence how features are distributed everywhere in the world, as we suggested with HDTV. Laser discs provide three highly desirable advantages to producers: distribution costs can be reduced by a factor of 100 (about $1,500 for a film print versus $15 for a disc); laser discs can be encoded to prevent pirating; and laser disc prints don't deteriorate. Properly handled, the 100th use is as sharp and distortion-free as the first use. As we went to print, negotiations had not been finalized, but Robert J. Estony, communications manager for Ikegami Electronics (USA), told us that "the company is in discussions with Jameson Entertainment Corp., CA, to supply video projectors and scan converters."

A classic example of this area is MTV. The international language of music made it a good bet for global expansion and coproduction, but Music Television's first decade surprised even itself.

MTV was started in the USA in August 1981, by Warner Amex Satellite Entertainment Company. It grew rapidly and constantly, adapting almost chameleonlike to shifts in audience preference in music and styles. By 1991 the seemingly formless formula had built to an estimated 54 million USA viewers, and that was just home base.

Another 50 million viewers were watching in thirty-nine countries, courtesy of MTV Internacional (USA/Latin America), MTV Brazil, MTV Europe, MTV Australia, and MTV Japan. And these are not carbon copies of each other. In each country MTV is coproduced to match its target audience. There is overlapping in many instances because stars like Sting and Madonna are global, not local, superstars. But on top of the core programming, each MTV outlet produces unique programming aimed at its unique audience.

For example, MTV Europe, based in London, announced that it has topped the 20 million subscriber mark. MTV had sales branches in London, Paris, Milan, Amsterdam, Stockholm, Munich, and Athens. It has more than 100 employees, representing fourteen nationalities. It was a coproduction, a partnership between Viacom International and Maxwell Entertainment Group.

What's in store for MTV? The aim is for the world: JSB in Japan . . . Gostelradio. . . Lenceltel, a jointure of Leningrad City Council and Rutter-Dunn Communications of Columbus, Ohio. Reportedly talks have been held with outlets in the Baltic nations of Latvia, Lithuania, and Estonia, with stations in Italy and Turkey. Coproductions all, no doubt.

Increasingly worldwide concerns—the environment, economics, entertainment—are drawing global companies into mutual projects. Coproduction will be a major trend in the future.

Chapter 20

Trends in Equipment

In a March 1990 *Scientific American* article ("The Road to the Global Village"), Karen Wright offered a "Global Village Phrasebook." In it she defined a human as "An analog processing and storage device with a bandwidth of about 50 bits per second," which excels at "pattern recognition" but is "notoriously slow at sequential calculations." And yet it is this 50-bit-per-second analog mind that has brought us to computers and fiber optics that can process and transmit billions, sometimes trillions, of bits of information per second.

The pattern and timing of the discoveries and inventions that made these advances possible also reflect the pace of the electronic/photonic revolution. And make no mistake, we are in a very rapid revolution. Our ancestors experienced the Agricultural Revolution, which shifted humans from being nomadic hunters to being farmers. That revolution lasted some thousands of years. The Agricultural Revolution was followed by the Industrial Revolution, which shifted population from the farms to the factories and cities. At different times in different countries, the Industrial Revolution spanned about a hundred years. In just *half* a century, the Electronic/Photonic Revolution—combined with the predicted doubling of the world's population by 2050—will alter human life on planet earth as nothing has altered it before. And it will take place at bewildering speed.

If we are to survive and thrive through this new revolution, it is important to understand the nature of the beast. One way we can do that is by keeping abreast of the new trends in communication. In this chapter you will meet the Video Toaster, Video Windows, microrobots—everything but the Kitchen Sync, right? No, it's here too. Don't let anyone tell you there can't be humor in a revolution.

Throughout the decade communication equipment will get smaller, more interactive, faster, cheaper. It *will* invade the home. Get ready.

Downsizing: Smaller Is Better

Downsizing in electronic and photonic equipment has been a trend for decades. Thanks to CCDs, cameras have become smaller and more portable. Other equipment has followed the trend.

Satellite Technology

Suitcase Satellites Suitcase-sized portable satellite communication systems have been in existence for some time. Many were used effectively during the 1991 Persian Gulf War. A trend? Magnavox claims to have the world's smallest, lightest, least expensive satellite terminal. The smallest model weighs 47 pounds, includes a collapsible umbrella antenna, and offers telephone, data, fax, and telex communications capabilities. There are also larger, heavier models. Microelectronics Technology Inc. also manufactures suitcase satellite uplinks, on Taiwan, an island just under 14,000 square miles in size. It's estimated that 20 percent of the world's PCs come from Taiwan. In 1990 a Taiwanese company, MITAC, acquired Wyse Technology, a USA manufacturer of computer terminals. The equipment is not alone in downsizing.

LEOs: Low Earth-Orbiting Satellites   While most of the world's focus has been on the geosynchronous Clarke belt, some, for example Motorola, Inc. and friends, have been looking elsewhere. They envision a satellite-based system of personal communication. By orbiting sixty-six small, smart satellites in low earth orbit (418 nautical miles high) and networking them together as a switched digital communication system, using the principles of cellular diversity, IRIDIUM is designed to provide continuous line-of-sight coverage *from any point to any point on earth* (see Figure 20–1). No need to know where anyone is—just dial the individual's number and be connected, instantly. (In 1992 AT&T introduced 700 numbers, which assign permanent individual numbers.) Service will be available on a country-by-country basis as negotiated with individual governments and/or individual telephone companies. The system is designed to be entirely digital, with 8 kilohertz bandwidth available for each voice channel.

And IRIDIUM will link with another downsize leader. Hewlett-Packard's 11-ounce HP95LX (Figure 20–2) was the first palmtop computer to merge PC power with built-in software, including Lotus 1-2-3, a world time clock, calculator, file manager, and so on. Beginning in 1992 it was designed to contain long-range wireless communications capability. Utilizing IRIDIUM's radio frequency paging technology, the palmtop can receive information regardless of the user's location.

In June 1992, TongaSat announced plans to launch a series of low-earth-orbiting satellites by mid-1995, aiming to establish a commercial mobile satellite system ahead of Motorola's planned debut. Geared most toward providing developing countries with telephony/data/paging services, TongaSat's announcement was, nevertheless, aimed at potential participants worldwide. (TongaSat has also reportedly been negotiating with Glavkosmos, the commercial space marketing representative for the CIS, for transponder space on a Gorizont satellite to provide telecommunications services along the Pacific Rim.)

All of this activity may have been spurred by the fact that in just the two years from 1990 to 1992 world subscribership to cellular telephone services more than doubled (7.8 to 15.8 million).

Micromachine Technology

Imagine a "smart pill" so tiny that hospital patients can swallow it, a pill which contains a silicon thermometer plus the electronics to broadcast instant temperature readings to a recording device . . . or a silicon light bulb thin enough to fit on a hypodermic needle . . . or an eyelash-wide silicon motor that rotates 500 times per minute on a tiny broadcaster that bank tellers can toss in with holdup money. These microtools exist because experimenters attached computer chips (microprocessors) to wheels, to arms, to legs. As computer and TV technology fuse, microcameras will deliver point of view positions we never dreamed of.

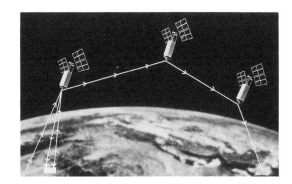

Figure 20–1   *Iridium* is number 77 on the periodic table of elements. The system was originally conceived using 77 sats. Increasing the number of beams per satellite permitted the cut to 66. Other companies and other countries are experimenting with similar systems.

Courtesy Motorola, Inc.

Figure 20–2 In addition to representing the downsizing technology of the 1990s, this eleven-ounce Hewlett-Packard palmtop computer began offering the technical capability to link with Motorola's IRIDIUM (see Figure 20–1) to provide instant worldwide availability of data.

Courtesy Hewlett-Packard Company.

Tiny microcameras are a logical extension, awaiting only acceptable video quality, of the Helmet-cam used by the World League of American Football (WLAF). This camera, 2 inches long and 9/16 inch wide, and the 2 pound battery pack/transmitter, were strapped to the back of the Orlander Thunder's quarterback in the WLAF's 1991 Orlando–San Antonio match on the USA cable network. Largely developed by Aerial Video Systems of Burbank, California, the Helmet-cam gave us one more point of view camera. But compared to a microprocessor microcamera, 2 inch equipment is gigantic.

Is this extreme downsizing just wishful thinking? Not in Japan. Under that country's National Research and Development Program (started 1966, usually called the Largescale Project), industry, government, and educational institutions cooperate in research and development of new technology "deemed important and urgent for the national economy" (Japan's Ministry of International Trade and Industry; MITI). In 1991 MITI added micromachine technology to the program. The project is planned to extend through the 1990s, with a cost approaching 25 billion yen (about USA$200 million). Japan's preliminary focus will be on industry (Micro Electronic Mechanical Systems using integrated circuit technology) and medical applications (biotechnology).

One proposed project is a microactive catheter for what you might call multimedia medical use. It would feature a headlight and lens— for sending back video pictures via optic fiber—a scooper for removing polyps, a scalpel that could double as a needle for injecting medicine, a tiny basket to withdraw snipped polyps or other tissue for testing, and a tiny motor to operate the surgical equipment. Flexibly guided by built-in sensors, the tiny catheter (1/5 inch thick) could be swallowed painlessly. The surgeon would work from a video terminal.

MITI's report indicates that similar research is going on in Germany and the Netherlands, as well as in the USA. Many scientists think micro is the future. So do some globalcasters.

Videoconferencing Uses Compression to Downsize

Three major USA manufacturers of videoconferencing equipment are Compression Labs Inc. (CLI) of San Jose, California; PictureTel of Peabody, Massachusetts; and Video Telecom Corporation of Austin, Texas. Both CLI and PictureTel have set goals of invading the consumer market using PC/multimedia compression techniques. The downsized price goal for a color motion picture system is USA$1,500, far from the $20,000 to $40,000 cost of current systems and a world away from the $80,000 it cost in 1980 to set up a videoconferencing room. The $1,500 model will doubtless resemble a picture-phone more than a business videoconference room outfitted with, say, an Axcess remote control system (by AMX Corporation, Dallas, Texas) that provides individual menu controls over lighting, multiple cameras (including zoom focus), multiple monitor screens, and audio levels.

PictureTel has forged links with Intel to develop new video compression algorithms to shrink costs and with MCI to link PictureTel's equipment with MCI's fiber optic system. In April 1992 AT&T offered a new computer chip, developed in cooperation with CLI, that it believed inexpensive enough ($400) to bring interactive live motion video— desktop videoconferencing—to PCs worldwide.

As noted elsewhere, Vyvx NVN offers portable broadcast-quality videoconferencing units compatible with their switched fiber optic DS-3 service. (Quality—especially motion quality—varies from system to system. The DS-3 transmission rate is 45 million bits per second; the rate for most other systems using phone lines, called T-1, is 1.5 million bits per second.)

Figure 20–3 shows perhaps the most interesting advance in videoconferencing: Bellcore's Video Window. Bellcore is short for Bell Communications Research, the research/engineering arm of the seven Baby Bells. At its research center in Morristown, New Jersey, you can see—or, more accurately, experience—the eight foot-high by three foot-wide Video Window. In this case the Video Window is connected via fiber optic cable to another Bellcore Lab about 50 miles away. The life size of the people in the Video Window, the hi-fi microphones/speakers, the multiple cameras involved combine to give a sense of togetherness the lack of which has made most communication experts rate videoconferencing well above audio-only communication but below in-person meetings. The Video Window isn't pressing-the-flesh, but maybe down the line, if you add a little virtual reality to the system, who knows? Bellcore's Video Windows are available for lease.

Fortunately for everyone considering use of videoconferencing, there is strong worldwide momentum to avoid a standards shootout like that over HDTV. In December 1990, the International Consultative Committee on Telephony and Telegraph (CCITT) approved a new set of global standards for videoconferencing to do away with varying codec standards that complicate the interfacing of video and phone lines.

The standard defines standards for decoding video transmitted at 64 kilobits per second (or multiples of 64 kilobits per second—the current ISDN level). To eliminate the hassle caused by multiple incompatible television standards, the CCITT ruling also aims to establish a global standard for videoconferencing. Called the Common Intermediate Format (CIF), it calls for resolution of 288 lines/352 pixels at thirty frames per second. (A lower resolution format reduces the numbers to 144/176 for smaller screen video.) *Videography* (June 1991, pp. 38–41) says the CCITT standard (H.261) is based on video compression algorithms similar to the JPEG/MPEG standards and that the MPEG working group is considering a modified version of H.261 for their proposed motion video standard.

It is possible that the rapidly decreasing costs and size of videoconferencing equipment—and of rental charges for use of established equipment—could lead to use of this technique to cut down on some costs of surveying inject sites for globalcasts. It could never replace on-site inspection, but preliminary and follow-up meetings could be much more productive with video attached. In much the same way the video camera linked my control room with Tokyo for *Earth 90,* videoconferencing could improve communication between a globalcast's executive director and inject directors when time or scheduling prohibits their meeting in person.

Downsizing the Television Studio

Out with desktop publishing, in with desktop production. The claim is that the name Video Toaster, from NewTek Inc., was a ruse to spook competitors during the development phase. Actually it's multimedia technology that can turn a private house or office into a miniature television studio.

Figure 20–3 The nearly life-sized images shown in this Bellcore Video Window demonstration foreshadow both interactive TV imagery and, ultimately, global videotelephony. Expanded fiber optic cabling and new legislation in many countries, including the USA, will have to precede such global interactivity.

Courtesy Bellcore.

The Video Toaster provides a switcher, an expansion board that plugs into a Commodore Amiga computer. Add software on a set of floppy disks and you end up with a DVE program offering 132 special effects. The software programming reportedly has more than ten times the number of command lines found in a 30 thousand line word processing program. The semiconductor chips employ RISC (reduced instruction set computing) technology.

Because Video Toaster has a switcher, users can move between and among video sources. There's a frame grabber to freeze and store images for manipulation, including 3-D animation. A character generator can insert data, including titles. How professional is it? Video produced by professionals using the Video Toaster has appeared on USA networks.

If you use a Macintosh computer, you can connect it to a Video Toaster workstation using Sundance Technology Group's Sundance/Toaster Video Production System. For additional timebase corrections and synchronizing functions with Video Toaster, you can turn to Digital Creation's Kitchen Sync.

Other companies are offering products, with less colorful names, specifically designed for multimedia uses, among them a system developed jointly by Interactive Media Technologies and Intelligent Resources Integrated Systems, Inc. These machines mark one more giant step on the way to ultimate convergence. The breakthrough will come the day the word *computer* disappears and people buy the multimedia communications device for what it does, not what it is. Microsoft's Chairman, William Gates, signalled one concrete move to convergence at the March 1992 International Conference on Multimedia and CD-ROM. By 1993 all of Microsoft's computer applications, he announced, would be available on CDs. And in 1993 palm-sized "Personal Communicators" appeared on the market, combining a laptop computer, a cellular phone and fax machine—with screens on which you write with a pen. IBM's PC-2 and Apple's Newton were among the entries.

Downsizing Television Screens

Sony's Watchman has sold in the millions. Check any sports stadium. The LCD screens, of course, made these miniature TV sets possible, and now other technologies are invading the field. A television audience won't necessarily be at home anymore. One of the newest Sony models is cube-shaped with a detachable screen that doubles as a camcorder. NEC's TurboExpress game player offers an optional tuner that turns the machine into a TV set. Size? Both have 2.7-inch screens.

Sony's CD-XA, dubbed Book Man, is a 2.2-pound hand-held interactive unit priced at about $1,000. Its 4.5-inch screen scans only fifteen frames per second. It's a stand-alone, but can be plugged into a TV set.

Another Japanese product, the Wave Chaser Helmet, boasts a tiny antenna that—in motion—can track satellite signals—DBS on the move.

Gradually—and probably that's the only way the world really works—the hardware reflecting the multimedia is converging. Hardware. Software. Wiring. VidTalk-StereoFax.

Remote Equipment: Technology to Go

Remote Camera Equipment: Triax

Remote operations have been a mainstay of many elements of globalcasting, news and sports in particular. New equipment is now beginning to turn old out-of-studio disadvantages into advantages. We've already noted CCD cameras' greater sensitivity to low-light areas. Add to that advances like those of Telemetrics Inc. of Hawthorne, New Jersey. Triax cable is about one-tenth the weight and cost of multicore cable, and it can be repaired in the field. Figure 20–4 shows a Triax Adapter mounted on a standard battery clip of an ENG camera. The TM-883W system is a Triax/Wireless connected camera control system. It is switchable between RF link and triax modes and features single cable link, up to one mile to tower. It's also compatible with the Pan & Tilt system.

Remote Repair: A Fax Fix

Serious remote repair service began, logically, with satellite manufacturers. No matter how much time he has on his hands, the Maytag repairman can't drop by a satellite. Computer manufacturers—heavily involved in the satellite systems—picked up on remote diagnosis systems. Soon other technologies joined the trend and found themselves able to serve customers more quickly and to save on on-site service calls. If you're on a remote and suffer fax failure, it can be reassuring to know there's remote diagnosis at the other end of a telephone.

Since the mid-1980s, Pitney-Bowes facsimile machines (manufactured by Matsushita), have been designed to facilitate remote servicing. The company offers round-the-clock diagnosis at its National Diagnostic Center in Melbourne, Florida. The problem fax is linked by phone to a matching machine in Florida, and the customer sends through a predesigned electronic status report plus a record of the last twenty transactions on the troubled machine. If the customer's machine passes muster, separate phone lines are used to check the customer's phone lines, machines with which the customer has been sending and receiving. It's a satellite-inspired trend that will extend through the 1990s.

Interactive Television

Viewers Become Directors, Writers, and Participants

The word *interactive* has almost too many meanings in our digital world. Electronic Speed Systems Corporation teamed with Structural Graphics to design an Absolut Vodka magazine ad to wish readers Merry Christmas and Happy Hannukah in three languages. Specialists predict that as the tiny computer chips develop, ads will call subscribers by name, question them, and provide buttons to register responses. If a magazine ad can do that, think what a television set could do! Think what it's already done, because television preceded the ad chips.

In the 1970s, in Columbus, Ohio, Warner Cable Corporation's Qube System not only offered viewers a choice among thirty channels, it also provided the first interactive TV in the USA. Each subscriber home was supplied an electronic selector with buttons linked to Qube's computer. Viewers could respond to on-screen questions by pressing a button. During an electronic town meeting, viewers were able to vote on issues. Qube's system didn't survive, but it pioneered.

Figure 20–5 shows an example of current interactive TV: ACTV (not to be confused with the HDTV standard), on New York cable, which allows even young viewers to select program alternatives via remote control. In addition, viewers can choose camera angles during ballgames.

Figure 20–4 This Triax adapter is mounted on a standard battery-clip of an electronic news gathering (ENG) camera. Triax interconnected camera control systems process signals transmitted between the camera and the base station into frequency- and pulse-coded multiplexed carriers. These carriers are combined with power and applied to the center conductor and inner shield of the Triax cable for transmission to and from the camera.

Wireless links are used for remotes that require complete mobility, such as floor cameras at political conventions or the Academy Award ceremonies.

Courtesy Telemetrics Inc.

Figure 20–5 Children of the 1990s will grow up with remote-controlled interactive video, reflected in games like Nintendo and this ACTV system, being demonstrated here to four-year-old Casey Leigh by ACTV chairman Dr. Michael J. Freeman. Note the numerical choices appearing on screen left.

Courtesy ACTV, Inc.

NTN, in San Diego, California, offers QB-1 (QB for quarterback) that lets viewers predict the next play. Diamond-Ball offers the same challenge in baseball.

In 1990, in Montreal, Canada, Le Groupe Videotron's Videoway service began giving its subscribers a hand-held remote control to aim at their Videoway terminal to pick the shots they want to see on the screen. A Videotron subsidiary, Tele-Metropole, telecast NHL hockey games, and Videoway viewers had the option of choosing a wide shot, either of two close-up cameras, or a 5 second instant replay. There was a monthly fee for this privilege of directing the show. If it's a trend and it grows, Directors Guild of America trainees could come on board with a lot more experience than in the past.

Explore Technology, in Scottsdale, Arizona, offers the Instant Video Receiver, which can download a full-length feature film in a matter of seconds. With advanced technology it's expected that viewers will be able to program the system to download selected items only. (You could also order the machine to, for example, tape anything with Elvis Presley in it.)

CD-I (for interactive) discs already offer films with alternative story lines, giving the viewer power to "write" the plot of the story, and in theaters equipped with electronic selectors on each seat, Interfilm (Controlled Entropy Entertainment) has exhibited its short feature *It's Your Man*. Audiences vote multiple times to determine the story outcome. With majority rule, the theater scene is often less like an intimate experience in darkened privacy and more like a midnight showing of *The Rocky Horror Picture Show* with viewers shouting to persuade others to vote their way. Shot on film, the feature was transferred to random-access laserdisc for projection. TV will copy. Today these technologies are floating around independently. The survivors will merge, and it'll be VidTalk-StereoFax time.

Next I discuss two current film techniques not yet directly linked to global television: three-dimensional (3-D) and motion simulators. Then I'll show you how the converging roads all lead home—and to the home viewer.

Three-Dimensional Simulators: In-Theater/On-Television

In Canada, Imax Systems offers Imax Solido, existing theatrical 3-D technology that replaces the 1950s-style red and green eyeglasses with battery-powered goggles featuring liquid crystal lenses. The system involves dual projectors, of course, and a huge concave Omnimax screen. (Imax screens are forty to fifty feet tall; Omnimax uses the same system as Imax, but with wider camera lenses to compensate for the wrap-around screen. Negative film area is almost nine times that of standard 1:66:1 35-millimeter film.) Audience seats are tiered and angled so the viewer is reclined, looking almost as much up as forward—surrounded by video.

But it's the goggles technique I'd like you to reflect on. If you have not yet read about optical computers—about AT&T's digital optical processor—in the section on photons, read it and compare the mirror technology there with the liquid crystal technology here.

Computer synchronized to the dual projectors, the lenses in the goggles are controlled by infrared light signals from the projection booth. These signals open and close the left lens and right lens of the goggles, alternately, twenty-four times per second, tricking the eyes into simultaneously seeing motion pictures and 3-D pictures. The open and closing of the lenses is accomplished by controlled electrical current switching back and forth. The electrical impulse turns the LCD lens from transparent to opaque, shutting off vision momentarily.

Will 3DO bypass these not-quite-virtual-reality goggles or will they make it to TV viewing? Maybe. Maybe not. *Some* form of 3-D will.

Three-Dimensional Simulators: In-Home Television Viewing

If research being done independently by Japan Broadcasting and Nippon Telegraph and Telephone Corporation pays off, the 3-D goggles won't even be needed—they'll go the way of 3-D glasses. Japan Broadcasting is testing a fifty-inch color screen—compatible with HDTV, of course—containing hundreds of tiny double-convex lenses placed in vertical ridges. In addition to filming with the customary two cameras to produce images the human eye can translate into 3-D, Japan Broadcasting is experimenting with four cameras, and possibly eight, to achieve a nearly holographic effect. Greater video compression will have to be developed to handle the huge amounts of data that would have to be transmitted. One problem is that to get the full 3-D effect, viewers have to sit 15 feet away from the screen. A possible solution is to use mirrors to shrink the distance. Nippon Telegraph and Telephone Corporation is doing its tests on a fifteen-inch liquid-crystal diode screen, which would allow much closer viewing. Neither set of experimenters sees immediate application to either terrestrial or satellite telecasting, but the research pushes forward.

So does other 3-D research continue, some involving the prediction I made in *Omni* (January 1987, pp. 36–40): a set of goggles with separate screens for each eye, not unlike Omnimax. There may be home video breakthroughs also, with double-lens camcorders carrying 3-D experimentation into the home.

Dynamic Motion Simulator (DMS) in Films

What about motion simulators and TV? Based on the concepts in aviation flight simulators, ride simulators have become famous at theme parks. Universal Studio's "Back to the Future" and Disney-MGM's "Star Wars" rides are two of the most famous. The simulators use digital technology to coordinate film action with hydraulically activated seats. Computer-synchronized, the action in the film and in the seats creates psychophysical sensations so realistic they produce shrieks of delight/fear/excitement.

Founded by Douglas Trumbull, responsible for the special effects in *2001: A Space Odyssey*, Showscan—whose high-resolution film (sixty frames per second) we discussed earlier—has been a leader in the simulator field, starting with "Merlin's Magic Motion Machine" at Excalibur Hotel/Casino in Las Vegas, Nevada. These simulators aim not just at the eye or the seat. Their aim is also at the inner ear, where your sense of balance lies.

Motion Simulators in Home Television

How will this motion simulation travel by television into the viewer's home? Let me suggest the paths, first with NHK's Telemusic system, second with Secondary Audio Programming (SAP).

In a television studio, picture a piano that has pressure sensors on its keys. The pianist sits and the concert is televised. The pressure sensors generate a digital data stream, which is broadcast simultaneously with the concert video. NHK's Telemusic technology beams by satellite to a viewer's home, where the computerized workings inside a Disklavier player piano convert the digital data from the telecast and turn the living room player piano into an instant concert piano. You could call it reverse interactive television.

But remember, this is a digital data stream. If it can make a piano play, why can't it activate an interactive in-home motion simulator in addition to the remotely activated player piano? There is no reason why it can't, except the home equipment hasn't been developed and because of the cost to the home viewer. The technology is there. And self-playing pianos and simulated rides aren't all that we'll be sending home in future global TV. One day the almost overwhelmingly vast amounts of data needed to motivate virtual reality will be compressed and mastered into equipment that will allow home TV viewers to live in the same scenes with their favorite television stars, from soap operas to films. That technology will not be available until well after the 1990s, but it will come. In the meantime, a lot of other data will go home via television, too. In fact, a lot of it has already started.

SAP: Secondary Audio Programming

All future television sets sold in the USA will have a built-in decoder that makes text available on the vertical blanking intervals (VBIs) in the video transmission. Some stations began well before 1992 to use VBIs to offer closed-caption programming for deaf viewers. VBIs have also been used to provide stock market quotations, news and sports headlines, community announcements, and so on. That's video. What about secondary audio channels?

Any television set offering stereo sound has the option of secondary audio programming (SAP) by pressing a button marked SAP, Track Two, or MTS (for multiple-channel television sound)—provided the station being watched is transmitting SAP. Separate transmission equipment has to be installed by each station.

The most important function for SAP is to serve blind television viewers, and many PBS stations use SAP for this purpose, providing explanatory commentary interwoven with the regular program audio. It's a little like sitting beside a kibbitzer in a movie theater, whispering comments in between film dialogue. It doesn't interrupt dialogue. It adds commentary. SAP can also provide alternative tracking, as is done with the Spanish version of *The MacNeil/Lehrer Report* on many PBS stations.

Less than 15 percent of US television stations report using SAP, but that percentage is sure to grow as newer television sets are sold. The future shift will be not just to a 16:9 aspect ratio or an HDTV standard, it will also be to multiple audio tracking . . . and interactivity in the form of motion simulation . . . and 3-D and . . .

There's a Photon in Your Future: Holography!

In 1990 AT&T scientists at Bell Laboratories in Holmdel, New Jersey, unveiled the result of five years' work: the world's first successful digital optical processor. In essence, they created a computer that substitutes light for all of the wires: a wireless computer. The memory storage in this AT&T processor wasn't large enough to qualify it as a computer, but the pattern worked. Because optics can handle a lot of light beams at the same time, Holmdel scientists envision optical processors processing more than one thousand times as much data as their electronic counterparts. The

Holmdel demonstration processor operated at one million cycles per second—less than most PCs. The potential of an optical computer, however, is several hundred million cycles per second, faster than many supercomputers.

Instead of silicon microprocessor chips, AT&T's digital optical processor combines laser light with mirrors. Instead of using on or off electrical current to produce ones and zeros, the mirrors either reflect or absorb light. Light particles—photons, not electrons—transmit the digital data, at the speed of light.

Because optical systems can transmit large amounts of data on different channels simultaneously, optical processors have built-in parallelism—the technique used in the Connection Machine supercomputers and announced by IBM for its 1990s computers. To accomplish the optical processing, the AT&T scientists had to develop new components. One of these is S-SEED (symmetric self-electro-optic effect device), an optical switch with a potential speed of one billion operations per second. Each switch contains two mirrors and measures 5 microns square. (A micron is about 1/100th the diameter of a human hair.) There are thirty-two S-SEEDs on each of four "arrays" within the processor.

Each of these arrays contains two 10 million watt modulated laser diodes that are divided into many separate beams to provide communication between the arrays. The lasers emit in the near infrared (850 nanometers).

The four arrays are separated by what they call lenses and masks, which serve the same function as connective wiring between logic gates in an electronic processor. The scientists discovered that if you expose a layer of gallium arsenide (a semiconducting material) to a burst from one of the laser beams, the layer becomes transparent—briefly. So they covered the mirrors in the optical system with gallium arsenide (the mask). By directing controlled bursts of laser beams, they could manipulate the mirrors. Hit with a laser beam, a mirror becomes momentarily transparent and reflects the next beam. A mirror that isn't hit with the laser beam keeps its mask of gallium arsenide in place and does *not* reflect. An "on" S-SEED represents a one. An "off" S-SEED represents a zero. And just like that you're ready to do digital at the speed of light. Input/output (I/O) functions can be accomplished by either optical fibers or laser beams transmitted in free space.

Obviously still in the early experimental stages, optical processors/computers will eventually make it possible to communicate so much data so quickly that graphics and effects could be created that cannot now even be dreamed of. Even attaching optical connections to electronic computers—a probably interim step on the way to optical computers—will expand digital capabilities and speed almost beyond imagination. Once more, the technology will outrace the content. What will we do with this vast power to communicate—globally, instantaneously, visually, graphically, verbally, endlessly?

Holography: Electronic Holographic Video

One thing we may do with optical computers is to create holographic television. Many who have viewed just plain MUSE-HDTV have felt a sense of 3-D and experiments are ongoing in that area. But in an interview with *Popular Science*, (January 1991, pp. 84–87) MIT professor Stephen Benton said the current compression techniques don't appear adequate to transmit the full holographic effect.

The basic theory of holography was the work of UK professor Dennis Gabor in the 1940s, but holography didn't become practical until the 1960s, when lasers came into being. (Professor Gabor won the Nobel Prize for Physics in 1971 for his holography theory.) Two University of Michigan scientists added a concept called the Leith–Upatnieks beam-splitting system, which increased holographic uses. A laser beam is divided in two by a beam splitter. One beam lights up the object, and that light is reflected onto film. The second beam also strikes the film so that the two beams interfere with each other. The resulting interference patterns build the holographic image. When laser light is projected through the film image, the divergent light rays recreate the in-depth outline of the original object.

Recording these images electronically requires more digital storage space than a regular video image, and it is this need for massive bits of data that keeps holographic television theoretical—just as it will slow the arrival of home virtual reality. In fact, Professor Benton's research employs a Thinking Machine CM2 Supercomputer, with 16 thousand microprocessors, capable of one billion FLOPS per second. Currently, holographic interferometry is used to measure stress, as in aircraft engines. Acoustic holography substitutes sound waves for light waves and allows physicians to "see" tissue (differentiating healthy from nonhealthy tissue) that x-ray images do not show.

In 1987, when *Omni* kindly invited me to contribute to "14 Great Minds Predict the Future," I indicated that holographic images *will* appear on TV screens, if not in the 1990s, soon thereafter. Optical computers will speed the day they will appear.

Smart Houses/Smart People

I referred earlier to "smart houses," models in which many of the multimedia we've discussed have already converged. Such models exist now. You can visit them. Cost and lack of cabling so far have prohibited mass construction, but the delay is a matter of time and money, not possibility.

USA's "smart homes" have a double coaxial cable system, providing two lines into every jack in the home, plus a LAN (Local Area Network) as a communications control that links all of the appliances. Smart Homes also have two electrical subsystems. One provides 120-volt alternating current. The second (A 24 volt direct current) is back-up, to provide an uninterruptable power supply.

This combination of power supplies, local area network, and phone lines offers the ability to control all of a home's appliances electronically as well as to program them in advance—lights to turn on or off, heat to rise or lower, and so on—and the ability to control them by remote, using the phone lines. Just call home to turn up the heat or the air conditioning, to start dinner cooking, to turn lights on early because it clouded up.

Invoking the powers of fuzzy logic, the Japanese have planned their Smart Home so people will be able to call home to regulate appliances, but they won't have to. There are sensors all around the house to monitor the temperature, the humidity, the air flow, even carbon dioxide levels. The sensors are connected to the appliances—the way sensors on Times Square in New York regulate the brightness of Sony's Jumbitron. If it's a great day, with clear skies and warm temperatures, the sensors will open the atrium windows and, at the same time, shut off the HVAC (heating, ventilation, and air conditioning). If it turns cold, or starts to rain, the sensors will order the windows to close and the HVAC to resume activity to bring the room back to optimum conditions.

How sensitive are these supersmart sensors? They even know if no one's home, or if friends arrive, or what they're doing. They sense human presence and the heat created by human activity—and adjust the appliances accordingly. If someone decides to read and turns on a bright light (which produces heat), the sensor will adjust the room temperature downward. Turn out the light to watch TV, the sensor will order HVAC to raise the temperature to compensate for the drop in room temperature.

And since this book is about globalcasting, we can't overlook Japan's plans for television coverage in this home of the future. The University of Tokyo planners suggest thirty-three television monitors, all of which are hooked up to video signals from seven cameras installed within the house, from however many VCRs are installed, from however many laser disc players are installed, and from television, satellite, and cable systems. With a little urging, I'm sure they could add facilities for the Telemusic Player Piano and an in-home motion simulator.

One final touch, mentioned earlier: in this home of tomorrow, toilet training takes on a whole new meaning. The computerized commode will check and report the user's pulse, check and report the user's blood pressure, and complete and report a basic urinalysis. If the commode computer is programmed to phone a computer in a doctor's office, no one may ever have to make an appointment or sit in a waiting room again. Trained toilets in this hygienic house probably won't make it to a globalcast either, but with convergence in the air, nothing is impossible.

Chapter 21

Jobs: A Look at the Future

I don't have to tell you that television is one of the most competitive areas in the job-seeking market. When I started in TV, there were virtually no television courses. There was just barely television. In fact, I wanted to go back to college to learn TV, but the college I approached wanted me not as a student but as a teacher. Now college courses are virtually mandatory for just an entry-level job. (The American Film Institute published a new edition of *Guide to College Courses in Film and Television,* covering both undergraduate and graduate courses.) And the number of jobs in television stations is constantly dwindling. (For a discussion by some of television's top directors and producers on how they got their first jobs in television, see my earlier book, *Live TV.*)

Ampex's Automated Cassette System and other such systems can be programmed to run stations round-the-clock. In New Zealand, when deregulation led to the creation of a third channel (two are state-owned), the new station's master control room contained a Grass Valley M-21P Master Control Switcher that permits one person to run the entire operation. Robot and remote control cameras have replaced some workers and will replace more. Technology makes jobs in television easier and often more exciting, but fewer in number, a special challenge for women and blacks as they try to make up for the years they were excluded from jobs in the industry. (A Radio-Television News Directors Association (RTNDA) study showed the percentage of women holding jobs as TV news directors grew from 1 percent in 1972 to 8 percent in 1980 to 18 percent in 1988. While the percentage increased, the salary level did not reach the male plateau. The average 1988 salary for female news directors was $26,000, versus $35,000 for male news directors.)

While the jobs are shrinking and applications are becoming even more competitive, the jobs themselves are changing. Technical directors who used to punch camera switches now program complex video effects. Cameras that once were dollied-in and dollied-back now zoom to cover the distance—and the zoom is electric, not manual. Cameras don't have viewfinders, they have monitors. Microphones are miniaturized and frequently wireless. Teleprompter went from huge typewriters and rolls of yellow paper to computerized instant-access keyboards. (When I taped President Bush in the White House in 1990, just days before the Persian Gulf crisis, he asked for minor changes in the copy I'd prepared for his welcoming greetings for *The 1990 Goodwill Games.* It took only a few keystrokes to rearrange the copy.) Fewer programs are aired live today, and many that are taped have work done in post, as producers and directors face computerized editing. News went from film to tape to electronic news-gathering to instant editing and one-person crews. New equipment comes on the market in a steady stream, not once a year or once a decade as in the past. Like Alice in Wonderland, workers in digitalized, miniaturized, computerized television have to run not to get ahead, but just to stay in place.

While the number of jobs at television stations and networks decreases, however, jobs in "television" are opening up in different areas. Corporate TV has been a thriving industry for decades. Making music videos has become a badge of honor among many. And the number of made-for-TV movies has escalated fantastically, with personnel floating relatively freely between the video fields. The debuts expected in DBS should swell these numbers even further. And organizations outside the corporate and music and film worlds are trying to capture and use the power video offers. Broad Street Productions created an eighteen-minute video for the New York Legal Aid Society to be distributed to law schools to recruit lawyers. In Aptos, California, Turn Up The Volume Inc. began production in 1991 of a rock 'n' roll video magazine called, fittingly, *Turn Up The Volume.* Distributed by Brentwood Home Video, the videomag began on a bimonthly schedule, with plans to advance to a monthly schedule. Its outlets reflected the multimedia theme: musical chains, video rental chains, and book stores. Can drug stores and food chains be far behind? *Billboard* (June 22, 1991, pp. 39–40) noted that prior to the start of *Turn Up The Volume,* similar "publications" had already been launched for country music, rap, and heavy metal. Intimate interviews, behind-the-scenes coverage, and talent development are some of the subject areas stressed in the videomags. As the pioneers in video magazines succeed, it's inconceivable that other print magazines will not follow. *Playboy,* of course, has been involved in video one way or another for years. Convergence in yet another guise. And jobs.

And convergence itself has created a new job category. More science than television, the area still reflects on and incorporates video and multimediatechnology. The field is called imaging science, and it combines elements of computer graphics, HDTV, satellite photography, and the academic fields of physics, mathematics, and chemistry. (Several schools offer courses in imaging science, including Massachusetts Institute of Technology, Rochester Institute of Technology, and the University of Arizona.) Electronic/photonic printing, medical diagnosis, and remote sensing, used in studying satellite images to analyze pollution or crop or drought conditions, are a few areas of imaging science specialization.

Via Satellite's second annual salary survey (December 1991, pp. 24–30) reported that more than 60 percent of satellite industry workers with five to fourteen years of experience surveyed in the USA made $50,000 or more per year. As in television, women workers unfairly earn less.

Most businesses and industries, including education, are as hostage to the convergence of the multimedia as are video and computer industries. Theme parks have used multimedia video to capture peoples' imagination. Video games are capturing the imagination of whole generations of youngsters, half of whom drop out of school because it's dull.

In his book *Organizational Communication,** Gerald M. Goldhaber defines organizational communication as a method of interacting to cope with uncertainty, uncertainty being the difference between information

---

*Goldhaber, Gerald M. *Organizational Communication.* Dubuque, Iowa: Wm. C. Brown Company, 1983.

available and information needed. Through long, expensive, often riotous years, business learned the lessons Goldhaber catalogs in his text. The best way for a company to succeed, to improve productivity, is to persuade individual workers that their personal goals match the company's goals, that if they help the company to succeed they will succeed. If the company and workers can pull that off, everybody wins. Companies have learned—also the hard way—that the quickest and easiest way to achieve that is to communicate the goals to the workers. How? Corporate videos. Videoconferencing. Many national companies (especially automobile manufacturers) have regular national videoconference networks to their dealers. Multimedia to the rescue. It may not be network TV. It may not be globalcasting—at least not yet. It is a lot of jobs, a lot of important jobs.

If you are aiming for a career in global television and are planning your education, I offer this advice: Multiply your skills as much as you can. Increasingly in all international businesses, language skills are critical. They certainly are in globalcasting. A knowledge of the mechanics of television can help advance you, too. Even if you get into television below ground level, if you have supplied yourself with multiple skills and manifest those skills in a dedicated fashion, word of mouth will spread your reputation and give you a chance for advancement. Being willing to work and offering multiple skills can open doors for you.

To provide a thumb-nail sketch of job futures, let me summarize the skills I think you will need to be part of global television in the 1990s:

Traditional production skills: These go virtually without saying. The basics remain the same. Get any job you can to learn and master them. The theory and philosophy can come from school. The know-how comes from on-the-job training. Be an intern. Be an unpaid volunteer. Be a secretary. Be anything. Get a job and learn all the jobs around you.

People skills: You have to be a self-starter who can work without constant supervision, but you also have to be able to interface with everyone. Patience and persuasion under fire. You have to make yourself stand out among the team members—but you have to be a team member. Learn diplomacy.

Computer skills: Whether you want to be a technician and work with the converging sciences or a production person dealing with programming, the material and mechanics with which you'll work will be either computerized or computer-connected. The more systems and software you are familiar with, the better.

Language skills: I've stressed these for global television because this category tends to set globalcasting apart from local and/or national TV. Other nations have learned this lesson much better—and sooner—than the USA. If you are a student in the USA, you may have to go out of your way to master multiple languages. Go out of your way. Spanish, Japanese, French, and Russian are probably the core. All languages are useful.

Creative skills: These may be the hardest to define, but they may provide "the edge" that separates you from the crowd. You have to accept enough of the past—and the authority represented by it—to get a job and be a team member. You also need constant inquisitiveness. Don't just learn what works, learn *why* it works, how it works. Always wonder: "Can it work better? How?" "What If?" You won't find many—if any—classes that teach creative thinking. Reading, watching, asking questions, *thinking* about the business—these are the elements that can push you past a job into a career.

Finally, you need desire. You have to decide what job, what career you want and fight to attain it. Find out what the demands of the job are, what skills are needed, what sacrifices you will have to make. If you're willing to pay those prices, nurture your desire until it's greater than the next person's. If you won't do battle for a job, if you won't go the extra mile, work the extra hour, run the extra errand, someone else will. Make sure there is never someone else who wants the job more than you do, who is better prepared than you are, who will work harder and more creatively than you will. Research the company or network you want to work for. Study annual reports. Hit the library for past history. Check magazines and trade papers for future plans. Determine how your strengths can match the company's goals, then put that in the job objective section of your resume. Tailor the resume *just* for that company, *just* for that job. Forget 400 resumes scattered to the world. No one will come looking for you. You have to go looking for the job. And you have to keep calling and writing. I refer you again to *Live TV*. See how often persistence paid off for today's top professionals. The meek may inherit the earth. They won't get jobs in global television. For those jobs you'll have to fight.

For the forseeable future, if you want to stay on top of developments in globalcasting—or any other major field of endeavor—your reading has to extend beyond books to periodic publications. Unless you read *Variety* every day or every week, unless you read *Broadcasting, Billboard, Hollywood Reporter, Omni, Videography, Satellite TV Week,* and *PC Magazine* (or similar publications)—unless you read these magazines and more, you cannot possibly stay abreast of the changes taking place. Of course there are useful, important books. *Global Television* is designed to be one of them. But as basic as books are to orient you to a science or a business, books take a long time to become books. Until *all* publications end up on CD-ROM or some similar technology, with updates being delivered to you on a steady basis, you will need current magazines to give you the edge you need to succeed, whether it's in globalcasting, genetic engineering, or population control or keeping track of how many dresses Vanna White wore while turning letters on *Wheel of Fortune*.

Part Six

A Final Perspective

Chapter **22**

Looking at a Crystal Ball

Virtual Reality

I've only hinted at virtual reality, a combination of Eyephone Goggles with a tiny screen in front of each eye, that lets you see moving images in three dimensions. Sensors connected with the goggles tell a silicon graphics computer which way your head is facing. If you move to the left, the computer compensates for that movement. The 3-D image you're watching shifts to the right, as it would if you turned your head in the "real world." The computer generates a new image every 1/20th of a second.

Put on a DataGlove, you can extend a computer-generated hand into your virtual world, to pick up a ball, to paint a picture, to perform a virtual operation on a virtual patient who will not die if you operate incorrectly.

Put on a DataSuit and you can have a computer-generated self enter the virtual world. If you take information from the body scan of a patient and feed it into a computer to create a virtual world of that patient's brain, you can walk inside the patient's brain to plan the surgery. Link yourself via optical fiber phone lines with another doctor on another continent, and the two of you can walk together inside the patient's virtual brain. Think of the medical miracles! And think of an episode of *Global Doctors*.

Property Rights/Privacy Rights

I've also only hinted at the effects of the electronic/photonic revolution on property rights. Software is not like hard copy. If you lend this book to someone to read, they have to return it to you in order for you to use it again. But you can give away a copy of software as often as you want and still retain the software. Property laws and economic laws are being rewritten. And they're not being rewritten by authors or store owners or lawyers or politicians. They're being rewritten by cable manufacturers.

In May 1993 Blockbuster Entertainment and IBM announced a system to give customers instant (six-minute) access to any album, film, game, downloaded from a computer-server—bypassing the need to stock (or manufacture) CDs, cassettes, etc.—an interim step en route to downloading at home. In Japan, for years you've been able to use your telephone to order and receive software—no video store, no book dealer. Just you and your telephone and a credit card. Convenience. But who owns what? Who controls what?

As the world shifts, however gradually, into a planet of instant communication on personal communication networks (PCNs), will existing broadcast entities adjust or disappear? Will individuals become broadcast entities? If our PCNs circle the globe, will Bruce Springsteen need a recording company to record/distribute his music, or will he just set up a 900 phone number by which each individual can download each new album? But if that's the case, how does anyone know Springsteen has a new album? Or what it sounds like?

Will the new technologies allow the developing nations to catch up? Or will boundary-leaping signals merely provide new methods of exploitation? Globalization is here. It can't be stopped. Can it be directed? How? By whom? By people who can't agree on a single television standard?

Copyrights, privacy rights, security rights—where and how do these fit into a photonic world that controls calculation and communication at the speed of light? As satellite signals and computer data flow across oceans and borders, what will define sovereignty of information? The expansion of DBS is combining with the miniaturization of satellite receivers to a point at which they eventually will not be visible externally. Will some countries still try to control this form of information? Is is necessarily true that if we open up communication among all nations that humankind will prosper—or will just certain segments of humankind prosper?

And if fiber optic cables are not liable to tapping as are copper wire cables, the storage capacity and instant accessibility of computerized files has created a different source of information to be tapped. Credit company files, police files, FBI files, Interpol files, medical files, heaven knows what other files are instantly accessible to anyone with the right password. And a properly programmed computer can unearth virtually any password, as has been proven by invasion of "secret" military files time and time again.

And if a sharp shudder struck the scientific community when a student-inspired computer virus invaded computer systems and began destroying computer programs all over the USA, a different but related, shock wave hit USA airlines when AT&T's lines went down in September 1991, knocking out not only long distance phone service, but also the communication system for the airports' air traffic controllers. In 1990 AT&T had software meltdown problems. In 1991 the planes were grounded for hours. When many major companies were surveyed, they said they weren't concerned. They'd already prepared a fallback position. They maintain lines with two long distance phone companies. Just as I protected the essential feed from the Vatican during *Prayer For World Peace* by renting transponder space on two separate satellites, so the business world backstops itself with double lines. But the more dependent we become on high-tech communication systems, the more vulnerable we will be to invasion, disruption, catastrophe.

What Kind of Global Village?

And as many major globalcasts have shown, our lives are linked. Catastrophes are seldom isolated in the 1990s. Worldwide ecology specials have repeated again and again that beef patties in one country affect the rain forests in another country. The population of Brazil affects the population of China affects the population of every country on every continent affects the predicted lifespan of the earth and its inhabitants. Heavy stuff. But globalcasting hasn't been shy. Like the United Nations, globalcasting thinks worldwide. A different way to think. A global village.

The UN reminds the individual member nations at least annually that the world's population is expanding at the fastest rate in history. The estimated birth rate is 250 thousand people each day (exactly the estimated number of television sets produced each day). But if the world population doubles, by 2050 will our communication-economic-ecological system match that growth, or will the world sink deeper and deeper into violently clashing developed/developing nations while the sun's rays rush through an ever-widening ozone layer, allowing uncontrolled cancer to reverse the population growth?

Let me recall a story of early hope outlined by Arthur C. Clarke in his book *Voice Across the Sea.* He has retained this title for part of his fascinating book *How the World Was One: Beyond the Global Village.* ** In a chapter of his earlier book, Clarke detailed India's Satellite Instructional Television Experiment (SITE), a cooperative venture between India and NASA. An early venture into DBS, an Applications Technology Satellite F (ATS-F) was launched with enough equipment to send a signal that could be received by a ten-foot antenna dish anywhere in India. The two goals of the Indian government were family planning education and agricultural productivity—goals remote from holographic statues and concert hall concerts. Villages were equipped with signal-receiving equipment. Hundreds of villagers crowded around the television sets. How successful was the experiment? Did the modern technology help the villagers? SITE began in 1969 and ran in the early 1970s, twenty years ago. Also twenty years ago, in 1971, a cyclone devastated Bangladesh. The cyclone was tracked by the world's weather satellites, but there was no communication system adequate to get that early-warning information from the satellite to the villagers of Bangladesh. Another storm, twenty years later, in 1991, was tracked by satellites, but there was still no communication system to warn the villagers. Thousands of Bangladesh citizens were killed or made homeless. Was this part of a global village?

I have no crystal ball. I can easily equate economic growth in the EC, Eastern Europe, the CIS, Asia, and South America, with added opportunity to sell more television sets and more VCRs, more computers and more holographic images cum symphonic sound. I can make the same equation with China's population shift, from 85 percent on farms twenty years ago to more than half in cities by the end of the 1990s. I have trouble equating those "advances" with the other end of the scale. Which end will be the dog, which the tail? Electronically, photonically, we are becoming one world—on paper, on disk. The news has not yet reached the villages.

The "Other" Radio Waves

The satellites took us to space. If you read *How the World Was One: Beyond the Global Village,* you will find Clarke detailing the history of global telephony and of the satellites, in addition to touching on many of the earthly subjects we have dealt with here—fiber optic cable, DBS, computers. But you'll also find that his interest in space goes beyond the satellites.

Scientists estimate that more than 90 percent of the known universe remains invisible. If you dig deep into their research, you'll discover the Big Bang theory of creation, followed by theories of cold dark matter and hot dark matter. This proposal says that the invisible portions of the universe contain *heavy* particles, including neutrinos that may have some mass after all. (Neutrinos are relatives of photons, used in the light transmission in computers and via fiber optic cables.) Clarke talks a lot about

---

*Clarke, Arthur C. *Voice Across the Sea*. New York: Harper & Row, 1974.

**Clarke, Arthur C. *How the World Was One: Beyond the Global Village*. New York: Bantam Books, 1992:109–145.

neutrinos for the simple reason that they are potential carriers of communication signals—just like radio waves.

Even when flawed, the Hubble telescope revealed images of space never before seen. Hubble was the first of four observatory missions on NASA's agenda. The second, already in orbit, is the Gamma Ray Observatory, placed there by the successful April 1991 Atlantis mission. The findings from the Gamma Ray Observatory instruments will be compared with those from Hubble for cross-referencing.

The method of studying gamma rays is significant, too. Well out of the electromagnetic area of visible light, gamma rays pass through telescopes and are recorded only when they interact with selected substances, such as crystals. It is the result of this interaction that is recorded as evidence of the gamma rays. Data from the telescope sensors are then analyzed and the strength of the rays measured. The study of neutrinos is necessarily similar to this.

If gamma rays race through telescope lenses, our old friends neutrinos do even better. With no charge and no proven mass, neutrinos make Superman look like a beginner. Neutrinos can not only race through telescopes, they can romp right through planets—including earth. To try to locate particles and their source(s), detectors have been built in the USA, Japan, and the Soviet Republics. Because of the penetrating and elusive nature of subatomic particles, these detectors are placed underground and underwater, to filter out unwanted particles. Tanks of water are placed in mines or lakes.

The newest detector is a neutrino telescope financed by the international consortium Deep Underwater Muon and Neutrino Detector (DUMAND). DUMAND is located off Hawaii, under three miles of Pacific Ocean water, aiming its detectors *down* at the ocean floor, to record neutrinos zapping their uninhibited way through the planet. It's a little like an old USA joke about digging a hole through the earth and ending up in China. The neutrinos do away with the digging.

The detectors are for both audio and video, sound and light waves. When a muon (a charged particle) speeds through water, it creates a wave that can be detected in the form of light and sometimes in the form of a sound wave. Sight and sound. Radio and television of the future?

The "DUMAND array," as it is called, is a cluster of 1,000-foot-long cables attached to the ocean floor and held vertical in the water by a series of hollow spheres, each containing a light sensor. These detectors and their hydrophonic partners will channel their data, via computer, through fiber optic cable to a land-based lab. The expectation is that if a muon wave passes through the cluster of cables, it will be detected on more than one cable, thus revealing the direction of its path, leading potentially to its source.

Not in the 1990s, but possibly beyond, globalcasters may be dealing with panels of computer-switched neutrinos.

What We Don't Know

We know the questions. We know the technologies. We don't know yet what the technologies will reveal—or when they will reveal it. Many scientist's dreams are being cut off by budget cuts. In a single decade the USA went from being the world's greatest creditor nation to being the world's greatest debtor nation. The two German nations were reunited. The USSR dissolved itself into multiple nations. The European Community is aiming for a whole new life. Asia and Australia are voices that insist on being heard. South America and Africa hover on the brink of rediscovering themselves. Worldwide, the number of refugees doubled from 1980 to 1990. Politicians and philosophers have only begun to tackle the massive moral questions raised by discoveries about DNA. Individual nations pass laws to control genetic engineering, as if geographical borders could contain genetics, as if our village were not global. Just as genetic engineering will increasingly force the religious and political worlds to reassess the very basis of their data and criteria, so will electronic engineering increasingly force the same kind of reassessment by the communications industry, the banking industry, lawyers, philosophers, and politicians.

We are facing twenty-first century reevaluation of what we are, where we live, and how we live. And television—specifically globalcasting—will reflect these pressures of reassessment. As we approach the new century, you can get good odds on both ends of the bet as to whether the human race will soar into space or sink lemminglike into a sea of slime. Two things are certain: Whether we humans discover or create great new societies among the galaxies or drown in our own decay, the event will be reported by a dedicated ENG camera operator, giving all to report the news. And the way we use global television will be a factor—perhaps *the* factor—in the future direction of the human race.

Afterword

The Future: The Weapon of the Mind

The Age of the Image . . . MTV . . . Thirty-second commercials with 100 cuts . . . Teleplomacy . . . Television wars . . . Satellites photographing destruction of the rain forests, "the lungs of the earth". . .

Like so many kids in a high school biology class, we seem to have taken the world apart only to discover that—like a frog—we had to kill it to do the dissection. As we approach the twenty-first century, we seem, as individuals—and ever so gradually as nations—to be driven to put the world back together again.

We are far from McLuhan's global village. Our fiber optics reach out, but not yet far enough. Our satellites soar, but only for the affluent, not yet for everybody. Whole nations remain beyond the fringe area. But the promise, the magic, is that we have the tools—if we hurry—to resurrect the world. The United Nations gave us ten years to save the planet. This book has outlined a thousand tools of communication that can help us achieve a global village. Globalcasting can be *the* tool, if we use it right.

This is what has driven me throughout my career—the need to sow seeds of something greater to come, the need to strive to make life better. I think down deep this is what sustains us all, this striving always to something better, for each next generation. We keep pushing ahead for the chance that human understanding will match our technical and scientific knowledge. The next great breakthrough in our world will come when the philosophical and religious sciences are forced to catch up with the physical sciences. Those of us in communications have to be ready to communicate this. Globalcasting offers an incredible opportunity to unify the world, a greater opportunity than the world has ever had before. Globalcasting can leap tall buildings and bounce across borders and send messages to political leaders and bury martyrs and walk on the moon and sing nations together and pray nations together and race nations together.

Tomorrow's television child will be attuned to up-to-the-minute global news and will be showered by a blitz of interactive entertainment. The future will be less passive. The Vidtalk-Stereofax machines may let people work at home. Voice recognition computers may let us vote from home. The circuited globe of the future will provide us with the technological speed to monitor all that we hold to be important. We face a race between our information and our intelligence. Globalcasting can at least provide the information. Viewers will have to provide the intelligence.

We have seen assassinated leaders buried on global television. We have seen princes married and queens coronated. We have seen World Series

273

and Super Bowls. We have seen a war fought, live, on global television. The only limit to the power of globalcasting is the limit to the power of the human mind. The only control over the power of globalcasting is the sense of self-discipline within the human mind.

In this book I have tried to ask the questions you will have to ask yourself. The answers will have to come from you, in each specific circumstance. No one could foresee the global impact of *Live Aid*. Not even as I linked three continents and three cultures through the combined magic of music and satellites could I imagine the impact that globalcast would have, or that later I would be able to link five continents for *Prayer For World Peace*. No one could foresee how satellite television transmission would help shape the events in Iraq and Eastern Europe and the USSR.

In one sense I wish I could drop Reese's Pieces for you to follow—as some friends did for E.T. But that would take away part of the challenge. No one can offer you a crystal ball. And if you found one, you'd probably smash it to bits, life would be so predictably boring. What I hope I have done is to point toward some of the critical questions you will have to answer.

Globalcasting is a powerful force now and will be a powerful force in the future. It can pave the way to opening closed societies. It can help reshape the world. It can be exciting as hell. But never forget that as exciting as it is, it places on you a responsibility. My message—from page one of this book—has been that the main weapon in the war of communication isn't technology but the human mind. Ideas are more powerful than bullets or bombs or broadcasting towers.

As you become involved in globalcasting, I want you to share the thrills and excitement I have felt being able to link millions of people in prayer, millions of people in a race to help others, millions of people in thrilling to music. I want you to share all that, and more. I want you to share the sense of power and responsibility, a sense you'll have to define and redefine with every program, every assignment. The future is globalcasting. The future is yours. Welcome to the future.

Appendix: Global Television Standards

Country	Language	Population	Television Standard	Currency
Afghanistan	Farsi/Pashtu	16,200,000	PAL	Afghani
Algeria	Arabic/French	25,714,000	PAL B	Dinar
Angola	Portuguese	8,600,000	PAL	Kwanza
Antigua	English	75,000	NTSC M	Dollar
Argentina	Spanish	32,000,000	PAL N	Peso
Aruba	Dutch	60,000	NTSC M	Florin
Australia	English	17,000,000	PAL	Dollar
Austria	German	7,600,000	PAL	Schilling
Bahamas	English	260,000	NTSC M	Dollar
Bahrain	Arabic/English	500,000	PAL B&G	Dinar
Bangladesh	Bengali	117,000,000	PAL B	Taka
Barbados	English	270,000	NTSC M	Dollar
Belgium	Flemish/French/German	9,900,000	PAL B&H	Franc
Belize	English/Spanish	200,000	Monochrome	Dollar
Benin	French	4,800,000	SECAM K	Franc
Bermuda	English	70,000	NTSC M	Dollar
Bolivia	Spanish	7,000,000	NTSC	Peso
Brazil	Portuguese	150,000,000	PAL M	Cruzeiro
Brunei	Malay/English	370,000	PAL B	Dollar
Bulgaria	Bulgarian	9,000,000	SECAM D&K	Lev
Burkina Faso	French/Regional	8,900,000	SECAM	Franc
Burma	See Myanmar			
Burundi	French/Swahili	5,600,000	Monochrome	Franc
Cameroon	French/English	11,800,000	PAL	Franc
Canada	English/French	27,000,000	NTSC	Dollar
Cape Verde	Portuguese	360,000	PAL	Escudo
Central African Republic	French/Sangho	2,870,000	SECAM	Franc
Chile	Spanish	13,000,000	NTSC	Peso
Colombia	Spanish	32,000,000	NTSC	Peso
Commonwealth of Independent States	Russian	290,000,000	SECAM	Ruble

(NOTE: CIS (USSR) now twelve separate Republics; Check each independently)

Country	Language	Population	Television Standard	Currency
Congo	French	2,300,000	SECAM	Franc
Costa Rica	Spanish	3,000,000	NTSC M	Colon
Côte d'Ivoire	French	11,870,000	SECAM	Franc
Cuba	Spanish	10,600,000	NTSC M	Peso
Curaçao	Dutch	280,000	NTSC M	Guilder
Cyprus	Greek/Turkish/English	685,000	PAL B&G	Pound
Czechoslovakia	Czech/Slovak	16,000,000	SECAM D&K	Koruna

(Separated January 1993. Check each republic separately.)

Country	Language	Population	Television Standard	Currency
Denmark	Danish	5,200,000	PAL B	Krone
Djibouti	French/Arabic	500,000	SECAM	Franc
Dominica	English/French	90,000	SECAM K	Dollar
Dominican Republic	Spanish	7,000,000	NTSC M	Peso
Ecuador	Spanish	10,500,000	NTSC	Sucre
Egypt	Arabic/French	55,000,000	SECAM	Pound
El Salvador	Spanish	6,000,000	NTSC M	Colon
Ethiopia	Amharic/English	50,800,000	Monochrome	Birr
Finland	Finnish	4,960,000	PAL B&G	Markka
France	French	56,000,000	SECAM D2-MAC	Franc
Gabon	French/Fang	1,100,000	SECAM	Franc
Gambia	English	850,000	Monochrome	Dalasi
Germany	German	77,800,000	(W) PAL B&G (E) SECAM	Deutsche mark
Ghana	Akan/English	5,000,000	PAL B	Cedi
Gibraltar	English/Spanish	30,000	PAL B	Pound
Greenland	Danish/English	50,000	PAL	Danish krone
Greenland (AFRTS)			NTSC	
Greece	Greek	10,200,000	SECAM	Drachma
Guadeloupe	French	670,000	SECAM	Franc
Guatemala	Spanish	8,000,000	NTSC M	Quetzal
Guiana	French	95,000	SECAM K	Franc
Guinea	French	7,100,000	SECAM	Syli
Guyana	English	750,000	NTSC	Dollar
Haiti	French/Creole	5,000,000	NTSC M	Gourde
Honduras	Spanish	4,500,000	NTSC M	Lempira
Hong Kong	English/Cantonese	6,000,000	PAL	Dollar
Hungary	Hungarian	10,600,000	SECAM D&K	Forint
Iceland	Icelandic	253,000	PAL B&G	Krona
India	Hindi/English	840,000,000	PAL B	Rupee
Indonesia	Indonesian	190,200,000	PAL B	Rupiah
Iran	Farsi/Turkish/Kurdish/Arabic	55,000,000	SECAM	Rial
Iraq	Arabic	18,500,000	SECAM	Dinar
Ireland	English	3,650,000	PAL 1	Pound
Israel	Hebrew/Yiddish/Arabic	4,500,000	PAL	Pound
Italy	Italian	57,500,000	PAL B&G	Lira
Ivory Coast (See Côte D'Ivoire)				
Jamaica	English	2,450,000	NTSC M	Dollar
Japan	Japanese	123,480,000	NTSC/HDTV	Yen
Jordan	Arabic/English	3,000,000	PAL	Dinar
Kenya	English/Swahili	25,000,000	PAL B	Shilling
Korea (N)	Korean	23,000,000	PAL D	Won
Korea (S)	Korean	43,450,000	NTSC	Won
Kuwait	Arabic/English	2,000,000	PAL	Dinar
Laos	Laotian	4,050,000	PAL	Kip
Lebanon	Arabic/French	3,500,000	SECAM	Pound
Liberia	English	2,500,000	PAL B	Dollar
Libya	Arabic/English/Italian	4,200,000	SECAM	Dinar
Luxembourg	French	368,000	SECAM/PAL	Franc
Macau	Portuguese/English/Chinese/Cantonese	340,000	PAL	Batacas
Madagascar	Malagasy/French	11,200,000	SECAM	Franc
Malaysia	Malay	17,000,000	PAL	Ringgit

Country	Language	Population	Television Standard	Currency
Mali	French	8,900,000	SECAM B	Franc
Malta	English	360,000	PAL B&H	Pound
Martinique	French	670,000	SECAM	Franc
Mauritania	French/Arabic	2,100,000	SECAM	Ougiya
Mauritius	English/French/Hindustani/ Creole	1,100,000	SECAM	Rupee
Mexico	Spanish	88,000,000	NTSC	Peso
Monaco	French	29,000	SECAM/PAL	Franc
Mongolia	Mongolian	2,100,000	SECAM	Tugrik
Morocco	Arabic/French/Berber	26,500,000	SECAM	Dirham
Mozambique	Portuguese	15,100,000	PAL G	Escudo
Myanmar	Burmese	41,200,000	NTSC	Kyat
Nepal	Nepali/English	19,000,000	PAL	Rupee
Netherlands	Dutch	14,800,000	PAL B&G	Guilder
Netherlands Antilles: See Curaçao				
New Zealand	English	3,380,000	PAL	Dollar
Nicaragua	Spanish	3,500,000	NTSC M	Cordoba
Niger	French	7,670,000	SECAM	Franc
Nigeria	English	118,500,000	PAL	Naira
Norway	Norwegian	4,200,000	PAL B&G	Krone
Oman	Arabic/English	1,300,000	PAL	Rial
Pakistan	Urdu/Punjabi/English	106,000,000	PAL	Rupee
Panama	Spanish	2,300,000	NTSC M	Balboa
Papua New Guinea	English	3,600,0000	PAL	Kina
Paraguay	Spanish	4,600,000	PAL N	Guarani
Peru	Spanish	21,900,000	NTSC M	Sol
Philippines	Pilipino/English	67,000,000	NTSC	Peso
Poland	Polish	38,200,000	SECAM D&K	Zloty
Polynesia	French/Tahitian	195,000	SECAM	Franc
Portugal	Portuguese	10,300,000	PAL B&G	Escudo
Puerto Rico	Spanish	4,000,000	NTSC M	Dollar
Qatar	Arabic/Farsi	450,000	PAL	Riyal
Reunion	French	580,000	SECAM	Franc
Romania	Romanian	23,500,000	PAL D&K	Leu
Saint Kitts and Nevis	English	40,000	NTSC C	Dollar
Saint Lucia	English	150,000	SECAM K	Dollar
Saudi Arabia	Arabic	16,500,000	SECAM	Riyal
Senegal	French	7,500,000	SECAM	Franc
Seychelles	French/English	70,000	PAL	Rupee
Sierra Leone	English	4,000,000	PAL	Leone
Singapore	English/Malay/Mandarin/Tamil	2,700,000	PAL B	Dollar
South Africa	Afrikaans/English	39,000,000	PAL	Rand
Spain	Spanish	40,000,000	PAL B&G	Peseta
Sri Lanka	Sinhala/English	17,170,000	PAL B	Rupee
Sudan	Arabic	24,900,000	PAL	Pound
Suriname	Dutch/English	410,000	NTSC M	Guilder
Swaziland	English	780,000	PAL	Lilangeni
Sweden	Swedish	8,420,000	PAL	Krona
Switzerland	French/German/Italian	6,600,000	PAL B&G	Franc
Syria	Arabic	12,450,000	SECAM	Pound
Taiwan	Mandarin	20,400,000	NTSC	Dollar
Tanzania	Swahili/English	26,000,000	PAL	Shilling
Thailand	Thai	55,000,000	PAL B&M	Baht
Togo	French	3,500,000	SECAM	Franc

Country	Language	Population	Television Standard	Currency
Trinidad and Tobago	English	1,250,000	NTSC M	Dollar
Tunisia	Arabic/French	8,000,000	SECAM	Dinar
Turkey	Turkish	56,000,000	PAL	Lira
Uganda	English	17,000,000	PAL	Shilling
United Arab Emirates	Arabic/Farsi/English	1,800,000	PAL	Dirham
United Kingdom	English	56,740,000	PAL	Pound
United States of America	English	250,000,000	NTSC	Dollar
Uruguay	Spanish	3,000,000	PAL N	Peso
Venezuela	Spanish	19,800,000	NTSC M	Bolivar
Vietnam	Vietnamese	68,000,000	SECAM	Dong
Yemen	Arabic	12,000,000	PAL/SECAM	Riyal
Yugoslavia	Serbo-croat	23,700,000	PAL	Dinar
(Status of Croatia undetermined as this book went to press)				
Zaire	French	35,000,000	SECAM	Zaire
Zambia	English	8,000,000	PAL	Kwacha
Zimbabwe	English	10,000,000	PAL	Dollar

Glossary

A

ABC (1) USA: American Broadcasting Company, a terrestrial, commercial network; (2) Australia: Australian Broadcasting Corporation, government-funded (see *SBS*)

Active-Matrix Video display technique developed in Japan; combines liquid crystal with a large semiconductor with over a million and a half transistors, each controlling a single pixel on a display screen (see *LCD*)

Actuator Motor to aim a dish antenna for satellite reception

Ad hoc network A group of independent stations joined temporarily to carry a single production

A/D (Analog to Digital) Process/device to convert signals

Advanced Communications Technology Satellite Lightning-fast NASA comsats

Advanced Television Test Center Inc. (ATTC) Private, nonprofit corporation set up by USA television industry to examine options for new USA terrestrial transmission standard for advanced television (ATV) service, including HDTV; members include Capital Cities/ABC, CBS Inc., NBC Inc., PBS, Electronic Industries Association (EIA), Association of Independent Television Stations (INTV), Association for Maximum Service Television (MSTV), and National Association of Broadcasters (NAB); reports to Advisory Committee on Advanced Television Systems

Advisory Committee on Advanced Television Systems USA industry committee set up to test EDTV and HDTV systems and report to FCC for 1993 decision on advanced system compatible with terrestrial broadcasting (see *Advanced Television Test Center, HDTV*)

Algorithm (Euclid's Algorithm) Any special way of solving a mathematical problem

Alias A type of image distortion connected with signal sampling, correctable via antialiasing; process that averages borderline pixels to smooth out the connection (see *Digital Compression*)

AM See *Amplitude Modulation*

AMSC American Satellite Co. domestic satellites (USA)

Amplitude Measurement of the variation in the height of a wave in the wavelength of an electromagnetic signal; in radio waves, corresponds to volume (see *Frequency*)

Amplitude Modulation (AM) (1) Radio signal carrier wave constant in frequency, varying in amplitude (intensity); (2) A broadcast system using this technique (see *Frequency Modulation*)

Analog Computer Works on principle of measuring, not counting; operates on continuously varying data represented as physical quantities; electronic analog computers work on voltages in lieu of numbers (see *Digital Computer*)

Analog Signals   Variable, continuous waveforms, as opposed to digital signals (data expressed in discrete binary form, discontinuous in time)

Antenna   Parabolic dish to collect, concentrate, focus electromagnetic energy from a satellite; any device that collects, concentrates, focuses electromagnetic energy

AOR (Atlantic Ocean Region)   One of four major satellite coverage areas (see also *A-PR, IOR, POR, Intelsat*)

A-PR (Asia-Pacific Region)   One of four major satellite coverage areas

Arab States Broadcasting Union (ASBU)   Twenty-two member coalition established in 1969

Arabsat   Satellite used by ASBU since 1987

Arianespace   Consortium of thirty-six European aerospace/electronic companies, thirteen banks, the French Center of Space Studies (Centre National d'Études Spatiales), world's largest commercial satellite launcher, from French Guiana in South America

ASCII (American Standard Code for Information Interchange)   A computer code for communicating data

ASC-1   Contel-ASC hybrid communication satellite (US)

Asia Pacific Broadcasting Union   Established in 1964; thirty-seven members include Australia, India, Japan, Philippines, N. Zealand, S. Korea.

AsiaSat   Privately owned (Hong Kong conglomerate) satellite

Aspect ratio   Ratio of screen width to height

ATM (Asynchronous transfer mode)   Superfast switching technology using uniform electronic packets of digital "shorthand" to route interactive data from consumer to/from server (see *Video Server*)

Atom switch   Method of controlling electrical current by manipulating a single atom of the element xenon, a rare, colorless, inert gaseous chemical element present in minute quantities in air, used in photographic flash lamps, high-intensity arc lamps for motion picture projection, lasers, vacuum tubes; process involves "scanning tunneling microscope" with sharp tungsten probe. An electrical difference between the probe and a flat nickel crystal causes a single atom of xenon to jump probe-to-crystal; reversing polarity sends the xenon atom back to the probe, creating the off/on formulation needed to operate digital computer functions, potentially faster than any existing system.

Audio Subcarrier   Auxiliary carrier, linked to main carrier, that can carry different audio signal (see *Subcarrier*)

Aurora   Alascom (Alaska) domestic communication satellites

Autocue (British)   See *Teleprompter*

AVHRR (Advanced Very-High-Resolution Radiometer)   Data-gathering device used on environmental satellites to detect sea-surface temperatures, identify snow and ice, and image in both visual and infrared bands; located on NOAA polar-orbiting satellites

Avid/1 Media Composer   Digital editing system, one of many available, to manipulate film and video interactively; can playback and edit at thirty frames per second; has CD sound; praised for ease of use

AWIPS (Automated Weather Information Processing Systems/Advanced Weather Interactive Processing Systems)   Computer station linked to data networks integrating ground data and satellite signal data

Azimuth   (1) In electronics, the angle at which audio/video is recorded to tape; by varying the angle on record heads, two separate tracks can be recorded on the same portion of the tape without interfering with each other; (2) In astronomy, the arc of the horizon, measured in degrees; (3) In satellite reception, one of two coordinates (with elevation) used to line up an antenna dish

B

Backhaul Feed   Signal transmitted by cable or satellite uplink/downlink from source to second terrestrial location for addition of new audio/video before transmission to viewers

Bandwidth   Measurement of the capacity of a signal-carrying channel. For analog signals bandwidth is expressed in terms of frequency. Higher-frequency responses equate with higher image resolution. Bandwidth expresses the difference in Hertz between the lowest and highest frequencies of the channel. For digital signals, bandwidth is expressed in terms of bits per second (bit rate). A channel with larger capacity can transmit more than one signal (multiplexing)

BASIC (Beginner's All-Purpose Symbolic Instructional Code)   Computer program language similar to but less complex than FORTRAN; in popular use on mini- and microcomputers

Basic Cable   System to transmit television signals to subscribers by way of a wired network

Baud Rate   A measure of transmission speed. Phone modems in computers currently operate at 1,200 or 2,400 Baud (bits per second); compare to the transmission rate over glass fiber optic cable (300 megahertz and up)

Beam-Matrix System   Replaces electron gun with grid pattern to control pixels on television picture tube, making thin tubes possible

Beam Splitter   (1) In television video, an optical device to split light signal into RGB components; (2) In holography, a system invented by Leith-Upatnieks to split laser beam, one lighting an object, one striking the recording film

Betacam/Betacam SP (Beta SP)   Broadcast quality half-inch videotape

Betamax   One-half inch VCR consumer-type format videotape

Binary Number   In digital information, a number represented by zeros and ones, with zero representing "off" and one representing "on"

Bird   Slang for satellite

Bit   Contraction for binary digit, referring to the representation of data/characters via ones and zeros; a single digit (see *Byte*)

B-MAC (Multiplexed Analog Component)   Television transmission standard used in Australia for satellite transmission, in USA for private satellite transmission, for production/transmission

Bpi   Bits per inch (see *Density*)

Brilliant Pebbles   Portion of US Strategic Defense Initiative (SDI) proposing to deploy series of small homing rockets to track/destroy incoming missiles by colliding with them

BSkyB (British Sky Broadcasting)   British DBS system formed November 1990 by merger of Rupert Murdoch's Sky Television and British Satellite Broadcasting (see also *Sky Cable*)

Buffer   In computers, auxiliary data storage area to store data temporarily (see *CPU*)

Buran   Russian for *blizzard*; Bus for planned USSR shuttle (reusable) fleet, orbited unmanned 1988, possibly doomed by budget cuts

Bus   (1) In television audio, the circuit that takes signals from multiple audio sources and feeds them to one source; in video, the bank of buttons for multiple video sources on a switcher; (2) In communication cabling, the device connecting cable from the central office (typically fiber-to-curb) to multiple curbside pedestals (now typically copper wire, in the future fiber-to-home); (3) In spacecraft, a satellite delivery system carrying the payload

Buy Rate   Percentage of (pay-per-view) homes ordering a telecast event

Byte   Computer term for fixed number of bits representing a character; most common byte size is eight bits

C

CAD (Computer-Aided Design)   Use of computer images/graphics to design products, in many cases replacing costly and time-consuming construction of models, often allowing creation of test situations impossible without computer's ability

CAE (Computer-Aided Engineering)   Computer program that allows engineering designs to simulate new designs on a computer and test them without having to build actual models, much used for designing circuit boards for optoelectronic circuits

Cathode Ray Tube (CRT)   Vacuum tube in which electron stream is focused by electron gun on fluorescent screen, in television set, computer, oscilloscope, etc.

CATV (Community Antenna Television)   Forerunner of the cable industry; rural communities unable to receive direct television signals erected large antennas to pull in the signal and wired their homes to the antenna

C-Band   Band of frequency (3.7 to 4.2 gigahertz) currently the most used by communication satellites (see *K-Band*)

CCD (Charge-Coupled Device)   Type of metal-oxide semiconductor (chip) used in film to translate images into digital one/zero (on/off) language to be enhanced with computer graphics, etc., then etched back onto film with a laser scanner; replaces tubes in miniaturized television cameras for ENG, other hand-helds, and sensor cameras on satellites; also used in telecines

CCIR (Comité Consultatif International de Radio)   International Radio Consultative Committee

CCITT (Comité Consultatif International de Téléphone et Télégraphe):   International Committee on Telephony and Telegraphy; sets communication standards

C-Cube Microsystems   A leader in the development of digital video compression (CL550-A Image Compressor Processor); the chair of the MPEG working group is the Director of Research at C-Cube. Figures G-1, G-2, and G-3 show varying degrees of compression; as a measure of the effectiveness of digital compression, they speak for themselves.

CD/CD-Graphics/CD-I/CD-ROM/CDTV/CDV/Computer CD/DAT/DCC/ MD/Photo CD (see *3DO*)

   CD (Compact Disc)   In 1982 when the CD was introduced by Philips of the Netherlands, everyone sensed that the shift in recorded music was revolutionary. Here was pure sound, optically read so that there was no wear on the disc. Suddenly the audio tape cassette, which had begun knocking out the LP, was itself in danger of being replaced. The LP had virtually killed off the 45 rpm, and the eight-track cassette was history, and these formats had filled attics with old 78 rpm shellac recordings. Will the simple CD survive the revolution it started? Survival until the mid-1990s is guaranteed because the "others" will only be arriving and—unless some projected new compatibility becomes reality—fighting each other for survival.

   CD-E   Recordable/erasable CD announced by Philips September 1992, due to market 1996.

   CD-Graphics   Songs of a different sort lie hidden inside JVC's CD-graphic system. JVC's portable RCGX-7 offered the first CD-graphic hardware in June 1990 (in the Asian market only). The blank space in

Figures G-1, G-2, G-3   Let your own eyes be the judge! These three scenes provided by C-Cube Microsystems show different degrees of video compression: 1:1 (Figure 1) 14:1 (Figure 2); and 66:1 (Figure 3).

Courtesy C-Cube.

the CD subcode stores graphic data, which can include such items as still pictures and karaoke lyrics. Will non-Asian CD buyers come out of their bars and showers with a yen to sing in their living rooms to home-style karaoke? (see also *CD-ROM*)

CD-I (Compact Disc Interactive)   Invented by Philips Electronics N.V. (Philips, Magnavox, PolyGram), codeveloped with Sony Corporation and Matsushita; combines sound, video, graphics, animation, and text on one five-inch silver CD disc that looks like an audio-only CD. The CD-I player is attached to any television set and stereo system. Figure G-4 shows that, unlike audio-only CDs, CD-I involves the viewer/listener, who uses a "thumbstick" remote control, by holding a continuous dialogue, asking questions, proposing actions, offering options, giving the viewer/listener the power of decision and action.

CD-I was established with global standards to ensure compatibility with all CD-I discs and players. All CD-I players are compatible with existing audio CDs, CD-Graphics, Photo-CDs (the Mini Disc is *not* compatible with these other existing technologies). Philips Consumer Electronics Company introduced Philips CD1910, the first consumer CD-I player available in the USA market.

CD-I processes data from a five-inch, 650-megabyte optical disc that allows up to 250 thousand pages of typed text, more than 7,000 photo-quality images, seventy-two minutes of full-screen digital video with CD-quality sound, full-motion animation, nineteen hours of speech, video capabilities of more than 16 million color variations, and four planes of visual effects.

CD-I is engineered with four types of sound: CD digital audio, high-fidelity, midfidelity, and speech. It can handle multiple-language commentary, up to sixteen parallel speech level audio tracks, with a maximum capacity of nineteen hours. The video is based on MPEG specs. The software is menu-driven, and it has main memory of 1 megabyte and contains an M68070 microprocessor. In addition to thumbstick remote control, CD-I operates with trackball accessories. Other controllers are optional.

CD-ROM (CD-Read Only Memory)   In computers, ROM is information built into the hardware. Users can read it but cannot write to it. RAM (random access memory) is where the computer software goes and is accessible. CD-ROM technology was first used as a mass storage peripheral for personal computers, basically in the form of data text. Because text requires vastly less storage capacity than video, a vast amount of copy can be recorded on a CD-ROM with about 680 megabyte capacity. Huge dictionaries, sets of encyclopedias, directories, and so on are available on CD-ROM.

CDTV (Commodore Dynamic Total Vision)   A combination of the Commodore Amiga computer and a CD player, CDTV provides interactive audio/video that is compatible with audio CDs and CD-Graphics (but not with Philips CD-I discs, which CDTV preceded).

CDV (CD Video)   Digital compression puts up to seventy-two minutes of film on a five-inch CD-sized disc; by 1999 full-length features are expected on the same size disc.

Computer CDs   If the Apple/IBM jointure doesn't develop a computer with storage on compatible CD discs, someone will, probably within the 1990s.

DAT versus Digital Duplication   The consensus is that digital audio tape will survive as a tool used by professional studios recording and editing music, not as a consumer system. Why not? DAT players will not play analog cassettes. DAT technology apes videotape technology. DAT's digital data is stored on the tape in a helical pattern (like the stripes on a barber pole or a candy cane). Both the tape and the head move. Analog tape has data stored along linear tracks, and the player head is stationary. (DCC has players compatible with both the new technology and analog cassettes.) Further, DAT cassettes came on the market at a higher price, partly because they have to be duplicated in real time; duplicating a two-hour cassette requires two hours.

DCC (Digital Compact Cassette)   One advantage of DCC is that its players will also play existing analog cassettes, both being the same size and both being recorded on linear tracks and read by a single stationary head. This is accomplished using digital compression. Like both consumer DAT and Sony's Mini Disc, DCC will incorporate the Serial Copy Management System (SCMS) to prevent unauthorized duplication. The system permits only one digital copy to be made and prevents further copies.

Mini Disc   Sony's Mini Disc (mini because of its 2.5-inch size) was released in late 1992. The Mini Disc concept offered two advantages over DCC: it was the first recordable/erasable optical disc system and it is a portable, battery-operated unit. The Mini Disc employs magnetic optical discs with a Sony-developed overwrite technology that allows simultaneous erase/rewrite using only one-third the power of standard magnetooptical discs. Sony's digital compression technique allows the Mini Disc to hold as much data as an audio CD in one-fifth the space, up to seventy-four minutes of audio data on the 2.5-inch discs, with random access. The discs come in a plastic cover much like that on 3.5-inch computer disks. Like DCC and consumer DAT, the Mini Discs incorporate the Serial Copy Management System.

Photo-CD   The Kodak/Philips Photo CD puts up to one hundred 35-millimeter videos onto a CD, which puts them onto a television screen with random access. The system reportedly has a built-in zoom capability that lets viewers expand any selected portion of the digital image for full-screen viewing, with no loss of quality. Video can be cropped as desired and printed via a videoprinter (see *Desktop Imaging*). It is viewable via CD-I players.

CD (Compact Disc)   Metal disc used to record music/data digitally; read via laser beam (see *Mini Disc*)

CD-I (CD-Interactive)   CD combining sound, video, graphics, and animation on a five-inch CD disc; player is attached to a television set

CD-ROM (Read Only Memory)   A CD used to store text (that may be supplemented with video and voice), including books, magazines, photos, and other print media

Channel   Portion of bandwidth used for a single communications link; NTSC television signals use a 6-megahertz band.

Character Generator   Electronic device to put letters/words on a screen

Chip   Electronic memory device; a semiconductor onto which an integrated circuit has been imprinted, or an imprinted integrated circuit itself

Chrominance   Color portion of composite video signal

Chryon   Commercial brand of character generator

Cinema Digital Sound   Audio technology developed by Eastman Kodak Co. and Optical Radiation Corp.; places six discrete channels of digital audio onto film, five full-bandwidth channels, one subwoofer channel; provides CD-quality audio in-theater and saves processing time; applied during the image-printing process

Cladding   Process of bonding layers of metallic alloys; in fiber optics, the layer outside the fiber core, configured to reflect the laser light signals back into the core at speeds resulting in a readable signal at the signal's destination (see *Core*)

Clarke Belt   Orbital belt named for Arthur C. Clarke, who conceived it, approximately 22,300 miles above the equator; satellites placed in orbit here, rotating at a speed matching the earth's rotation, remain stationary relative to earth (called geosynchronous)

CLASS (Custom Local Area Signaling System 7)   Computer-based system installed by Baby Bells in USA to route calls; also referred to as SS7

Clean Feed   On a show with multiple feeds going out, feed with no commercials, graphics, or other video that might interfere with local broadcast of the show (see *Dirty Feed*)

C-MAC (Multiple Analog Component)   Television transmission standard used in Norway for satellite transmission (see *B-MAC*, *D-MAC*)

Coaxial cable   Cable used to transmit high-frequency electronic/light signals

COBE (Cosmic Background Explorer)   Unmanned, solar-powered NASA spacecraft (16 feet tall, 28-foot span) orbiting 560 miles above earth to record radiation in space, aiming to support/refute the Big Bang theory of creation of the universe

COBOL (Common Business-Oriented Language)   Computer language used in business applications

Codec (Code/Decode)   Device that converts analog signals (voice, music, video) to digital form for transmission via digital media, then back to analog in receivers; compression for transmission uses a mathematical equation (called an algorithm) to digitalize the analog video signal and reduce its bandwidth requirement (see *Sampling, A/D, D/A*)

Color Bars   Television signal consisting of bands of color and sections of black-and-white areas, used as a reference signal in television production

Common Carrier   Organization providing regulated telephone, telegraph, telex, and data communications services

Component Video   Color and brightness signals kept separate, not encoded (mixed) as in composite video (see *D-1*)

Composite Video   Video signal in which luminance and chrominance have been encoded (mixed), as in NTSC, PAL, SECAM (see *Component Video*)

Compression   See *Digital Compression*

COMSAT (Communications Satellite Corporation)   Private company established as a monopoly representing the USA to Intelsat and Inmarsat; now one of several companies providing satellite links to international television transmission (see *Intelsat, Inmarsat*)

Comstar   Comsat communication satellites (USA)

Connection Machine   Supercomputer system that links 64 thousand parallel processors, produced by Thinking Machines Corporation

Contelsat   Contel-ASC communication satellites (USA); Contelsat-1 is a hybrid satellite

CONUS (Contiguous United States)   Satellite footprint acronym

Coord Director (Coordinating Director)   On a show with multiple feeds (commercial, noncommercial, and so on), the director for any one of these feeds, which the executive director coordinates with the basic "clean" feed

Copyright Royalty Tribunal (CRT)   USA government organization that rules on copyright issues

Core   Inner portion of fiber optic cable, composed of the optical fiber itself, surrounded by cladding to return any zigzagging light signals to the core (see *Cladding*)

Cross-Strapping   Interconnecting different bands on a single hybrid satellite—uplinking on one band and downlinking on the other

CPU (Central Processing Unit)   Device in central computer containing the main storage area and arithmetic/logic unit controlling the operation of the computer unit; usually a series of terminals is connected to a central computer (mainframe) to input/output

CRT   See *Cathode Ray Tube, Copyright Royalty Tribunal*

Cyberspace   Coined by Autodesk Co.; see *Virtual Reality*

D

D/A (Digital to Analog)   Process/device to convert signals

DAB (Digital Audio Broadcasting)   Broadcast system with less interference than AM or FM analog systems, mostly delivered by satellite to cable television systems to subscribers for monthly fee, currently more prevalent in Europe than in USA due to EBU's Eureka Project; multiplexing allows sixteen DAB CD-quality radio channels in the same spectrum space as sixteen FM radio channels; DAB is not compatible with AM or FM receivers

DAT (Digital Audio Tape)

Data Base (Data Bank)   Centralized storage of computer data that can be accessed by remote terminals

Data Discman   Sony CD-ROM player; plays music CDs and stores data (encyclopedias, court transcripts, newspapers); weighs 1 pound, has a pop-up LCD screen and headset

DBS (Direct Broadcast by Satellite)   Programming from broadcaster direct to home antenna dish; in the USA one of two FCC classifications for communication satellites: DBS covers only direct public television service, fixed service satellites covers all other satellite services (see *DTH*)

DCC (Digital Compact Cassette)   Digitally recorded cassettes, compatible with analog cassettes

Decoder (Descrambler)   Unscrambles encoded video signals made indecipherable except to subscribers with decoders, to prevent pirating of signals

Degrees Kelvin   See *Kelvin Degrees*

Density   The number of characters in one inch of magnetic tape or on a computer disk track, usually expressed as bits per inch (bpi)

Desktop Imaging   In 1990 Polaroid introduced its Digital Palette CI-3000, a universal desktop computer film recorder, a plug-and-play system that produced a 35-millimeter color slide instantly off IBM and IBM-compatible computer screens, in-house, in under ten minutes. Imaging options include slides, small format transparencies and prints, four × five-inch color prints, three × four-inch self-developing AutoFilm color and black-and-white prints. In 1991 came the Digital Palette CI-5000 film recorder (see Figure G-5), which operates in both IBM and Macintosh environments and can be interfaced to connect both an IBM (or compatible) and a Macintosh II or SE-30 computer to the same film recorder. If you can get an image onto a computer screen, you can get it on the air in minutes. Javelin's Video Color Printer JP2600 delivers high-definition video prints in ninety seconds from any video source. Autosynchronous, it automatically scans/stores composite or component RGB video signal frequencies at horizontal rates between 15 and 32 kilohertz, with vertical rates between 45 and 70 kilohertz. A serial (RS 232) port is provided for control via computer.

Digital Compression   Process for shrinking images for transmission or editing (unshrinking them via decompression) (see *JPEG, MPEG*)

Digital Computer   Computer that uses numbers to perform logical and numerical functions (see *Analog Computer*)

Digital Effects   Figure G-6 shows the Ampex ADO (Ampex Digital Optics) 100 Digital Effects System (one- and two-channel systems, online and offline effects storage), a low-cost system for smaller broadcast, postproduction, corporate/industrial facilities. ADO 2000 is for broadcast and postproduction (thirty preprogrammed effects with no more than two key strokes), upgradable to ADO 3000 for high-end postproduction (one- to four-channel systems, user-configurable). Figure G-7 shows the ADAPT digital layering device, significant because it operates in conjunction with existing analog switchers. Also available is the ACR-225 D-2 format composite digital automated cassette system, for round-the-clock on-air commercial/program-segment playback (capacity 256 32-minute cassettes).

Digital Sound Field Processing   Figure G-8 diagrams Yamaha's Digital Sound Field Processor (DSP-A1000) that lets television viewers/listeners turn their homes into Anaheim Stadium, the Roxy Theater, a New York disco, a Tokyo disco, a jazz club, a concert hall, and so on. It combines Dolby Pro Logic Surround (Directional Enhancement) and Yamaha's DSP with data stored in the DSP-LSI chip. Other companies have developed similar amplifiers, including Sony and Onkyo. Toshiba has incorporated Audio Designer Bob Carver's Sonic Holography processor in some of its television sets. Sony has used Hughes Aircraft Company's Sound Retrieval System in its top-line television sets.

Digital Video Interaction   Experimental technique that merges computer with VCR, at SRI-David Sarnoff Research Lab in Princeton, New Jersey

Digital Video Interactive (DVI)   Intel's digital imaging codec system for still and motion images

Figure G-5   This Polaroid Digital Palette CI-5000 Computer Film Recorder operates in both IBM and Apple Macintosh environments and can produce 35-millimeter slides, prints, or small-format overhead transparencies in minutes.

Courtesy Polaroid Corporation.

Figure G-6   This Ampex ADO 100 Digital Effects System, designed for small broadcast, postproduction, and corporate facilities, offers a wide range of two- and three-dimensional effects. Larger systems range up to ADO 3000 for high-end postproduction.

Courtesy Ampex Corporation.

Figure G-7 This Ampex ADAPT composite digital layering device is one of a series of products developed as transitional tools between analog production—with which it can interface—and the ultimately expected all-digital postproduction.

Courtesy Ampex Corporation.

Figure G-8 Yamaha's DSP-A1000 (Dynamic Sound Field Processing A/V Integrated Amplifier) can input and control multiple sources. Pressing an on-screen key on the remote superimposes the current DSP program name and parameter settings on the video monitor. Sound processors range from disco and jazz club to churches and concert halls.

Courtesy Yamaha Electronics Corp.

Diode Device with two terminals, with low resistance in one direction and high resistance in the other, used to regulate voltage and to convert alternating current to direct current; a light-emitting diode (LED) produces light when current passes through it

Dirty Feed On a show with multiple feeds going out, a feed containing identifying graphics, commercials, and so on that might clash with local broadcasting restrictions (see *Clean Feed*)

Dish (Dish Antenna) Slang for parabolic microwave antenna, used to receive satellite signals

D-MAC (D-2MAC) (Multiplexed Analog Component) Transitional European transmission standard, more advanced than PAL but less advanced than HD-MAC

D-1/D-2 (1) In video editing, D-1 is a *component* digital format for editing tape/film, sometimes called 4:2:2; D-2 is a *composite* system with the RGB signals combined/encoded into a single signal. D-2 is analog-compatible and less expensive; D-1 has close to double the data quality. (2) Intelsat assignation for small earth stations in its Vista service, begun in 1983, to serve remote areas lacking communications facilities; a standard D-1 is a single small station, standard D-2 is a domestic hub or master earth station serving multiple D-1s.

Doppler Effect Named for Christian Doppler, Austrian physicist, the apparent change in frequency of sound/light waves, varying with the relation of the source and the observer

DOS (Disk Operating System) Computer program stored in ROM or RAM portion of computer memory; contains information to provide basic computer functions

Downconverter Circuit to drop high-frequency signals to a lower portion of the electromagnetic spectrum (see *LNA, LNB*)

Downlink Earth station receiving signal from a satellite, used as both noun and verb (see *Uplink*)

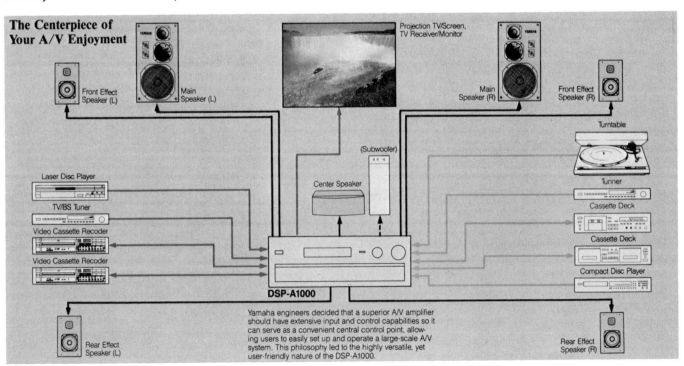

The Centerpiece of Your A/V Enjoyment

Front Effect Speaker (L)
Main Speaker (L)
Projection TV/Screen, TV Receiver/Monitor
Main Speaker (R)
Front Effect Speaker (R)
Turntable
(Subwoofer)
Tunner
Laser Disc Player
Center Speaker
Cassette Deck
TV/BS Tuner
Cassette Deck
Video Cassette Recoder
Video Cassette Recoder
Compact Disc Player
DSP-A1000
Rear Effect Speaker (L)
Rear Effect Speaker (R)

Yamaha engineers decided that a superior A/V amplifier should have extensive input and control capabilities so it can serve as a convenient central control point, allowing users to easily set up and operate a large-scale A/V system. This philosophy led to the highly versatile, yet user-friendly nature of the DSP-A1000.

D-RAM (Dynamic Random-Access Memory) Memory chip used in computers; in 1991 a maximum of 4 megabits was available, but an existing 64-megabit prototype is expected by 1995

DS-3 (Digital Signal-3) DS-0 is the quantum unit of fiber optic transmission capacity, a channel with a bandwidth of 64 kilobits per second; DS-1 is a channel with a capacity of 24 DS-0s (1.54 Mbps); DS-3 is a channel with a capacity of 28 DS-1s (45 Mbps). DS-3s are used for transmitting broadcast quality video signals.

DTH (Direct-to-Home) Same as DBS

Dual Feedhorn On a satellite antenna, a feedhorn that can receive both C-Band and K-Band signals

Dynamic Motion Simulator (DMS) Computer system coordinating film action with hydraulically activated theater seats to involve viewing audience in action

E

Earth Station Ground-based equipment used to communicate with satellites; also called ground stations

EBU (European Broadcast Union) Organization representing thirty-nine member broadcasters in thirty-two countries in Western Europe; absorbed its Eastern European counterpart, the International Radio and Television Organization (OIRT), in 1993 (see *OIRT*)

EC (1) The European Community, an economic coalition of twelve European nations (the number may expand) (2) The European Commission, which helped establish the European Community

EDL (Edit Decision List) List of edits produced during off-line, non-linear editing as a guide to final edit

EDP (Electronic Data Processing) Processing of data by electronic digital computer(s)

EDTV (Enhanced or Extended Definition Television) Transitional sets between 1991 standard sets and HDTV-compatible sets

Effects Bus Switcher bus used for DVE effects, and so on

EIA/CEG Electronic Industries Association's Consumer Electronics Group

EIRP (Effective or Equivalent Isotropic Radiated Power) The measure of the power of transmission from an earth station's antenna, measured in decibels above one watt of power (dbw)

Electromagnetic Pulse Shock wave resulting from nuclear explosion, powerful enough to disrupt electronic communications (see *Emergency Broadcast System*)

Electromagnetic Spectrum Full range of wavelengths (frequencies) of electromagnetic radiation, including (from low to high) radio, microwave, infrared, visible light, ultraviolet, x-ray, gamma ray, cosmic ray, all of which travel at 186,000 miles per second. The radio frequency is best suited for communications. Satellite communications can use the microwave portion of the spectrum.

Electronic Cottage Futurists' view of everyone's working at home, interconnected via computer networks (see *Telecommuting*)

Electronic Still Store (ESS) Electronic device to store still video frames and recall them with random access

Emergency Broadcast System (EBS) Radio stations that are theoretically protected from destruction, especially by an electromagnetic pulse resulting from a nuclear explosion (see *Electromagnetic Pulse*)

Encryption Scrambling a signal to avoid piracy by changing or varying the bitstream to make it random and undecipherable without a decoder

End User Ultimate recipient of a television signal

ENG (Electronic News Gathering) Use of portable camera for fast, mobile news coverage, usually sent by microwave or satellite back to broadcast source (see *SNG*)

ERIS (Exoatmospheric Reentry Vehicle Interceptor System) Anti-missile device, largely self-guiding, to intercept missiles faster than SCUD missiles; has ground-based supercooled infrared sensor to detect heat radiated from a warhead

ESS See *Electronic Still Store*

Ethernet Local area network (LAN) standard for interconnecting (interfacing) multiple computers or to link computers to video equipment

Eurospace Nonprofit industrial group, founded 1961, based in Paris

Eurovision Program exchange system of European Broadcasting Union

Eutelsat (European Telecommunications Satellite Organization) Satellite network serving the EC (see *Intelsat*)

Exosphere Outer layer of earth's atmosphere; contains helium, hydrogen, radioactive particles, and bands of radiation (see *Mesosphere, Thermosphere*)

F

Fax (1) Facilities: all equipment necessary for a show. (2) Facsimile: electronic transmission/reproduction of data

Feedhorn Wide-beamed antenna that collects signals from the reflector in a satellite antenna system

Fiber Distributed Data Interface A standard for glass fiber optic cable

Fiber Optic Cables Cables that carry data as pulses of light, usually laser, gradually replacing copper wire for telephone, computer, and television cables. Glass fiber optics have a standard (fiber distributed data interface) carrying 100 megabits per second. Some phone companies have run tests to a trillion bits. Cable consists of hair-thin fiber core covered with reflective cladding, acting like a mirror to reflect back zig-zagging light signals. Plastic fiber optics are unbreakable, but still limited to 10 to 16 megabits per second, the same rate as copper wire. Fiber optics cannot be tapped without detection, as copper wire can.

Field (1) In computers, a group of characters. (2) In television, converting video standards involves frames and fields. The NTSC standard for scanning calls for 525 lines of pixels at thirty frames per second, sufficiently fast that the human eye "sees" a complete picture. Each frame is a composite of two alternate-line scannings. The odd-numbered lines are scanned, then the even-numbered lines. The odd lines alone, or the even lines alone, are called a field. Thus there are two fields per frame. In NTSC, with thirty frames per second, there are sixty fields per second. In PAL/SECAM there are 625 lines at twenty-five frames per second and fifty fields per second. Some converters go only from PAL/SECAM to NTSC. More expensive models are bidirectional. Some equipment that converts NTSC video that originated on film (twenty-four frames per second) can produce erratic movement called "judder." (see *Interlaced Scan, Progressive Scan*)

Financial Interest and Syndication Rule (Finsyn) FCC ruling that prohibited networks from owning/syndicating their programming

FLIPS (Fuzzy Logic Inferences per Second) What fuzzy logic chips make in fuzzy logic computer programs (see *Fuzzy Logic*)

FLOPS (Floating Point Operations) The number of mathematical calculations a computer can perform involving numbers that are not integers (whole numbers of zero) (see *Teraflop, FLIPS*)

Flowchart Visual reproduction of logical steps to be taken in a computer program, in a job, in a commercial, and so on

Flyaway   Portable satellite uplink

FM   See *Frequency Modulation*

Font Operator   Person responsible for the graphic look of a telecast by inputting text data from sources such as statisticians and recalling them for viewer identification

Footprints   Beam patterns showing the coverage and strength of a satellite signal; important in determining transponder rental for transmission, the earth equipment needed for home reception

FORTRAN (Formula Translator)   High-level computer programming language, most widely used for scientific and mathematic programs

4:2:2   International standard (CCIR-601) for digital component signals; the first number refers to relative resolution of luminance and the other two to relative resolution of chrominance signals (see *D-1*)

Four-wire   A set of dedicated phone lines (PLs) that do not go through the standard switching system of any company or country; consists of two pairs of open phone lines, one pair in each direction. On a news or sports remote, a four-wire allows a producer to talk into the correspondent's earpiece while the correspondent uses the other line to report on the air, speaking live to the television audience; made famous outside the industry by CNN's use during the start of the 1991 Persian Gulf War.

Frame   Two interlaced fields (see *Field, Interlace*)

Frame Grabber   Device to capture/record still video images in real time (1/30 second)

Frame Store   See *Electronic Still Store*

Frequency   Measure of the rate of waves in electromagnetic wavelengths; i.e., the number of waves that passes a given point, measured in cycles per second (cps), usually expressed as Hertz (Hz)—1,000 cps is 1 kilohertz, 1 million cps is 1 megahertz (see *Modulation*)

Frequency Division Multiplexing (FDM)   Division of a bandwidth into subdivisions, each with enough bandwidth to carry one voice or data connection. Multiplexing mixes multiple signals and transmits them as a single signal, to be separated again at reception. FDM separates signals by assigning each to a different portion of the frequency. Opposite of time division multiplexing.

Frequency Modulation (FM)   (1) Radio signal carrier wave constant in amplitude, varying in frequency; (2) A broadcast system using this technique (see *Amplitude Modulation*)

F-Sat (Frugal Satellite)   Mid-size satellite bus designed by Lockheed

FSS (Fixed Service Satellites)   One of two FCC classifications for USA communication satellites; covers all satellite service except DBS (see *DBS*)

Fuzzy Logic   Artificial intelligence, as in computer programming, that recognizes imprecise commands, used to control household appliances, television sets, and so on; created in the USA in the 1960s by Dr. Lofit A. Zadeh, at the University of California, not developed commercially until Japan used it in consumer products such as electronic gear, camcorders, and television sets

G

Gain   Ratio of output current/power to input; knob or fader used to control gain is called poteniometer (pot)

Galaxy   Hughes Corporation domestic communications satellites (USA)

Gateway   (1) In satellites, an earth station uplink providing access to signal transmission or an interchange between two satellite networks; (2) in computers, an electronic gate offering access to otherwise inaccessible data system(s)

GATT (General Agreement on Tariffs and Trade)   International pact designed to reduce global barriers; often misused to establish national quotas

GEOS (Geodynamic Experimental Ocean Satellite)   Not to be confused with GOES

Geosynchronous   Satellite orbiting in the Clarke Belt at speeds matching the earth's rotation

Genlock   Device used to lock synchronizing sources to allow switching from source to source without picture roll

Gigabit   One billion bits

GIGO   Garbage In, Garbage Out

GIS (Geographic Information Systems)   Software/hardware systems for electronic map technology, combining computerized maps with business/scientific/transportation data, and so on

Global Beam   Satellite footprint covering all visible earth surface

Global Positioning System (GPS)   Satellite-based location system using three satellites to triangulate geographical position(s), four for holographic purposes, developed by US Air Force Space Systems Division in 1973; needs twenty-one satellites for round-the-clock positioning, twenty-one had been launched by December 1992; twenty-four are planned by 1994

GMT (Greenwich Mean Time)   At Greenwich, England, used as universal standard time; now called Universal Coordinated Time.

GOES (Geostationary Operational Environmental Satellite)   Part of the USA weather observation system, a series of geostationary meteorological satellites; futures planned by NASA are called GOES-NEXT

Gorizont   (Russian for *horizon*) USSR/CIS fixed satellite system, used for USA/USSR(CIS) Hot Line and USSR/CIS link to Inmarsat

Groupe Spéciale Mobile (GSM)   European Community's all-digital cellular system in eighteen countries, with time-division multiple access technology; not backward-compatible to analog as is USA's.

H

HDCD (High Definition Compatible Digital)   Improved digital recording/playback system developed by Pacific Microsonics Inc. of California. Twenty-bit recording process increases initial sampling rate to several hundred thousand per second (as opposed to current standard 44,000), then compresses the signal via a new algorithm. Encoding during recording improves playback on a CD with no decoder but reaches maximum improvement on players with decoders.

HDTV (High-Definition Television)   Broadcast standard for which no single definition is yet complete because of still competing standards. All systems substitute a picture ratio of 16:9 for the older 4:3 ratio, with double or more NTSC 525 lines and much sharper picture. The oldest HDTV, Japan's MUSE-E (Hi-Vision), transmits 1125 lines at a field-scanning rate of sixty (interlaced); developed by NHK/Sony of Japan for DBS, not for terrestrial broadcasting.

HD-MAC (High-Definition Multiplexed Analog Component)   Television transmission standard proposed for the EC in 1995/6

Headend   Electronic control center for cable television system

Helmet-cam   Two-inch-long POV camera used for television action coverage

Hertz   Named for German physicist Heinrich R. Hertz; electromagnetic frequency representing the electrical waves completed in one second (1 kilohertz = 1,000 Hertz; 1 megahertz = 1 million Hertz; 1 gigahertz = 1 billion Hertz).

Himawari   Japanese geostationary meteorological satellite, launched in 1977

Hirachi Fiba   Optical fiber in Japan

Hi-Vision   Japanese term used for NHK's MUSE-HDTV

Hologram   A three-dimensional image on a high-resolution photographic plate exposed with laser light; acoustic holograms, which substitute sound waves for light waves, are used in medicine

Holography   Lens-less photographic method using laser light to produce three-dimensional images; splits laser beam in two, uses mirror image, records on photographic plate; light shined through the plate produces a "virtual image" (see *Virtual Reality*)

Holographic Interferometry   Technique employing an interferometer to measure wavelengths of light; used, for example, to seek fatigue in airplane engines

Home Satellite Television System   A parabolic antenna combined with a feedhorn, a low-noise amplifier (LNA), a downconverter (or an LNB combining an LNA with a down converter), a satellite receiver (inside the house), and a television set

H.261 (CCITT Recommendation H.261)   Motion video compression standard created for ISDN, currently accommodates multiple signal channels of 64 kilobits per second

HuMaNet (Human Machine Network)   A Bell Laboratories PC-based speech-recognition system that controls lighting (facility control), audio/video sources (database control), and telephone lines (ISDN control) in a conference room

HUT (Homes Using Television)   A percentage reported in television rating services

Hybrid Satellite   Satellite with both C-Band and Ku-Band transponders

Hydrosphere   Water portion of earth, including water vapor in air

I

Ibero-American Television Organization (Organizacion de la Television Ibero Americana-OTI)   Broadcast group founded in 1971, headquartered in Mexico

IBS (Intelsat Business Service)   Global digital service with speeds up to 8 megabits per second, used for videoconferencing, audioconferencing, and facsimile; begun 1983; sometimes called International Business Service

IC   See *Integrated Circuit*

Icon   On computer screen, a pictoral representation of a function in lieu of a printed word (e.g., a wastebasket for delete)

Iconoscope   Early television camera pickup tube, patented by Vladimir Zworykin in 1924, requiring vast amounts of light to register acceptable video (see *Image Orthicon, Vidicon*)

IDB Communications   Commercial common carrier in Culver City, California, awarded the right to offer public access telephony service to the USSR/CIS via Intersputnik

IDR   Intelsat's digital equivalent of its analog frequency division multiplex/frequency modulation (FDM/FM); offers ISDN quality with data rates from 64 kilobits to 45 megabits per second

IF (Intermediate Frequency)   See *Downconverter*

IFB (Interrupted Feedback)   Intercom system allowing director to give directions to on-air talent; carries program audio

IFPA   International recording trade group working to establish worldwide subcode to identify tracks of recordings (see *ISRC*)

IKI The Space Research Institute in Moscow, whose marketing interests in the USA are represented by The Space Studies Institute in Princeton, New Jersey

Image enhancement Electronic manipulation of data to improve visible effects

Image-Orthicon (I-O) First practical television pickup tube; replaced iconoscope and preceded vidicon

Infomercials Program-length advertisements masquerading as entertainment

Infotainment (1) In television jargon, program surrounding information with entertainment to attract a larger audience than a straight documentary might draw; (2) in computers, multimedia programs incorporating educational data

Infrared Portion of the electromagnetic wavelength spectrum critical in satellite and light/heat sensors to visualize images beyond human sight levels

Infrared Astronomical Satellite Satellite fitted with infrared-sensitive sensors to map the universe, as opposed to earth observation satellites

Inmarsat (International Maritime Satellite Organization) Global satellite communication network run by forty-eight-member nations, headquartered in London, serving ships, remote oil rigs, remote landsites, and so on, and providing data, facsimile, telephony, and telex transmission; founded in 1979, it uses its own satellites and leases transponder space from Intelsat and the European Space Agency; uses both C-Band and L-Band for shore-to-ship and ship-to-shore transmission; the US representative to Inmarsat is Comsat

Insat Indian geostationary meteorological satellite system, first launched 1982; operates in C-Band and S-Band

Institute of Space and Astronomical Science Division of Japan's Education Ministry, one of nine government agencies with space programs

Integrated Circuit Electronic circuit with multiple interconnected amplifiers and circuit devices formed on a single semiconductor chip (see *Microchip*)

Intelnet Digital service from Intelsat for distribution and collection of data from remote areas, based on VSATs, available for domestic and international applications

Intelsat (International Telecommunications Satellite Organization) A not-for-profit commercial international cooperative headquartered in Washington, D.C. that owns and operates a global satellite system providing telecommunications services connecting the entire world; founded in 1964, membership is now more than 100 nations and links more than 170 nations; Comsat is the USA representative to Intelsat

Interface (1) A device that connects two units or parts of a computer system; (2) the actual connection

Interlace Scan Combines two picture fields to produce a full video frame (see *NTSC, Progressive Scan*)

Interactive Television Computer-compressed system allowing television viewer to select camera shots or variable endings to access different programs; two-way television, linked via fiber optic cable or satellite, connecting non-studio participant (including home viewer) to participate in live telecast

International Gateway Origin of global satellite feed; an uplink

Internet Four-node computer network started by US Defense Department in 1969, connecting more than 175 thousand computers, 7 thousand computer networks worldwide

Intersputnik   CIS(USSR)/Eastern European commercial satellite telecommunications organization using multiple satellites (see *IDB*)

Intervision (Intervidnyie)   Broadcast union/program exchange part of
OIRT

I/O   Input/Output

Ionosphere   Layer of earth's atmosphere 80 to 600 kilometers above
earth's surface; contains electrically charged particles via which radio
signals are transmitted

IOR (Indian Ocean Region)   One of four major satellite coverage areas
(see also *AOR, A-PR, POR, Intelsat*)

IPPV (Impulse Pay-Per-View)   See *Pay-per-View*

IRD (Integrated Receiver-Decoder)   Sophisticated home satellite receiver
incorporating VideoCipher II Plus decoder module; can move home dish
antenna by remote control; parental lock-out, VCR timer are among
IRD options

IRIDIUM   Global personal communications system using sixty-six
lightweight, low-orbiting satellites, from Motorola

ISDN (Integrated Services Digital Network)   (1) International standard
for high-speed digital information networks (information highways), with
a current minimum set at 64 kilobits per second; (2) fiber-optic telecommunications system designed to link homes and businesses to multiple
sources, such as telephones, television, and electronic newspapers,
capable of carrying both voice and data over two voice channels and one
data channel, varying from 16 kilobits per second to experimental rates
of 10 megabits per second; (3) the technology involved

ISO   (1) Isolated camera and/or the shots taken with that camera; (2)
International Standards Organization

ISRC (International Standard Recording Code)   Digital coding technique
used in the EC to identify music video, in Japan to identify both video
and audio tracks, and proposed as a global standard to facilitate distribution of royalties for public performance and home recording via interactive services such as DAB, DBS, cable, and phone lines

ITU (International Telecommunication Union)   Specialized agency of the
United Nations, currently with 163 member nations, that sets standards
for equipment and techniques used for global communications, including
allocating orbital positions for satellites, registering global use of
frequencies, and sponsoring World Administrative Radio Conferences to
discuss these issues; under the UN since 1947, but the oldest inter-
governmental organization, originally established in 1865 after the
invention of the telegraph; charged with promoting/offering technical
assistance to developing countries in the field of telecommunications

ITVA (International Television Association)   Organization for nonbroadcast video professionals with 9,000 members in 103 chapters plus 3,500
in fourteen affiliated international groups in 1991; headquartered in
Irving, Texas

J

Japan Satellite Broadcasting (JSB)   Private DBS corporation that began
broadcasting April 1991 via WOWOW Home Theater Channel, in a joint
venture with New York-based Reiss Media Enterprises; JSB has more
than 260 investing companies, including earth station companies (19
percent), newspapers (16 percent), banks, railways, realtors, and electronics companies

Joystick   Hand operated control used to position wipe patterns on a
screen and control effects in computer video games

JPEG (Joint Photographic Experts Group) International standards group for CCITT and ISO that sets standards for compressing digital images (see *MPEG*)

Judder see *Field*

Jumbitron Giant 23 x 32-foot television screen in Times Square, New York, composed of 560 individual picture tubes, by Sony; programmed with component video using Trinilite picture elements; programming recorded on optical video discs, brightness controlled by sensor on building exterior; larger Jumbitron, in Toronto Skydome, has 40 x 120-foot screen.

K

Ka-Band Satellite transmission bandwidth from 19 to 30 gigahertz frequency range; higher than Ku-Band

Karasync From KG-Digital, synchronizes an auxiliary soundtrack with a film. The film producer has to deliver a separate digital audio track made from the same master that recorded the film's optical track. The two tracks get in sync within seconds, and Karasync fades from optical to digital playback. If a splice or other disturbance occurs, Karasync switches back to optical until synchronization is reestablished, then returns to digital. The system then marks the frame loss in its memory and anticipates the next playback. Digital feed offers CD quality, permits playback in multiple language tracks. One print in a multiplex theater can show same film in half a dozen languages.

Karaoke ("Empty Orchestra") Videodisk player hooked up with television screen/speakers and a microphone; a laser beam reads both video and audio but feeds audio through the microphone so singing voice can be mixed in (including enhancement)

K-Band Frequency spectrum including Ka-Band and Ku-Band, from 10.7 to 36 gigahertz (10.7 to 36 billion cycles per second)

Kelvin Degrees (Degrees Kelvin) Named for Baron Kelvin, British physicist, a scale for measuring temperature in degrees Celsius from absolute zero; used to measure color temperature(s) and to express noise temperature of an LNA, sometimes called degrees absolute

Keyhole Satellites (KH-11, Advanced KH-11) Low earth-orbit USA weather satellites

Keykode Numbers Kodak's system applying machine-readable edge numbers in bar code format to facilitate transfer to and from film and computer for editing

Kinescope (1) Recording technique used prior to videotape to make motion picture film directly off a television tube; (2) The resulting film

Knowbots Adjustable computer programs designed to prioritize incoming data

K-Prime Partners USA DBS system begun November 1990, owned by satellite operator GE Americom and nine cable companies

Ku-Band Frequency bandwidth reserved for satellites using microwave frequencies between 10.7 and 18 gigahertz (10.7 to 18 billion cycles per second); includes DBS satellites

L

Lacrosse Radar Imaging Satellite Remote sensor satellite capable of "seeing" through cloud cover

LAN (Local Area Network) System that links stand-alone computers and other electronic units in business offices and also found in "smart homes"

Landsat Series of USA earth observation satellites in polar orbit, begun 1972

LASER (Light Amplification by Stimulated Emission of Radiation)   Device used to create and amplify narrow, intense, focused beams of light by exciting atoms so that more are at higher energy levels than at lower energy levels. The process called stimulated emission occurs when a photon strikes an excited atom, causing it to emit a second photon aiming in the same direction as the original so that both photons can hit an excited atom, producing a rapid chain reaction and emitting a narrow, intense beam.

Last Mile   Telecommunications term for the final segment of a network that carries communication signals to/from the end user; can be fiber optic cable, coax cable, satellite, or microwave

L-Band   Bandwidth used by some satellites, 1 to 1.17 gigahertz

LCD (Liquid Crystal Display)   Screen readout used on computers, television, calculators, watches, and so on. Liquid crystal has component particles approaching the order of solid crystal, but being a liquid is easily modified by electromagnetic radiation, temperature, and mechanical stress.

LED (Light-Emitting Diode)   Semiconductor diode that emits light when voltage is applied, used for alphanumeric displays as on digital watches, small-screen television, portable computer screens and so on; consumes very little power (see *Diode*)

Letterbox   Method of televising wide-screen films, used more in Europe than the USA; in lieu of cutting off part of the film video to fit pre-HDTV 4:3 aspect ratio, full picture is shown with black band top and bottom

Light   Radiation visible to the human eye; electromagnetic wavelengths from 400 to 750 nanometers, between infrared and ultraviolet waves, transmitted at 186,000 miles per second

Lightsats   Lightweight satellites (circa 1,000 pounds), generally used for experimental purposes

Line Monitor   Monitor showing a picture going on air or onto videotape

Linear Editing   On-line editing, no random access (see *Nonlinear Editing*)

Lithography   Process that uses light to shrink chip circuitry onto silicon wafers (see *Integrated Circuit*)

L-1 Libration Point   Position approximately 1 million miles from earth at which satellites can be equally balanced—and so made "stationary"—between the earth and sun; needed to predict solar storms

LMDS (Local Multi-point Distribution System)   Microwave system transmitting on ultra high frequencies, capable of providing 40-plus channels of cable television (including two-way interactive), telephone services, videoconferencing, etc., to dish antenna under six inches; developed by Cellular Vision of New York Inc. (Freehold, New Jersey)

LNA (Low Noise Amplifier)   Device to receive/amplify weak signals reflected by an antenna (not to be confused with LAN) (see *LNB*)

LNB (Low Noise Block Downconverter)   Incorporates LNA and downconverter in a single housing; many new home satellite systems can be used only with an LNB

Lock Box   In coproduction, all money for production put in single account requiring multiple signatures

Low Earth Orbit   Satellite orbit 350 to 600 miles (550 to 950 kilometers) above the earth's surface, used by earth observation/meteorological satellites for remote image sensing

Low Wavelength Light   Light with a wavelength of 0.19 micron, difficult to produce

Luminance   Brightness portion of composite video signal (as in NTSC, PAL, SECAM); monochrome component of color television (see *Chrominance*)

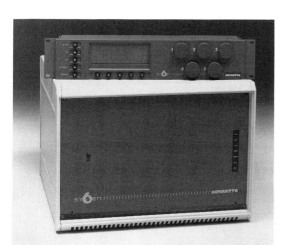

Figure G-9 Ultimatte System-6 can interface with an optional external disk drive that can be used to store or load all System-6 settings, including the time-code list. It can also interface with an IBM/IBM compatible computer to emulate the System-6 remote control. All of the control functions performed by the System-6 remote can be performed via a series of menus on the computer screen.

Courtesy Ultimate Corp.

M

MAC (Multiple Analog Components)   Television transmission standard with varying types of sound/data channels (A-MAC is little used) (see *B-MAC, C-MAC, D-MAC, HD-MAC*)

Magnetosphere   The region around a planet in which its magnetic field is stronger than the magnetic field beyond it (in interplanetary space); extends 3 million-plus miles from earth

MAPS (Multiple Array Processor System)   Advanced computer program, one of five supercomputers (one in Italy; one in Japan; three in the USA) designed as massive parallel computers to test basic theories of physics

Master Control   Termination and distribution point for all incoming/outgoing video/audio signals on a telecast/globalcast

Matting   Electronically keying together video from two sources. Figure G-9 (Ultimatte System-6) is a good example of how far matting has progressed. For example, for the music video *Ebony and Ivory* Ultimatte System-6 was used to make it look as if Stevie Wonder and Paul McCartney, actually on two different continents, were playing the same piano. Patented Screen Correction is achieved by first recording the blue or green backing without the foreground subject but with the same lighting and camera setup to be used for the actual shot. If the setup is shot static, one clean frame of the backing is all that's needed, recorded on a component tape recorder or stored on component still store. If motion is involved, the entire move is recorded with just the backing, no foreground and synchronized with the camera move for the foreground element, which is played back as the shot is composited. Screen Correction requires an RGB signal comparable in resolution to the foreground RGB signal, so D-1 is recommended for motion control or Betacam-SP component analog. System-6 has internal memory capable of holding seventy-five files plus a built-in SMPTE/EBU time code reader, which can maintain a list of 400 events associated with time code. Also available is a high-definition version that incorporates all System-6 features, operates with 1125/60 or 1250/50 standards, can be programmed for "any subsequent standard." Ultimatte's ITA (Interface to Anything) component routing switcher lets one System-6 composite foregrounds from up to sixteen different sources. It can also be linked to a PC.

Mediacitra Indostar   DBS serving Indonesia via S-Band and C-Band with ten television channels, up to eight CD-quality radio channels.

Megabit   One million bits

Mesosphere   Earth's atmosphere from circa 30 mi (50 km) to 50 mi (80 km)

Meteosat   Series of European Space Agency geostationary meteorological satellites first launched 1977; polar-orbiting satellites

Microchip   Microscopically thin layers of chemicals (e.g., silicon and gallium arsenide) containing electronic components to run computers; also called chips or semiconductors (see *Microelectronics*)

Microelectronics   Process of squeezing components onto a computer chip; components include microscopically thin layers of silicon and gallium arsenide

Microgravity   The state of zero gravity experienced in space, significant for its affects on humans and possible affects on manufactured products

Micron   One millionth of a meter

Microrobots   Highly miniaturized robots still largely used experimentally in medical, industrial, and automotive equipment

Microwave(s) Frequencies above 1 billion Hertz (1 gigahertz) and up to 3,000 gigahertz; radio waves with wavelengths less than 20 centimeters are used to transmit video and audio data. Line of sight (or reflected line of sight) is required for transmission. Microwaves are concentrated into a beam for transmission, not *broad*cast (see *C-Band, Ku-Band*)

MIDI (Musical-Instrument Digital Interface) Computer program that synchronizes/integrates multiple electronic musical instruments; helped revolutionize the "sound" of recorded music in 1980s by integrating synthesized instrumentation and helped popularize electronic pianos. Some expensive factory stand-alones are Yamaha's Disklavier and Bosendorfer's concert grand, which has optical sensors scanning the position of every key, hammer, and pedal 800 times per second. A remote control controls volume, tempo, and key. "Retrofits" can be installed on existing pianos. All are MIDI-compatible.

Minidisc Computer optical disc (2.5 inches) that is recordable/erasable

MIR (Russian for *peace*); USSR/CIS Space Station, 100-feet long, launched 1986, of modular design, with docking port for Soyuz spacecraft to land with exchange cosmonauts; includes a unit to manufacture crystals and two Kvant modules for scientific experiments and can house six cosmonauts

MITI (Ministry of International Trade and Industry) Japanese governmental agency that prioritizes national research programs, one of Japan's nine agencies with space programs, also highly involved in microrobots

Modem (Modulator-Demodulator) Device to modulate a signal to make it compatible for transmission/reception via an otherwise noncompatible source (e.g., a phone modem changes computer digital signals into signals that can be transmitted via analog phone lines, then changes them back to digital read-out on a receiving computer)

Modulation In electronics, the process of varying the amplitude or frequency or phase of a radio wave for transmission; of adding audio/video data to a carrier wave; or of adjusting, via a switcher, the rate of amplitude at which the edges of a wipe pattern move

Molniya (Russian for *lightning*) Domestic USSR/CIS satellites in nongeosynchronous orbit that carries them over the Hudson Bay in Canada; each of the Soviet satellites telecasts to the CIS for six hours

Monochrome Literally, one color; television transmission in black-and-white with no other colors

Morphing (from *metamorphosis*) Jargon for digital manipulation of video images, especially human to nonhuman form, mostly in film

MOS Japanese earth observation satellites

Mosaic (1) Breaking video image into equal-sized squares of limited luminance/chrominance; (2) Series of overlapping earth images matched to produce a continuous picture

MOSKVA (Russian for *Moscow*) USSR/CIS small antenna radio/television satellite system

MPEG (Motion Picture Experts Group) International standards group that sets standards for compressing full-motion digital video (see *JPEG*)

MSO (Multiple-System Operator) Company owning/operating more than one cable television system

MSS Mobile Satellite Service

MTV (1) Music Television, a worldwide network of video music channels begun in the USA in 1981 and global by 1991; (2) Maygar TV, in Hungary

Multimedia Integration of animation, music, video, and voice (see *Video Toaster*)

Multiplexing (1) In electronics, using compression techniques to send multiple signals on a single transmission line or transmitting two separate audio signals on a single carrier to produce stereo broadcasting; (2) In film, showing multiple films simultaneously on multiple screens; (3) In cable television, airing the same programming on multiple channels

MUSE-E (Multiple Sub-Nyquist Sampling Encoding) Original form of NHK's HDTV

N

NAB (National Association of Broadcasters) US broadcast industry association

NACS (National Advisory Committee on Semiconductors) Committee established by US Congress as part of 1985 Omnibus Trade Act

Narrowcasting See *Niche*

NASA (National Aeronautics and Space Administration) US space agency, responsible for launching satellites

NASDA (National Space Development Agency) Japan's equivalent to NASA in the US; division of Japan's Ministry of Science and Technology

National Reconnaissance Office (NRO) Agency (combined US Air Force/CIA) in charge of spy satellites

NAVSTAR (Navigation System Using Time and Ranging) Spacecraft element of the US Air Force/Navy Global Positioning System (see *GPS*)

NESS (National Environmental Satellite Service) Part of the National Oceanic and Atmospheric Administration (NOAA), which, in turn, is part of the US Department of Commerce

Neutrino A subatomic particle, with mass approaching zero (fact in dispute) and no charge, that occurs only during the decay of other particles (see *Neutrino Detector*)

Neutrino Detector (DUMAND: Deep Underwater Muon and Neutrino Detector) An astronomical "telescope" suspended above the floor of the Pacific Ocean near Hawaii, financed by an international group including universities from Germany, Japan, and the USA and the US Department of Energy; muons and neutrinos are two of many particles that combine to form atoms, the basic unit from which matter is formed

Newstrack In India, a monthly video magazine formatted like *60 Minutes*, both rented and sold; similar Indian news tapes include *Business Plus*, and *Observer News Channel* (India has historically banned both private broadcasting and international satellite networks)

NEXRAD (Next-generation Radar) Ground-based US National Weather Service radar detectors, sometimes called doppler radar, for weather stations, airports, and military sites; nearly 200 in place by 1997. Computers convert radar signals into digital data, then into usable weather information for transmission to meteorologists.

NHK (Nippon Hoso Kyokai) Japan Broadcasting Corporation, founded in 1926

Niche Programming Programming aimed at specific, frequently limited, audiences (e.g., all sports, comedy channels, nostalgia); same as narrowcasting

Nintendo Entertainment/Education System The world of television—and education—ignore Nintendo at their peril. Where television networks and schools face increased dropouts, this system has captured the imagination of millions of young people. The Nintendo Entertainment System (NES) has the components to fit the convergence theory. The Game Pak contains a ROM chip. There's an RF modulator to convert the sound signal. The signal's similar to that produced by a cable TV station. The central processing unit (CPU) has power to match a PC.

NOAA (National Oceanic and Atmospheric Administration) (1) Part of the US Department of Commerce; (2) series of polar-orbiting weather satellites with AVHRR

Noise   Unwanted signal interference with reception; in satellite systems microwave spurious noise appears as black and white dots called "sparklies"

Nonlinear Editing   Off-line editing, editing out-of-sequence without disturbing final sequence (see *EDL*). Despite the fact that most episodic television shows use nonlinear editing, it is not that old. Nonlinear means you can edit out or edit in without disrupting the rest of the project. The work, done offline, can be manipulated, viewed, then edited back in. Its instant random access makes the digital computer storage comparable to a film editor's bins. You can grab/access anything at any time. Each scene is identified by a small picture of the first frame of the scene. Some systems also show the last scene. All systems produce an edit decision list (EDL). Nonlinear's fast and efficient. Here are a few of the systems available in post-production editing suites:

AVID 1/Media Composer   Uses Macintosh hardware with multiple monitors and video compression; can handle thirty frames per second recording; introduced in 1989, with new models in 1991 utilizing JPEG video compression and magnetic or optical disk drives; offers multitrack CD-quality audio mixing

BHP Touch Vision   PC-driven system, with the editor touching on-screen instructions to control images visible on multiple source monitors; can source from tape or disk; leased in Los Angeles for episodic television and television films; in 1991 introduced D/Vision based on Intel's DVI video compression board.

CMX 6000   Nonlinear editing started in the 1970s with the CMX 600, withdrawn from the market, redesigned, and reborn in 1987 as CMX 6000; sources from laserdisc and can produce either a negative cutter's list or a CMX-compatible disk for online conforming; requires pretransferring material to laserdisc via an Optical Disc Company (ODC) machine.

Ediflex   Script-oriented, with the shooting script on-screen; uses a light pen to activate footage; sourced from videocassette or laserdisc, viewable on up to twelve images simultaneously; available only by lease

Editdroid   First generation released in 1984 as one of first laserdisc systems, back in 1990, manufactured by LucasArts Editing Systems for sale or lease; sources from laserdics and two 3/4-inch playback machines and one 3/4-inch record deck; requires pretransfer to laserdisc via an ODC recorder; accessed via touchpad with KEM Knob Motion Control.

EMC2   PC-oriented, like AVID, Editing Machine Company's system, introduced in 1988, was the first to introduce removable, reusable digital optic disks for storage; one disk holds 1.5 to 4 hours of material; EMC2 generates EDLs compatible with all standard formats; edits based on images in lieu time code; uses JPEG video compression.

MONTAGE   First edition out in 1984, with Montage II in 1987 and Montage System III in 1991; a full digital, nonlinear system; uses multiple monitors that display both first and last frame on each clip; laser storyboard printer can turn out hard copy printouts of EDLs and head/tail picture label contents.

Video FX   Made by Digital F/X, uses Macintosh and Disk F/X, which uses Intel's DVI Technology for hardware-assisted compression, based on Mac II board developed by NewVideo

NTSC (National Television Standards Committee)   (1) One of the three main video standards used worldwide at the start of the 1990s (see *PAL, SECAM*); (2) an ad hoc group of representatives from the NAB and EIA to recommend standards and specifications; two NTSC committees were formed in the USA to advise the FCC. In 1941 the first committee recommended adoption of the RCA system, but using 525 scanning lines at a ratio of thirty frames (technically 29.97 per second. RCA had used 441 lines) (see *Field*). In 1953 the second committee, reconvened by the FCC to resolve the question of colorcasting, recommended replacing the previously approved CBS color wheel system, which was incompatible with existing black-and-white sets, with a compatible RCA system transmitted at rates of 525 scan lines, sixty frames per second. Technically only 483 of the 525 lines are seen on home receivers.

NTT   Nippon Telegraph and Telephone

Nyquist Frequency   See *Sampling*

Nyquist Rate   Minimum sampling frequency for correct digital reproduction of a video signal (see *Sampling*)

Nysernet (New York State Education and Research Network)   Nonprofit company created 1986, based in Troy, New York, that uses fiber optic cabling and computers to link subscribers

O

Offline Editing   Editing video, as opposed to film, is a two-step operation. Tape is on reels or cassettes and lacks CD or computer random access, therefore the first phase of video editing is offline: scenes are viewed, edited, and mentally positioned. An edit decision list (EDL) is prepared; then online editing puts the material in linear sequence, adding needed special effects. Nonlinear digital editing using laserdiscs or optical CDs is more expensive but faster (see *Nonlinear Editing, AVID*)

OIRT (Organisation Internationale de Radio/Télévision, International Radio and Television Organization)   Eastern European group representing thirteen member broadcast groups in six countries (Bulgaria, Czechoslovakia, Hungary, Poland, Romania, USSR), merged into the European Broadcast Union effective 1993 (see *EBU*)

Online Editing   Electronic editing going directly from original tape to program master

Optical Computer (Processor)   Experimental processor (created by AT&T Bell Labs) that uses laser beams (photons), not electrons, to carry information. Optical switching involves two sets of laser beams, a series of mirrors (instead of silicon transistors) coated with gallium arsenide and a semiconductor device that opens and closes like a venetian blind; this technology could raise computer power to speed of light (see *S-SEED*)

Optical Lithography   Optical method of using light to shrink chip-circuiting patterns on semiconductor chips, almost the reverse of photo enlargment. Laser light is used now; x-ray lithography is the ultimate goal (see *X-ray Lithography, Phase Shifting*)

P

PAL (Phase Alternation by Line)   One of the three main video standards in use worldwide at the start of the 1990s (see also *NTSC, SECAM*). PAL derived from NTSC, but avoided hue shift caused by phase errors in the transmission path by reversing the phase of the reference color bursts on alternate lines. Like SECAM, PAL is a 625-line system (as opposed to NTSC's 525-line system), at the rate of twenty-five frames per second. PAL is the approved standard in Australia, England, Germany, Spain, and more than thirty other countries. PAL-plus is the enhanced version, offering a 16:9 screen ratio, as opposed to 4:3, on new sets (see *D-MAC, Field, HDTV, SECAM*)

PASCAL   Named for the French mathematician/inventor Blaise Pascal, a general purpose, high-level computer language

Pay-per-View (PPV)   Programming, usually cable or DBS, for which subscribers pay on a one-time basis

PCN (Personal Communication Network)   Communication system that uses radio waves to provide wireless, digital information services, including telephone services

Pen-based Computers   Computers that employ pens on touch-sensitive screens. Available from Go Corp., the company that started the trend, Grid System, IBM, Microsoft Pen Windows, and NCR. Sony's Palmtop uses pen on touch-sensitive screen.

PET (Positron Emission Tomography)   Scans that let scientists see the human brain at work

Phase Shifting   Technique that extends the capability of optical lithography. Normal method passes light through stencil, similar to photo negative, with transparent and nontransparent sections. In phase shifting the mask is all transparent but varies in thickness of glass. Used by Japanese to develop their 64-megabit D-RAM chip due by 1995. By the end of the 1990s, expect 256-megabit D-RAM and a 1 billion bit chip early in the twenty-first century.

Photon   Measure of electromagnetic energy exhibiting both particle and wave behavior, with no mass and no charge but with momentum; carries the energy of electromagnetic waves, including visible light

Photonics   The use of the energy of light for communication, as the use of laser beams to transmit digital off/on signals via fiber optics, satellites, and radio waves and in optical computers/processors

Photovoltaic Cell   Semiconductor diode that converts sunlight into electrical current; used to provide electrical power on satellites

PIP (Picture-in-Picture)   Digital effect allowing multiple video sources on a single television screen

Pixel (Picture Element)   Smallest element of a video image, the "dots" that make up an electronic image, comparable to print dots in photo reproduction; term borrowed from computer language for photosensitive dots to describe the grid layer of a television camera picture tube; hundreds of pixels make up the lines that are scanned by the beam from an electron gun to be translated into electrical current that becomes the video signal for broadcasting

PL (Private Line)   Dedicated phone line not going through a switchboard

Pod   Group of commercials/announcements during a break in programming

Polarization   Way of increasing channel capacity on satellites by polarizing electromagnetic waves horizontally and vertically so both can be set on the same frequency, doubling the number of channels available

Polyvision   Flat panel display by Alpine Polyvision Inc. using solid film silk-screened onto a single glass panel

POR (Pacific Ocean Region)   One of four major satellite coverage areas (see also *AOR, A-PR, POR, Intelsat*)

PPV   See *Pay-per-View*

Premium Programming Service   Optional channels available to cable subscribers for an added fee; separate fee per channel called à la carte

Preview Bus Device with a row of buttons to input video to a preview monitor; separate fee per channel called à la carte

Preview Monitor Monitor showing video source other than the on-air picture, includes monitor with video the director plans as next shot

Progressive Scan The tube in a television camera contains elements sensitive to light. The camera lens focuses on the image before it. Light waves of that image (photons) go through a conductive layer of the tube. The photons are passed to a gridlike photoconductive layer that contains photo-sensitive dots (pixels) arranged in a series of lines with hundred of pixels per line. An electron gun in the tube aims its beam through a ring of deflective magnets, which deflect the beam into a scanning pattern that goes both back and forth and up and down. The nature of this scan pattern, in part, determines the transmission standard of a color television signal. *Interlaced scanning* scans alternate lines; the beam scans line one, is turned off by a blanking pulse (creating a vertical blanking interval), then is deflected back and down to line three. It scans all alternate lines, then goes back and scans all even lines. The odd scan plus the even scan separately are called fields. Combined, the two fields are called a frame. *Progressive scan* scans all lines, in sequence, simultaneously, resulting in a sharper picture than is produced with interlaced scanning (see *Field, Vertical Blanking Interval, VIRS, VITC*)

PTT (Ministry of Post, Telephone and Telegraph) Government bureau, in many countries, controlling rights to televise events and import/use broadcast equipment

Pulse Code Modulation Process using pulse amplitude modulation to convert analog signals to digital zeros and ones, sample them, and further digitize them for transmission

Q

Q-Sound Commercial sound enhancement system that simulates stereo-surround sound on musical recordings, developed in Canada by Archer Communications and used during the mixing process after all the tracks are laid down; can also be applied to stereo television, radio, and film

Quad Split Video effect that divides the screen into four different images that appear simultaneously

Quantization Limiting a quantity (signal) to multiples of an indivisible, fixed unit (e.g., eight pixels per unit in digital sampling); Scientific-Atlanta's HDB-MAC's compression uses vector quantization (VQ)

Quark (Word created by James Joyce in *Finnegan's Wake*) Hypothetical particle believed to be a basic unit in matter

Qube Interactive cable system operated in the 1970s in Columbus, Ohio

R

RADAR (Radio Detecting and Ranging) System/device sensitive to reflected radio waves by which it determines the existence, size, shape, and speed of the reflecting object; used in weather, military, environmental detectors

RBDS (Radio Broadcast Data Systems) USA version of EC's RDS, adds digital data, via inaudible subcarrier, to audio; can include commercials

RAI (Radiotelvisione Italiana) Public broadcast system based in Rome

RAM (Random-Access Memory) (1) In computers, data storage area into which you can input and retrieve information (i.e., usable memory space, as opposed to ROM); (2) in CD players, similar computerized devices, the ability to access data in nonlinear fashion, going to any portion at any time, not restricted to sequential access

RAPPORT Bell Labs' experimental system allowing people to "meet" on computer screen (computer conferencing)

RARC (Regional Administrative Radio Conference) Group that establishes ITU policies for a single region (see *WARC*)

Real Time Simultaneous collection/display of electronic data, no delay

Regatta Planned series of USSR/CIS solar sail satellites substituting a solar sail and series of rudders for rocket thrusters used to maneuver a satellite in orbit; proposed as L-1 satellite for solar storm warning

Repeater Device/technique to boost an electronic signal being transmitted over distance

Rewritable Videodisc Recorder (Figure G-10) Pioneer's Rewritable Videodisc Recorder, like the Mini Disc, is a dual-head design that allows simultaneous erasing and recording on a Magneto Optical disc; with thirty minutes of recording time per side, the disc is highly reliable due to noncontact pickup and is rewritable one million times; average random-access time is two seconds

RF (Radio Frequency) Mike Wireless microphone

RGB (Red-Green-Blue) The three primary color components mixed to produce composite color television

RIAA Record Industry Association of America

RISC (Reduced Instruction Set Computing) Super-fast computer technology introduced on IBM's RS/6000 work stations and elsewhere

ROM (Read-Only Memory) In computers, portion of memory permanently stored, to be used but not accessed by user

RS-232 Interface Standardized interface (by EIA) to connect a modem and computer

RTNDA Radio-Television News Directors Association

S

Sampling Part of the process of converting analog signals to digital signals. The converter (see *CODEC*) samples the analog signal at regular intervals. If samples are taken regularly enough, the digital signal will reproduce the analog exactly. The frequency of the sampling has to be at least twice the highest frequency in the analog signal being sampled (called the Nyquist frequency); also part of process of digital compression

S-Band Satellite downlinking in spectrum area of 2.6 gigahertz

SBCA Satellite Broadcasting and Communications Association

SBS (1) Special Broadcasting Service, a multicultural service of the Australian government that broadcasts in forty-plus languages; (2) Satellite Business System; a series of satellites—SBS-1, -2 and -6, operated by Hughes, SBS-3, operated by Comsat (MCI), and SBS-4 and -5, operated by IBM

SCPC (Single Channel per Carrier) Transmission format for radio via satellite; whereas conventional audio subcarriers can be received by any satellite antenna dish, the SCPC format can only be picked up by adding an SCPC receiver

Scrambled Signal Signal unintelligible without a decoder (see *Encryption*)

SCSI (Small Computer System Interface) A computer card that allows up to seven disk drives to be connected to an edit system for storage of source material, standard on Macintosh and some PCs

SECAM (Système Électronique Couleur avec Mémoire) One of the three main video standards in use worldwide at the start of the 1990s (see *NTSC, PAL*), a French standard also used in the USSR/CIS and Eastern European countries. Like PAL, SECAM scans 625 lines at a rate of twenty-five frames per second (as opposed to NTSC standard of 525/30). SECAM cannot accommodate some digital special effects (e.g., wipes) that PAL and NTSC use (see *Field, D-MAC, HDTV*)

Figure G-10 The video format for Pioneer's Rewritable Videodisc Recorder is time-compressed analog component. Time code is based on SMPTE. The unit weighs about 100 pounds. Dimensions are 17 1/8 inches wide, 12 5/8 inches high, 25 5/8 inches deep.

Courtesy Pioneer Communications of America.

Secondary Audio Programming (SAP) Audio broadcast via vertical blanking intervals of video transmission

Sematech A USA industry/government-funded consortium to provide research in semiconductor manufacturing

Semiconductor Substance with poor conductivity at low temperatures and improved conductivity with the addition of other substances or the application of heat, light, or voltage (e.g., gallium arsenide, silicon, germanium); term often used interchangeably with *chip*

Serial Port In computers, the I/O opening to which to attach an RS-232 interface to transmit data

Server See *Video Server*

SES (Société Européenne des Satellites) European Satellite Society

Signal-to-Noise Ratio (SNR) Comparison of signal strength versus unwanted noise, expressed in decibels

Sky Cable USA DBS system (Cablevision Systems, NBC, and News Corp.), not to be confused with Sky TV or British Sky Broadcasting

SkyPix Corporation USA pay-per-view DBS system

Sky Television DBS system in Britain, merged November 1990 with British Satellite Broadcasting to form British Sky Broadcasting (BSB)

Smart Card In the VideoCipher II Plus RS (Renewable Security) scrambling system: a replacement card that instantly changes the scrambling code to stop pirating

Smart Homes Homes in which electronics control/coordinate basic functions such as phone, television, temperature controls, lights, and appliances, in essence a LAN controlling all power systems, programmable in-person or by touch-tone phone; Smart Homes have been built in the USA by the National Association of Home Builders and in Japan by the University of Tokyo

Smart Skin Proposed flat antenna to be embedded in skin of, for example, a plane, allowing global communications via satellite; a system linked to an aircraft's electronics would provide a warning of missiles or enemy fire from ground or air

Smart Television Suggested term for combined television/computer

SMATV (Satellite Master Antenna Television) System that downlinks programming and distributes it to clients, such as hotels and motels

SMPTE Society of Motion Picture and Television Engineers

SMPTE Time Code Signals put on film or tape for editing, producing digital, clocklike readout (as often aired on FBI undercover tapes); named for group that established it

SNG (Satellite News Gathering) Electronic news gathering (ENG) using satellites to transmit signals. Downsized equipment, including portable uplinks, frequently put local TV stations on an equal footing with networks for nonlocal coverage, especially live.

Solar Storm see *Solar Wind*

Solar Wind Steady stream of particles from the sun; intermittent and unpredictable cool spots (sun spots) create flares that toss quantities of particles centrifugally towards earth, creating electromagnetic (solar) storms that disrupt power stations and communications

Soyuz Series of USSR/CIS space vehicles used, among other tasks, to supply the MIR space station with replacement cosmonauts

Space Bridge Satellite hookup involving two-way audio and two-way video, usually used for broadcast purposes or videoconferencing

Spacenet GTE hybrid communication satellites (USA); private commercial system in C-Band and Ku-Band serving business in video/audio/data, including broadcasters

Sparklies Microwave noise seen as black-and-white dots in satellite video, comparable to terrestrial "snow"

Spin Transistors Electrons in normal electric current spin randomly. Bellcore experimental spin polarization transfers the ordered spins in magnets to electrons via solid metal, creating on/off impulses; could banish moving parts, multiply capacity, miniaturize

SPOT French earth observation satellite(s), in polar orbit

Spot Beams Satellite antennas concentrating coverage on a limited geographical area, as opposed to larger zone beams or still larger hemispheric beams (see *Footprint*)

Spotnet Hybrid communication satellites (USA)

SRS (Sound Retrieval System) Hughes Aircraft Co. sound system promising 3-D sound from two ordinary loudspeakers

S-SEED (Symmetric Self-Electro-optic Effect Device) Optical switch, used in AT&T Bell Labs' optical computer/processor, in essence replacing wires with light, offering a potential speed of 1 billion operations per second (see *Optical Computer*)

Station-keeping Process of keeping a satellite in its assigned longitude/latitude position, via on-board thrusters

Steadicam Trade name for special bodybrace supporting a free-floating device to hold a camera steady while the operator moves

Still Store Digital technique for storing television frames on disc for random access recall for graphics use; also called electronic still store

Stratosphere Layer of earth's atmosphere from 10 to 50 kilometers above earth's surface

Subcarrier Separate frequency offset from main signal carrier, used to transmit television audio, a second audio channel, color information, and data

Sunspots See *Solar Wind*

Sun-synchronous Orbit Near-polar orbit, with satellite plane directed at the sun, making it ideal for earth observation

Superstation Local broadcast station that transmits its signal via satellite (or other means) to reach other stations/cable systems nationally

Svensat Joint Polish-Swedish firm, largest producer of satellite dishes in Eastern Europe

Switcher (1) Device used to select and control video sources, now largely digital and computerized; (2) Operator of the device

T

TBS (1) Turner Broadcasting System (USA); (2) TBS-Channel 6 (Japan)

T-Carrier Cable network capable of operating rates of 270 megabits per second

Telecine (Television/Cinematography) (1) Unit with film/slide projectors, dedicated film cameras/tape machines; (2) the room containing the unit(s); sometimes called film chain

Telco (Telecommunications) Generic term for a telecommunications company, usually "telephony" in its broadest sense

Telecommuting Working at home, linked to office by computer network (see *Electronic Cottage*)

Teleconferencing A form of videoconferencing electronically joining people at different locations via an uplink signal transmitted to multiple downlinks linked via telephone lines (see *Videoconferencing*)

Telemusic NHK technology that sends sync pulses via satellite video, activating digital player piano in television viewer's home to play music simultaneously with on-screen piano

Tele Piu Italy's first national pay television channel

Teleprompter Device mounted on a camera (or podium, etc.) so that personality can read script/speech while looking at the lens (audience); the British term is autocue

Telesat Canada Private company owned by government and common carriers, owner of Canada's Anik satellite system

Teletel Videotex network owned by French government linking more than 5 million French households, using Minitel terminals, supplied free by the government to access thousands of electronic information services, to make airline reservations, send private messages to and from users, and so on. Like Prodigy in the USA and similar systems in Great Britain and Germany, it uses phone lines. The lack of international standards may doom existing systems after EC unification.

Teletext Transmitting text/graphics via the blanking interval of the television signal; not interactive; requires a decoder (or, if advertiser-supplied, a teletext equipped television set) (see *Vertical Blanking Interval, Videotex*)

Telstar AT&T communication satellites; Telstar 301/401 is a hybrid

Teraflop Supercomputer Parallel computer system designed by fifteen universities and Thinking Machine Inc., capable of performing 1 trillion floating point operations (FLOPS) per second

Texture Mapping Technique in computer imaging to create surfaces such as wood grain; "bump mapping" creates textures such as velvet

Thermosphere Layer of atmosphere between the mesosphere and the exosphere (see *Exosphere, Mesosphere*)

3DO Multimedia player for interactive entertainment, has two unique graphic chips that produce 3-D color images; upgradable to digital decoding of compressed video from cable TV

Time Code Digital code added to tape for editing; produces time readout (hour:minute:second); developed by the Society of Motion Picture and Television Engineers (see *SMPTE*)

Time Compressor Device to replay tape faster or slower than recorded speed without perception of change

Time Division Multiplexing (TDM) Multiplexing mixes multiple signals and transmits them as a single signal, to be separated again at reception. TDM separates such signals by time, each signal having a turn in the bandwidth (see *Frequency Division Multiplexing*)

TIROS (Television and Infra-Red Observation Satellite) NOAA polar-orbiting weather satellite system, the first weather satellite, launched in 1960, which preceded the NOAA satellites with AVHRR

Transducer Device to convert one form of energy into another (e.g., microphone, which converts vibrations into a radio signal)

Transistor Solid-state electronic device of semiconductor substance to control current flow without using vacuum; functions like electron tube, but requires less power and lasts longer (see *Semiconductor*)

Transponder Channel on a satellite. Each satellite has multiple transponders, each of which can receive and send both audio and video signals. They receive uplink signals on one frequency, downlink on another, lower frequency to avoid interference. Satellite users rent or buy use of appropriate transponders (see *Hybrid, Tribrid*)

Trap Device to attach to a cable converter to turn a nonaddressable subscriber into an addressable subscriber for a specific (PPV) event

Tribrid Satellite Satellite with transponders for three different bands

Trinilites Picture elements, as in Sony Jumbitron, containing red, green, blue cells, needed to create a color image

Tropopause   Layer of earth's atmosphere between the troposphere and stratosphere, where air is calm and storm-free

Troposphere   Layer of earth's atmosphere from earth's surface up to 10 kilometers contains 75 percent of atmosphere's gases

TV Answer   VCR-sized, interactive system, by Hewlett-Packard

TVRO (Television Receive Only)   A dish antenna and satellite receiver with no uplink capability, sometimes called a private terminal (see *Earth Station*)

U

UCT (Universal Coordinated Time)   See *GMT*

UHF (Ultra-High-Frequency)   Bandwidth in 300 to 3,000 megahertz range, includes television channels 14 through 83

Uplink   (1) to send signals to a satellite; (2) an earth station with uplink capability (see *Downlink*)

Uplinker   Annual guide listing international television broadcasters alphabetically by continent/country, published by Uplinger Enterprises, Washington, D.C.

V

VCR (Video cassette recorder)   See *VRC*

VDT (Video Display Terminal)   A computer screen, stand-alone or part of a system

Vector   Physical quantity with both magnitude and direction (e.g., the path of a plane, a missile, a television signal); HDB-MAC uses vector quantization in its compression technique (see *Quantization*)

Vertical Blanking Interval   See *Progressive Scan*

Vertical Refresh Rate   Computer term for scanning rate of video signal

VESA (Video Electronics Standards Association)   Organization of computer graphic adapter manufacturers, monitor makers, and computer manufacturers; created in 1988 when super VGA screens first appeared. Established standards for super VGA signals (800 × 600 pixel resolution, vertical refresh rate of 72 Hertz), a standard compatible with 1,024 × 768 displays at 60 Hertz vertical refresh rate as well as with older rates

VGA (Video Graphic Adapter)   Board inserted in computer to permit display of graphic software on screen, now also super VGA

VHF (Very High Frequency)   Bandwidth in 30 to 300 megahertz range, including television channels 2 through 13; wavelength and transmission power make VHF television easier to receive than UHF

VHS (Video Home System)   Consumer video cassette format using half-inch tape

VideoCipher   Commercial scrambling system for satellite transmission, the current industry standard, from General Instrument

Videodisc   Like CDs, videodiscs contain permanently formatted digital information read optically by a reflected laser beam. Until recently, unlike videotape, videodiscs could not be erased and rerecorded. Like CD audio recordings, this laser-read information offers random-access capability that videotape lacks. Videodiscs can locate and play programming in any order, with instantaneous selectability and/or computer-programmed selectability. Interfaced with a computer, videodiscs become interactive. For television programming, especially news and sports, the basis of the Electronic Still Store, which also offers selective replay, freeze frame, and freeze search

Videoconferencing   Using video and audio links to hold a meeting among people at different locations, via satellite or cable links

Videojournalist   Solo reporter carrying his or her own equipment and no other crew members

Video printers   Devices that print on paper still frames grabbed from television/computer video (see *Desktop Imaging*)

Video Server   The "mother" computer that feeds on-demand digitally compressed audio/video product from storage to consumers (see *ATM*)

Videotex   Interactive system transmitting text and graphics via phone lines or cable; generally links home computer to news, banks, airline reservations, shopping and so on (see *Teletel*)

Video Toaster   System from NewTek Inc. using switcher, computer expansion board, and a set of floppy disks to turn a Commodore Amigo Computer into an in-home edit suite

Videoway   Interactive television system allowing the viewer to switch camera video from home; available to Le Group Videotron viewers in Montreal, Canada

Video Windows   Bellcore teleconferencing technology offering near life-size images, a "virtual presence"

Vidicon   Television camera pickup tube developed in the 1960s, cheaper and more durable than the image-orthicon tube; late models include Plumbicon (Philips N.V.), Leddicon (EEC), Vistacon (RCA) with a lead-oxide target, and Saticon models (Hitachi, RCA) with selenium arsenic tellurium target layer (see *Iconoscope, Image-Orthicon, CCD*)

VIRS (Vertical Interval Reference Signals)   Data inserted in the blanking intervals during line scanning in the NTSC standard; The blanking period during each field, odd and even, is scanned and used to insert time code, *teletext data,* and *test signals* (see *VITC*)

Virtual Environment   Term used by NASA (see *Virtual Reality*)

Virtual Reality   (VR)   Computer technology viewed through goggles (Eyephones) that place a television screen in front of each eye to create a 3-D image and the illusion, via electronic visuals, that parts or all of viewer's body (wearing wired DataGloves and/or DataSuit) are interacting with the computer-created images; has great potential in medicine, architectural designing, and video games; also called Artificial Reality, Cyberspace, Virtual Environment, and, in Japan, Intimate Presence; already used in commercial units in parks, malls

Virtual Vision   Bifocal-like visor; wearer can focus on either real-world or TV image. Cf. Sony's Visotron, UK's Goggle Vox vision-blocking goggles.

VisEurope   Dedicated satellite news system available to broadcasters in EC, part of Visnews

VITC (Vertical Interval Time Code)   A form of Vertical Interval Reference Signal inserted in the blanking period during line scanning on the NTSC standard, to locate frames; critical for editing (see *VIRS, SMPTE Time/Code*)

VLSI Chips   Very Large Scale Integrated Chips

VNR (Video News Release)   Promotional video by commercial sponsor prepared and offered in the form of a news item

VRC (Voice Recognition Code)   Used to program computers by voice

VSAT (Very Small Aperture Terminal)   Computer dish linked to computer system to send/receive (usually business) data via Ku-Band satellite

VSDA   Video Software Dealers Association, USA multimedia industry group

W

WAIS (Wide Area Information Service)   Computer network that uses Internet to interface thousands of public/private computer networks allowing them access via a Thinking Machine supercomputer network to information from multiple sources, such as the Library of Congress

WARC (World Administrative Radio Conference) International group that establishes global system specifications, carrying out rulings by the International Telecommunications Union (ITU) (see *RARC*)

Wavelength The horizontal distance between points on successive waves, measured in the direction of the wave

Window (1) Period of time during which a group (e.g., a television network or distributor) has exclusive rights to air or sell a program or product; (2) use of part of a television/computer display screen to show other data or video; (3) computer technology that integrates motion video from composite video sources (such as television) into a computer's component video; Windows is a trademark of Microsoft Corporation software

Wipe Video transition in which one image wipes across another, progressively replacing it

Worldnet Global TV network of US Information Office, comparable to Voice of America radio network

W-VHS JVC (Victor Company of Japan) analog version of VCR that records signal after it is decoded by the TV set, allowing taping of any digital/HDTV signal. Other companies (Toshiba, Hitachi, Sony, Goldstar) propose digital recording of compressed signal. All are compatible with existing VHS tapes.

WYSIWYG (What You See Is What You Get) Term used by computer programmers; what you see on-screen is what will print out on paper

X

X-Band Bandwidth spectrum range: uplink, 7.9 to 8.4 gigahertz, downlink, 7.25 to 7.75 gigahertz

X-ray Lithography Use of x-ray (wavelength 0.001 micron) to shrink chip circuitry onto silicon wafers, producing smaller circuits than are possible using optical lithography (wavelength 0.4 micron); more costly than optical lithography

Z

Zapping Changing channels by remote control to avoid commercials or dull programming

Zone-Plate Antenna A rollup satellite antenna—a series of concentric metalized rings printed on a plastic sheet or a glass pane to focus incoming microwave signals into a feedhorn and thus to a television receiver; British-made, 25 inches in diameter; can be installed on blind, window pane

Index

Page numbers followed by f denote figures.

313